Cancer Drug Delivery Systems Based on the Tumor Microenvironment

Yasuhiro Matsumura • David Tarin
Editors

Cancer Drug Delivery Systems Based on the Tumor Microenvironment

Springer

Editors
Yasuhiro Matsumura
Division of Developmental Therapeutics, Exploratory Oncology Research & Clinical Trial Center
National Cancer Center
Kashiwa, Japan

David Tarin
Moores UCSD Cancer Center and Department of Pathology
University of California San Diego
La Jolla, CA, USA

ISBN 978-4-431-56878-0　　ISBN 978-4-431-56880-3 (eBook)
https://doi.org/10.1007/978-4-431-56880-3

This Springer imprint is published by the registered company Springer Japan KK part of Springer Nature.
The registered company address is: Shiroyama Trust Tower, 4-3-1 Toranomon, Minato-ku, Tokyo 105-6005, Japan

Preface

This book describes a systematic approach to the treatment of solid malignant tumors, commonly called cancers, based upon the pathophysiological properties that they all share. It focuses on the use of cytotoxic agents to destroy malignant cells without harming their healthy counterparts and assembles in one place a wide body of information that can be used in different combinations to suit the particular features of specific tumors in different patients. Other important complementary methods, namely, surgery and radiation therapy, which are valuable in the treatment of cancer patients, are beyond the scope of this book, but the methods discussed here can provide powerful adjuvant or neoadjuvant treatment options to supplement other therapeutic modalities.

While much has already been written about the general topic of cancer chemotherapy, the specific advances described in this volume relate to the delivery, concentration, and retention of drugs and/or other bioactive agents, such as DNA, RNA, and viruses, in the tumor and their targeted release into and onto the cancer cells. This technology requires the assembly of a variety of natural and synthetic components described herein, collectively termed a drug delivery system **(DDS)** and is an essential part of a strategic approach to the effective use of medications in the control of human cancer.

The narrative describes the dynamic structural and physiological changes within a growing tumor. It also provides a comprehensive account of a wide array of organic, inorganic, physical, and biological entities such as micelles, liposomes, cleavable linkers, antibodies, vascular permeability effects, and other modalities, which can be assembled in different combinations to maximize a therapeutic effect, according to the specific properties of the neoplasm. In particular, this book draws attention to the non-neoplastic, stromal components of the tumor that are important, formerly under-recognized, participants in the neoplastic process and shows how stromal properties can be harnessed to enhance antitumor therapy.

The overall objective of cancer chemotherapy delivered via the vascular system is to deliver inhibitory or toxic agents to the malignant cells without harming healthy organs. The contents of this book demonstrate, via preclinical and clinical studies, that this can be achieved without compromising the cardinal rule of the compassionate physician, which is to first and foremost do no harm.

Kashiwa, Japan Yasuhiro Matsumura
La Jolla, CA, USA David Tarin

Contents

About the Editors and Contributors

Editors

Yasuhiro Matsumura, MD PhD
Director, Division of Developmental Therapeutics, Exploratory Oncology Research & Clinical Trial Center, National Cancer Center.
Visiting Professor of Graduate School of Frontier Sciences, University of Tokyo
Visiting Professor of Graduate School of Medicine, Keio University
Visiting Professor of Kansai Medical University

Formerly
Head, Department of Medicine, National Cancer Center Hospital.
Senior Clinical Research Scientist, Nuffield Department of Pathology John Radcliffe Hospital, Oxford University.
Assistant Professor, Department of Microbiology, Kumamoto University Medical School.

Over 35 years experience in
Laboratory cancer research in drug delivery systems, cancer-induced blood coagulation, molecular diagnosis, and monoclonal antibody development

David Tarin, MD, PhD, FRCPath
Professor of Pathology,
University of California, San Diego

Formerly
Director, Moores UCSD Cancer Center, San Diego, California, USA
Professor of Pathology, University of Oxford, John Radcliffe Hospital, Oxford, UK
Professor of Pathology, Royal Postgraduate Medical School, Hammermith Hospital, London, UK

Over 50 years experience in
Laboratory cancer research with special interests in metastasis, carcinogenesis and cancer-stromal interactions
General surgical pathology with special interest in cancer pathology

Contributors

Yuki Abe Oncology Research Laboratories II, Daiichi Sankyo Co., Ltd., Tokyo, Japan

Toshinori Agatsuma Oncology Research Laboratories I, Daiichi Sankyo Co., Ltd., Tokyo, Japan

Gilles J. A. Gallant Oncology Research & Development, Daiichi Sankyo Inc., Basking Ridge, NJ, USA

Hideyoshi Harashima Laboratory for Molecular Design of Pharmaceutics, Laboratory of Innovative Nanomedicine, Faculty of Pharmaceutical Sciences, Hokkaido University, Sapporo, Japan

Hiroto Hatakeyama Clinical Pharmacology and Pharmacometrics, Graduate School of Pharmaceutical Sciences, Chiba University, Chiba, Japan

Kyoko Hida Vascular Biology and Molecular Pathology, Graduate School of Dental Medicine, Hokkaido University, Sapporo, Japan

Yasuhiro Hida Department of Cardiovascular and Thoracic Surgery, Faculty of Medicine, Hokkaido University, Sapporo, Japan

Takahiro Jikoh Oncology Research & Development, Daiichi Sankyo Inc., Basking Ridge, NJ, USA

Ikramy A. Khalil Faculty of Pharmaceutical Sciences, Hokkaido University, Sapporo, Japan
Faculty of Pharmacy, Assiut University, Assiut, Egypt

Mitsunobu R. Kano Department of Pharmaceutical Biomedicine, Okayama University Graduate School of Medicine, Dentistry, and Pharmaceutical Sciences, Okayama, Japan
Department of Pharmaceutical Biomedicine, Okayama University Graduate School of Interdisciplinary Science and Engineering in Health Systems, Okayama, Japan

Yoshikatsu Koga Department of Strategic Programs, Exploratory Oncology Research & Clinical Trial Center, National Cancer Center, Kashiwa, Japan

Nako Maishi Vascular Biology and Molecular Pathology, Graduate School of Dental Medicine, Hokkaido University, Sapporo, Japan

Shino Manabe Ph.D. Cluster for Pioneering Research, RIKEN, Wako, Japan
Research Center for Pharmaceutical Development, Tohoku University, Sendai, Japan

Yasuhiro Matsumura Division of Developmental Therapeutics, Exploratory Oncology Research & Clinical Trial Center, National Cancer Center, Kashiwa, Japan

Takashi Nakada Oncology Research Laboratories I, Daiichi Sankyo Co., Ltd., Tokyo, Japan

Takashi Nakamura Laboratory for Molecular Design of Pharmaceutics, Laboratory of Innovative Nanomedicine, Faculty of Pharmaceutical Sciences, Hokkaido University, Sapporo, Japan

Nobuhiro Nishiyama Laboratory for Chemistry and Life Science, Institute of Innovative Research, Tokyo Institute of Technology, Yokohama, Japan

Naoto Oku Department of Medical Biochemistry, School of Pharmaceutical Sciences, University of Shizuoka, Shizuoka, Shizuoka, Japan
Laboratory of Bioanalytical Chemistry, Faculty of Pharma-Science, Teikyo University, Tokyo, Japan

Tsuneo Saga Department of Diagnostic Radiology, Kyoto University Hospital, Kyoto, Japan

Javad Shahidi Oncology Research & Development, Daiichi Sankyo Inc., Basking Ridge, NJ, USA

Kosuke Shimizu Department of Molecular Imaging, Institute for Medical Photonics Research, Preeminent Medical Photonics Education & Research Center, Hamamatsu University School of Medicine, Hamamatsu, Shizuoka, Japan
Department of Medical Biochemistry, School of Pharmaceutical Sciences, University of Shizuoka, Shizuoka, Shizuoka, Japan

Kiyoshi Sugihara Oncology Research Laboratories I, Daiichi Sankyo Co., Ltd., Tokyo, Japan

Hiroyoshi Y. Tanaka Department of Pharmaceutical Biomedicine, Okayama University Graduate School of Medicine, Dentistry and Pharmaceutical Sciences, Okayama, Japan

David Tarin Moores UCSD Cancer Center and Department of Pathology, University of California San Diego, La Jolla, CA, USA

Atsushi B. Tsuji Department of Molecular Imaging and Theranostics, National Institute of Radiological Sciences, National Institutes for Quantum and Radiological Science and Technology (QST-NIRS), Chiba, Japan

Ryo Tsumura Division of Developmental Therapeutics, Exploratory Oncology Research & Clinical Trial Center, National Cancer Center, Kashiwa, Japan

Masahiro Yasunaga Division of Developmental Therapeutics, Exploratory Oncology Research & Clinical Trial Center, National Cancer Center, Kashiwa, Japan

Part I
Cancer Pathophysiology

The Cancer Stroma and Its Relevance to Tumor Survival and Treatment

David Tarin

Abstract Current cancer research focuses mainly upon the cancer cells in malignant tumors and is providing an expanding database about aberrations in their genetic composition. However, solid tumors also contain non-cancerous host tissue, referred to as the stroma, which plays an active and indispensable role in tumor growth and influences the virulence of the neoplasm towards the host. So far, little attention has been given to this important component of the tumor although progress in cancer therapy has been glacially slow and there is a need to consider alternative approaches. Many cell types inhabit the stroma, and are loosely distributed amidst apparently inert fibrous and viscous matrix material, composed of complex polysaccharides, proteins and other molecules. Actually, all of these elements are in constant turnover, and haphazard deviations from control mechanisms regulating this process, incurred during tumorigenesis, cause unpredictable evolution in the collective properties of the whole community. This chapter provides pathologic observations and data on reciprocal interactions between these stromal and neoplastic cellular components of tumors and how they change during the course of the disease. Malignant progression depends upon derangement of complicated communications between different specialised lineages within the cellular society normally inhabiting the organ, and this enables rapid adaptation to changing circumstances. Opportunistic misuse of such communication networks enables tumor cells to recruit and incorporate adjacent non-neoplastic stroma into their midst and modify it, so that they may grow, infiltrate and parasitise the host. The absolute dependency of primary tumors and metastases on their diverse stromal components for survival and their insatiable need to continuously recruit more stroma to support expansion, renders them vulnerable to strategies capable of disrupting the cellular interactions needed for such recruitment. This dependency of cancer cells upon the stroma is of critical importance for cancer therapy research and proposed methods for turning this parasitic behaviour of tumors against them-

D. Tarin (✉)
Moores UCSD Cancer Center and Department of Pathology,
University of California San Diego, La Jolla, CA, USA
e-mail: dtarin@ucsd.edu

Y. Matsumura, D. Tarin (eds.), *Cancer Drug Delivery Systems Based on the Tumor Microenvironment*, https://doi.org/10.1007/978-4-431-56880-3_1

selves are suggested below. Research on drug delivery systems for oncotoxic agents has shown that the diffusion and active transport of molecules through the interstitial space of the stroma affects their bio-availability and therapeutic efficacy. This is an avenue worthy of exploration to enhance drug therapy. This book will also describe how the inert intercellular materials can be used as scaffolding for delivering, trapping and holding cytotoxic anticancer drugs in close proximity to cancer cells, thereby enhancing the destructive effects of these drugs upon the malignant cell population, without poisoning the host. Recognition that non-neoplastic cells of the stroma are not simply passive bystanders opens the further novel possibility of stimulating the living non-neoplastic components to deny support to the tumor cell population, thereby hindering its expansion. This new approach offers opportunities for bypassing difficulties diminishing effectiveness of current cancer therapies.

Keywords Basement membrane · Cell therapy · Collagen · Diffusion · Drug therapy · Electron microscopy · Exosomes · Fibrin · Intercellular space · Invasion · Metastasis · Microenvironment · Proteases · Scar · Stroma · Supra-molecular factors · Tumour stromal cell interactions · Tumour regression · Tumour progression · Vascular permeability

Introduction

This chapter provides a brief description of the structure and life cycle of tumors and their functional microenvironment, as a basis for the novel therapeutic advances described in subsequent chapters in this book. It focuses on the stroma, defined in detail below, because this component of solid tumors and of healthy tissue, provides an indispensable medium for transport and exchange of signalling molecules, nutrients, metabolic products and therapeutic agents and a structural framework within, or upon which, cells live. Knowledge of its structure and functions is therefore essential for understanding the basis of the therapeutic interventions described in this book. The stroma inevitably interacts with and provides scaffolding and life support for vital components (such as nephrons, liver epithelial cords, lung alveolar cells etc.) of most organs, including the bone marrow. Consequently events in the stroma have substantial persistent effects on the lives of tumor cell colonies and can be successfully utilised for therapeutic advantage in many ways described elsewhere in this book. The brain and the blood differ from other organs in having little or no stroma and require separate consideration. Their constituent cells are consequently adapted to live in specialized micro-environmental conditions and their malignant tumors behave differently from those of other organs. Brain tumors, for instance, do not metastasize to sites outside the central nervous system(CNS). However, it should be noted that haemorrhage and necrosis in brain tumors can result in the influx of macrophages, fibroblasts and blood vessels resulting in fibrin deposition and collagenous scar formation, as a healing response. This pathologically generated, ectopic, stroma provides the opportunity for unconventional stromal-based therapy of CNS tumors (see chapter "Principle of CAST Strategy" by Matsumura), but there is no similar phenomenon in the circulating blood. Consequently, the features and treatment of haematological malignancies are topics beyond the scope of this Chapter and this book.

Healthy Stroma

Before considering the structure and profound role of the stroma in neoplasia, it is appropriate to briefly consider the composition of healthy stromal tissue. It is derived from the mesoderm of the embryo and, in its most unspecialized form, is referred to as loose mesenchyme or loose connective tissue. All stroma is composed of living cells loosely distributed amongst inert intercellular materials, which they secrete and maintain. As can be seen in Fig. 1a and c, the intercellular space within the stroma is much more extensive than in other tissues such as epithelia (also shown in these pictures) or neural tissue of the brain and spinal cord. Stroma contains materials classified as the "formed elements of the connective tissue" that consist of collagen and elastin fibres embedded in a featureless, amorphous ground substance or matrix of gelatinous consistency. The fibres give the tissue tensile strength combined with elasticity and determine the macroscopic appearance of each organ. In different types of stroma the fibres can be densely packed in parallel bundles as in tendons or membranes (e.g. cranial dura), or loosely distributed in non-aligned orientation, as in loose connective tissue that, as its name implies, fills spaces between other tissue components and binds them together. Various specialisations of the stroma occur in different parts of the body. In some places the stromal cells are almost all filled with lipid (adipocytes) and the stroma is termed adipose tissue, whereas in other regions the fibres are precisely organized into lamellae which become calcified to form bone, within which the marrow resides. The marrow is also a form of stroma inhabited by various specialised cell types, including fibroblasts, adipose cells and haematological precursor cells. Blood vessels are also a specialised form of stromal component. Further discussion of the detailed composition of the specialised forms of stroma in different body locations is beyond the scope of this book. However, it should be emphasized that <u>all</u> types of stroma contribute actively to tumor formation and progression in the organ that they normally inhabit, sustain, nourish and provide with structural integrity.

The Tumor Microenvironment

The environment within and around a tumor is a highly heterogeneous, dynamically changing, zone within the otherwise stable, reciprocally interacting tissues of the body of the host. Histopathological analysis of tumor structure reveals that it is composed of many different living and inert components arranged in a variably

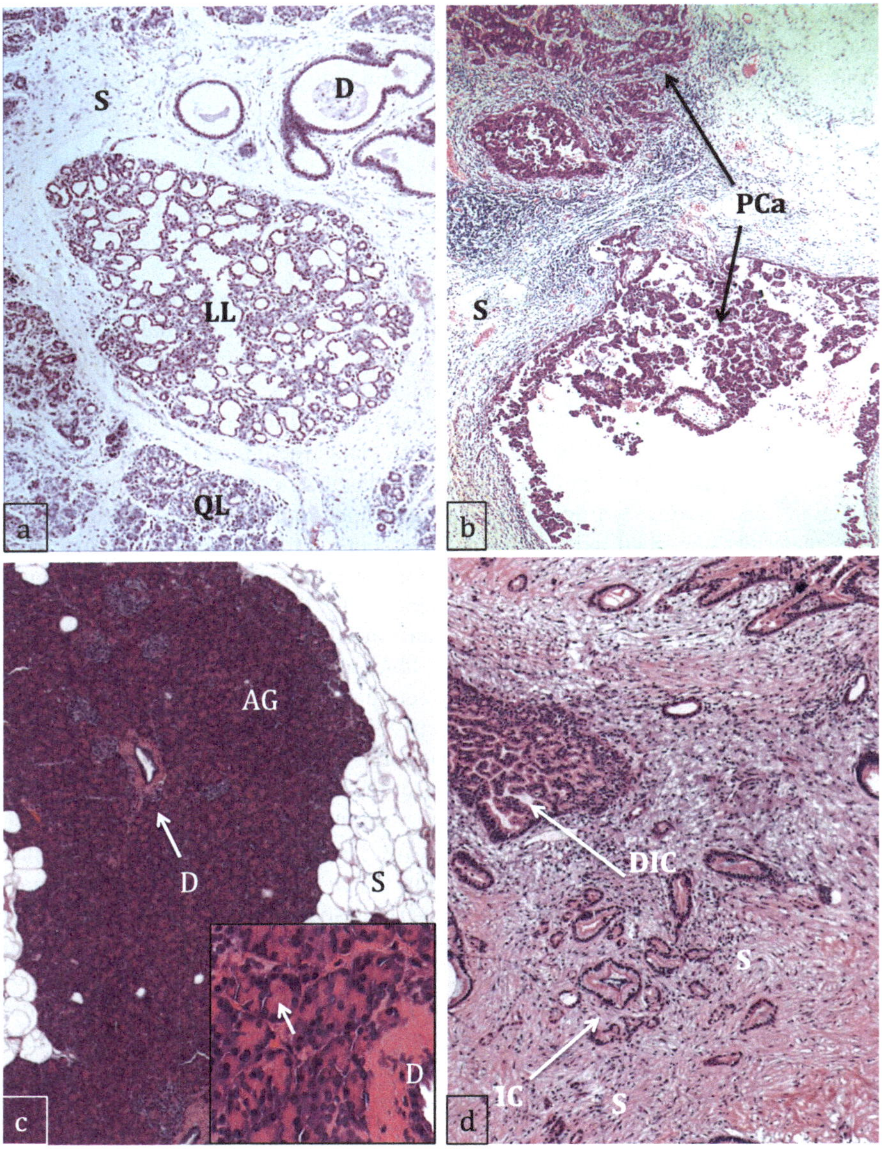

Fig. 1 (**a**) A photomicrograph of healthy breast tissue showing a lactating lobule (LL), a quiescent lobule (QL) and a breast duct (D) embedded in loose connective tissue stroma (S)
(**b**) Invasive and in situ papillary ductal carcinoma of the breast (PCa). The stroma (S) around the malignant cells is more cellular and fibrous than at the top right of the field of view, where looser connective tissue and adipose tissue are visible
(**c**) Healthy pancreas composed of tightly grouped acinar glands (AG) around a draining duct (D) and surrounded by loose mesenchymal adipose stroma (S). Inset magnified shows acinar gland (white arrow) with tiny central lumen adjacent to a duct (D)
(**d**) Invasive ductal carcinoma (IC) of the pancreas and in situ carcinoma (DIC) within a duct, embedded in dense fibrous and cellular stroma (S)

irregular resemblance to its tissue of origin. The degree of such resemblance depends upon the tumor's stage of development and its extent of internal functional disorganisation. This organoid configuration of cells and intercellular materials consists of malignant and healthy cell types interacting within a milieu of soluble and insoluble products that they have secreted and the composite entity creates the living tumor microenvironment. Functionally, the disorderly behaviour of tumor cells fails to integrate properly with the activities of the healthy cells of its organ of origin. The neoplastic cell community emits irregular streams of signalling molecules which can disrupt the functions of adjacent and distant organs [1]. It also responds erratically to regulatory messages from its surroundings. In summary, a tumor is not simply an expanding ball of multiplying cells. It is a malfunctioning living entity consisting of interwoven, interdependent, neoplastic and non-neoplastic components which overgrows, infiltrates and out-competes its neighbours for available resources. This chapter will review the evolving structure, and function of tumors in their life cycles and how these characteristics can be used against them for effective cancer therapy.

Structural Aspects

Tumor Cells

The histopathological composition of tumors varies greatly between tumors of the same origin and even between different regions within the same tumor (Fig. 1a–d), but there are usually sufficient similarities to their ancestral tissue structure to enable diagnostic identification of the organ of origin, in most cases. The tumor cell population is often, but not always, the dominant cell type in the neoplasm (see Figs. 1b, d and 3b). It may show substantial uniformity of cellular morphology (technically termed a monomorphic appearance) but more commonly the tumor cells show great variation in size, shape and staining characteristics, known as pleomorphism or anaplasia. The degree of uniformity of morphology of the tumor cells and their similarity to non-neoplastic counterparts of the same tissue of origin reflects the amount of disturbance in differentiation of the tumor cell phenotype and is a useful, but not completely reliable, prognostic indicator.

The Tumor Stroma

This consists of non-neoplastic cells within the tumor, together with inert intercellular materials such as collagen fibres, amorphous basement membranes and a morphologically featureless "matrix" in which the cells, fibres, and vessels are suspended (Figs. 1, 2 and 3). The distinguishing microscopic feature of neoplasia is disorderly distribution of all these components within the lesion. Healthy organs are composed of orderly combinations of basic tissue types including epithelia, connec-

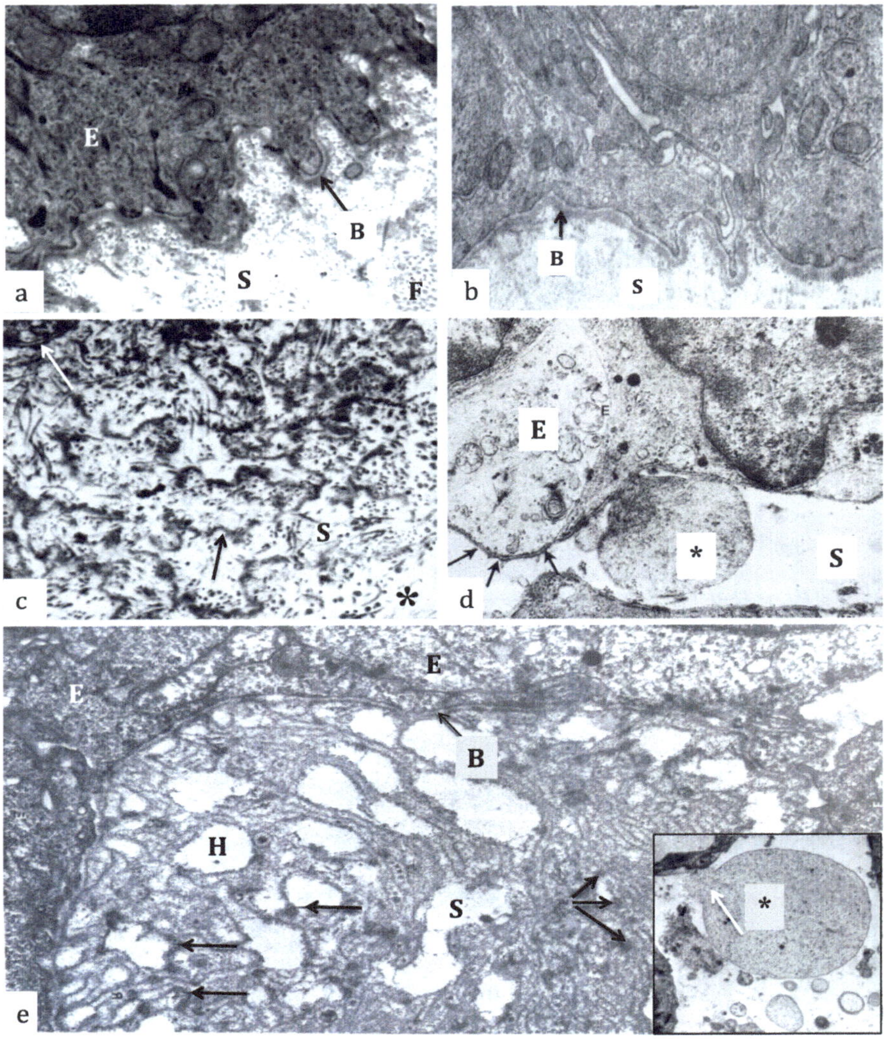

Fig. 2 (**a**) Transmission electron micrograph of the epithelial (E)-stromal (S) junction in skin showing the basement membrane (B), stroma (S) containing collagen fibres (F) composed of bundles of smaller fibrils seen in cross section. Note the regularity of fibril thickness, displayed in a fibre in the right bottom corner
(**b**) An electron micrograph of the epithelial stromal boundary around a breast duct lined by epithelium (E) and ensleeved in a basement membrane (B) separating it from the stroma (S)
(**c**) Early experimental skin carcinogenesis: several reduplications (black arrow) of the basement membrane are seen near the boundary between epithelium (upper left) and stroma (S) in this electron micrograph. A portion of the original basement membrane (white arrow) is still visible next to the base of the epithelial cell in the top left corner. Note the marked variation in collagen fibril thickness seen near the asterisk in the bottom right corner and throughout the stroma. Compare with (**a**)

tive tissue (stroma), blood vessels, nerves and other components. The boundary between the stroma and each different tissue type composing an organ is sharply delineated by a thin lamina of condensed material, best visualized with the electron microscope, called a basement membrane (Fig. 2a, b). The molecular composition of this structure varies in detail between different organs but characteristically contains collagen type 4 and other collagens in the dense layer and laminin in the lucent layer. In addition the basement membrane can contain other molecules, such as CD44 and collagen 17, which have tails that cross the limiting membranes of adjacent epithelial cells and anchor them to the basement membrane. Other constituents of this latter structure include fibronectin, elastin, proteoglycan and several other glycoproteins. The basement membrane acts both as a substratum for attachment of healthy epithelial cells and other cell types as well as forming a mechanical barrier to invasive cells and to other pathogens. Characteristic sequential changes are seen in the morphology of the boundary zone between healthy and neoplastic tissue during cancer formation (see below).

The non-neoplastic cell population within the stroma is scattered loosely within the fibres and matrix composing the intercellular material (Fig. 1b, d). These include many different cell types, known as cell lineages, including fibroblasts, smooth muscle, myofibroblasts, pericytes, mast cells and endothelial cells. Some of these were already resident in the region where the tumor originated and others were attracted or recruited into the developing neoplasm by molecular signals released by the malignant cells [2]. The topographic relationships between malignant and non-malignant cells are haphazard and change over time. In addition, a transient population of lymphocytes and other immune-related cells (easily seen in Fig. 1b) traffics through the lesion and, in most tumor types, does not stay to contribute to expansion of the lesion. However, a few tumor types, such as medullary carcinomas of the breast [3], Hodgkin's lymphomas, seminomas and some melanomas, characteristically contain large numbers of lymphoid cells. The stroma of healthy organs and of tumors exerts powerful invisible effects upon the behaviour of the cell populations that it supports and provides a valuable platform for new anti-cancer therapies described in this book.

Fig. 2 (continued) (**d**) Virally-induced mammary tumor of a mouse showing an epithelial carcinoma cell (E) extruding an exosome (asterisk) containing granular material into the stromal space (S) where collagen fibres have been digested away. A portion of basement membrane (arrows) is still visible at the epithelial stromal boundary
(**e**) Electron micrograph of the epithelial-stromal junction in a later stage mouse mammary carcinoma showing a portion of a neoplastic epithelial cell (E), stroma (S), reduplications of the basement membrane incorporating fragments (arrows) of older basement membranes and large digested holes (H) in the stroma, where collagen fibres have been completely removed in front of the invading carcinoma cells. (Inset shows an exosome (asterisk), which has been released from a carcinoma cell that is rupturing (arrow) and releasing contents into the stroma). This phenomenon is accompanied by the appearance of holes in the stroma shown in **e**

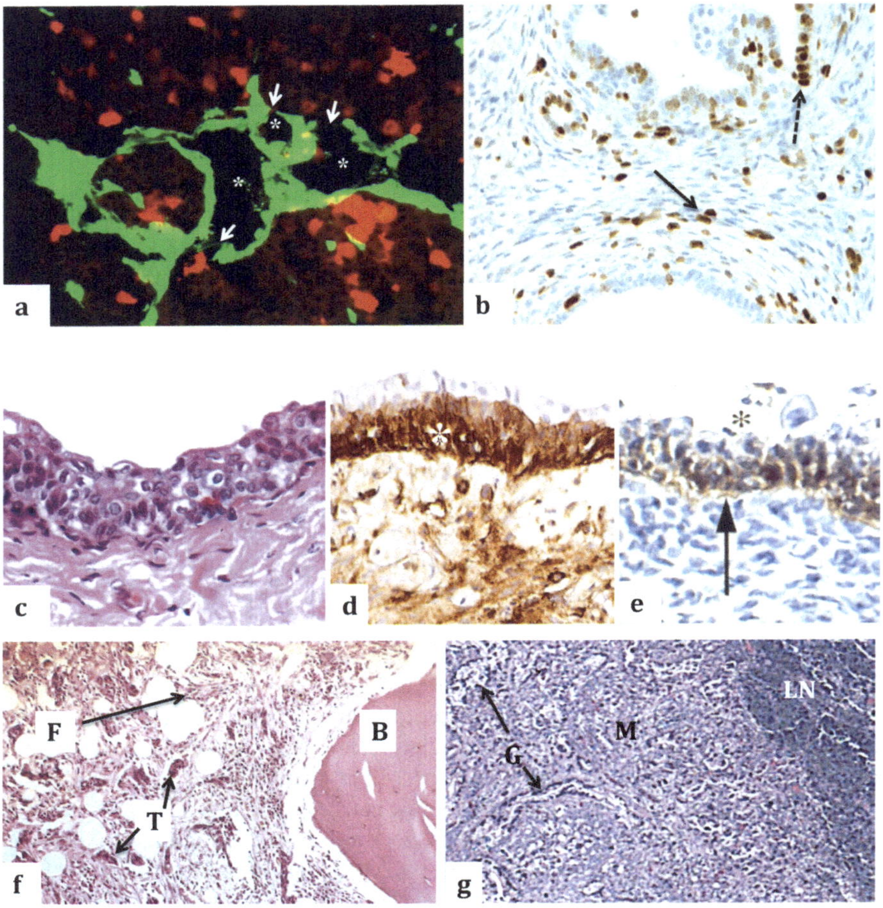

Fig. 3 (**a**) Histological section of a xenograft of a human melanoma labelled with red fluorescent protein growing in a green fluorescent mouse. The stromal blood vessels (asterisks), derived from the host, have large abnormal fenestrations (arrows) in their green endothelium, through which contents can leak into the surrounding stroma
(**b**) Section of a human pancreatic carcinoma xenograft growing in a mouse. The immunoperoxidase-labeled antibody staining for Ki67 antigen reveals that the desmoplastic stromal cells (unbroken arrow) are proliferating at least as much as the epithelial tumor cells (dashed arrow)
(**c**) Photomicrograph of human bladder mucosa with epithelium showing severe dysplasia. The lumen of the bladder is in the upper part of the field of view and the stroma containing tightly packed collagen fibres and scattered mesenchymal cells lies below the epithelium stained with haematoxylin and eosin
(**d**) Human bladder mucosa showing epithelial dysplasia and marked over-expression of CD44 proteins (asterisk) stained with antibody to CD44 standard form (CD44s)
(**e**) Healthy bladder mucosa stained with antibody to CD44s for comparison with **d**. The epithelium is less thick and less heavily stained. The asterisk marks the bladder lumen, into which some surface epithelial cells are desquamating
(**f**) Human bone marrow containing metastatic breast cancer cells (T) stimulating reactive desmoplastic hyperplasia of host mesenchymal fibroblasts (F). The apparent empty spaces are adipocytes. A spicule of non-neoplastic cancellous bone (B) is visible on the right
(**g**) Metastatic pancreatic carcinoma (M) growing in a lymph node (LN) and stimulating marked stromal cell hyperplasia. Some of the tumor cells are forming distorted glands and ducts (G)

Functional Aspects

The non-neoplastic stromal components of a tumor provide the scaffolding of matrix materials, fibres and blood vessels, which are essential for the survival, and growth of the tumor cell population. In healthy tissues, organs and organisms, the stroma is a source of growth factors, nutrients, oxygenation and structural scaffolding as well as a medium for diffusion of fluids, ions, hormones, gases (e.g. CO_2, NO, O_2) and waste products. In tumors, chaotic changes occur in these functions and the ability of the tumor cells to adapt to these different conditions determines the survival and rate of growth of the neoplasm or of different regions of the lesion relative to each other.

Tumors are extremely inhospitable places where lethal competition rages between tumor cells and normal cells and the conditions for orderly social interactions that operate in healthy, thriving, counterpart tissues, do not apply. Adaptability to unpredictable changing conditions determines which malignant cell clones dominate available resources at any given time and which die out. Changes in clonal composition among tumors has been demonstrated by genetic marking experiments conducted by Moffett et al. [4], Baban et al. [5] and others. The changing conditions are created by the tumor cells themselves and by an altered flux of signals and metabolic materials amongst all the living elements in the field of the developing neoplasm. This generation of clonal and genetic diversity by the unstable genomes within the ever-changing tumor cell population provides it with substantial scope to adapt to selection pressures exerted by conditions within its environment according to the principles of natural selection described by Darwin. Such changes in clonal composition can also be caused by cytotoxic therapy and explain how cancers can often recur, even after highly effective sub-total eradication (It should be noted that killing of 99% of the tumor cell population still leaves 10^4 living cells for every 10^6 living cells in the original, untreated, tumor – a substantial reservoir for its potential recovery).

One of the important changes in the tumor microenvironment, which occurs from an early stage in neoplastic development, is an increase in vascular permeability. This occurs because tumor capillary blood vessels are malformed, in possessing large fenestrations between endothelial cells lining the capillaries (Fig. 3a), which are not fully ensleeved with smooth muscle cells and pericytes. High molecular weight compounds can leak into the interstitial space of the tumor via these pores [6], in greater quantity than in normal tissues and are retained there for extended periods as initially demonstrated by the pioneering work of Matsumura and Maeda [7]. This phenomenon termed the enhanced permeability and retention (EPR) effect can be exploited for therapeutic purposes. EPR is a passive phenomenon which can be actively augmented by various measures [8] including hypertension, and by pharmacological measures to increase local vasoactive agents such as bradykinin and nitrous oxide. Its value can be enhanced much further by techniques to concentrate onco-therapeutic drugs in tumors and delay their clearance by binding them to intercellular components of the stroma [9, 10] and other methods described in this book.

Attention to the changes in the stroma of developing tumors began with histopathological studies as early as 1914 [11] but the pioneering work of JW Orr [12] subsequently provided definitive evidence that such stromal changes occurred long before the tumor was clearly recognizable. His further studies also provided excellent evidence that the changes in the stroma may have functional implications important for the advancement of the carcinogenic process [13–15]. Later studies by Tarin [16, 17] examined these changes at higher resolution, using the electron microscope and demonstrated that they were similar in tumors of different organs, induced by different aetiological agents. Moreover, they occurred in a specific sequence at different stages of the neoplastic process, beginning at the boundary between neoplastic and healthy tissues and were not simply incidental collateral damage inflicted by the carcinogenic agent [18]. This indicated that the changes were causally related to the neoplastic process. The next section will consider these changes in detail.

The Stroma in Developing and Established Neoplasms

The stroma in a growing tumor is a zone of dynamic activity. In early neoplasms, it shows gradually increasing cellularity and disorder of intercellular collagen, elastin and matrix components. Experimental and clinical histopathological studies show that in early carcinogenesis [12, 17, 19, 20] progressive disorganization of the stroma adjacent to the cancer cells is evident, even before epithelial dysplasia becomes detectable. Subsequent high resolution electron microscopic studies on the development of carcinomas in various organs of humans and laboratory animals [21–24] have reproducibly confirmed that the earliest changes in neoplasia consist of loosening and fragmentation of the basement membrane in the region of the primordial cancer, followed by reduplication of this laminar structure to form several incomplete layers (Fig. 2c, e). These reduplications appear to be repeated attempts to repair the membrane.

Concomitantly, the epithelial cells show pleomorphism, hyperplasia, disturbed orientation and loss of cohesion to each other, a change termed dysplasia. At this stage production of many CD44 protein isoforms in the primordial tumor cells becomes exaggerated. This is demonstrable by immunohistochemistry (Fig. 3c–e) and studies using reverse transcription PCR reveal overproduction of many aberrant RNA transcripts, some of which contain introns [25]. These histopathological and molecular changes become progressively more severe and the lesion is then regarded as carcinoma in situ. Eventually, the cancer cells push bulbous-ended pseudopodia with narrow necks through the fragmenting basement membrane (Fig. 2d) into the stroma. The pseudopodial necks constrict and sever to release vesicles, containing granular material, into the adjacent stroma. These vesicles, known as exosomes, rupture (Fig. 2e, inset) with simultaneous fraying and lysis of collagen fibres in the vicinity resulting in the formation of lattice-like holes in the intercellular materials of the stroma close to the tumor cells (Fig. 2e). The holes merge, creating large spaces into which the cancer cells migrate. This sequence of changes indicates that uncontrolled release of proteases is occurring and that this process facilitates local invasion (Fig. 2d, e and inset) by the cancer cells [16, 17, 20].

Later, the host stroma undergoes rapid multiplication of mesenchymal cells (Fig. 3b), often to a degree exceeding that of the tumor cells, and disorganized collagen fibres composed of fibrils of irregular thickness and orientation polymerise in the intercellular space. New deformed, leaky capillary vessels (Fig. 3a) grow into the area and establish a disorderly network. Crowding of mesenchymal cells (called desmoplasia) in such tumors (Figs. 1d and 3b) is often mirrored by similar changes in their metastatic deposits (Fig. 3f, g). In other tumors, where stromal cell proliferation and increased vascularity is not evident, large serpiginous areas of necrosis occur in the packed epithelial cell population, due to lack of stromal support. Larger tumors often contain extensive highly fibrous areas resembling dense scars containing both collagen and elastin (Fig. 1b, d) Exaggerated examples of this can be seen in pancreatic cancers and pulmonary "scar' cancers. Mature large cancers are often highly heterogeneous in patterns of cell differentiation, stromal cellularity and distribution of dense fibrous areas. These features reflect the highly labile, variable and unregulated nature of tumor-stromal cell interactions and the increased turnover of cells and intercellular materials in the developing neoplasm. Evidence of the important role of such interactions can be seen in the early phases of the development of metastases in the bones, lungs, bone marrow, lymph nodes (Fig. 3e, f), brain [26, 27] and other sites, where recruitment of adjacent host stroma, desmoplasia, and disorderly collagen fibre deposition are all common.

As the developing neoplasm proceeds through dysplastic and in-situ stages, to invasive malignancy, the changes in the stroma around the cancer cells intensify. This illustrates that the current focus on disturbances inside the genomes of the individual cancer cells, as the main cause of malignant behaviour, is too narrow a perspective, because important field changes are occurring in the immediate neighbourhood of the cells and are equally relevant for establishment of malignant tumor formation. For example, Dawe's [28] landmark studies with virus-induced head and neck cancer in rodents, in which infected and non-infected epithelium and mesenchyme were assembled in different combinations, demonstrated that the unit responding to the neoplastic stimulus is not simply the epithelial cell but an epithelial-mesenchymal complex. The experiments demonstrated that despite the presence of the virus causing genetic disturbance in the putative tumor cells, tumor formation could not occur without the involvement of the stroma. The experiments of Orr, Marchant, Billingham [13–15] and colleagues, involving exchange of epithelial and stromal components between carcinogen-treated and untreated skin provided further compelling evidence that the stroma plays an *active* role in epithelial carcinogenesis.

The morphological changes described in this section of this chapter cannot be seen at the molecular level but make major contributions to the course of events. Supra-molecular factors which determine the course of events in a specific neoplasm include (a) the identities of the cell populations which produce the molecules that exert important effects and (b) the specific locations in the tumor where they are synthesized and secreted. The effects of bioactive molecules are dependent upon spatio-temporal factors (i.e. where and when they are produced) as well as on their affinity, avidity, concentration and specificity for their targets. So, in order to understand the significance of changes in a disorganized and poorly coordinated, mixed, cellular population during tumor induction, growth, invasion and metastasis, and how they lead to independence from regulatory controls, it is *essential* to study events at the cellular and tissue level, as well as from the genomic and transcriptomic perspective. In short, events at the supra-cellular level determine the mode of implementation of the mechanisms of carcinogenesis initiated by aberrant molecular processes at the genetic level, because pathological processes at the molecular level can only be executed if they are feasible at the supra-cellular level.

Subsequent studies discussed below have revealed that reciprocal interactions between the stroma and the tumor cells occur within each growing neoplasm in unique and subtle, context-dependent ways. These interactions determine the behaviour and responses of all components in the tumor system developing in a given organ and result in local and remote clinical effects that impact upon the state of health of the whole patient or host organism. For example local effects of the neoplasm on the host can include invasion and damage of adjacent organs, whilst systemic manifestations can include metastasis, inappropriate hormone secretion or paraneoplastic syndromes [29–31]. It must be stressed that the changes that occur within each tumor *are unique to that growth* and even vary from one region to another within the neoplasm, as well as with time, This poses challenges to determining pathways and mechanisms of disease and to choosing optimum therapy for a specific patient (see section below on clinical implications).

The stroma continues to be of *critical* importance to the neoplastic cells during *both* primary tumor formation and metastasis. When a cell or a group of cells ceases to play its part in the regulated harmony of the whole organ or tissue and can replicate independently, its behaviour eventually disrupts the orchestrated balance of interactions between other cells in its locality. Some of these small colonies of dysfunctional cells evolve mechanisms to recruit adjacent normal stromal cells into their midst to provide them with support for their parasitic expansion. The tumor can then grow exponentially and, if interactions with neighbours change further, may invade local vascular channels and metastasise. Other times, recruitment of non-neoplastic stroma does not occur and the early tumor colony ceases to grow, or regresses or dies. This description of supra-cellular events determining the fate of tumor cell colonies is supported by recent reports on the outcomes of human breast cancer screening (mammographic) programs [32] discussed in the next section.

The Life History of Tumors

From the data presented above it is possible to construct a narrative of the cycle of events, which commonly occur in the parasitic interaction between a tumor and its host, with which it shares near-complete genetic identity.

Progression, Regression, Dormancy and Metastasis

Tumor cell colonies evolve over time and their molecular, structural and behavioural properties are affected by the stroma associated with them. The blend of stromal and cancerous cells and the relationships between them (the tumor system) create the specific characteristics of each tumor. Such properties, created by several million living units (cells) living together, disappear when the units are separated from each other and are called "emergent" properties". (This phenomenon can also explain why metastases from a tumor can sometimes have different characteristics, from the original primary tumor, because the cancer cells in the two growths are interacting with stromal cells from different organs).

The resulting malfunction of the combined "tumor system", is evident in the distorted organoid histological structure visible microscopically and in its patterns of behaviour in vivo. If the tumor cells and the host stromal cells proliferate cooperatively, the tumor grows and expands into adjacent non-neoplastic tissue, replacing it either by compression, or by infiltration and destruction. This process is termed tumor progression and may result in penetration of blood and lymph vessels and metastasis. Cooperative interactions with the host stroma require reciprocal signalling and if this is not achieved, local or widespread necrosis and haemorrhage will occur. If this is extensive, tumor growth may decrease or stop, causing it to become an inert nodule, or the neoplasm may regress and disappear. Alternatively, especially in metastasis formation, solitary tumor cells or small groups of cells may remain "dormant" or inactive in an ectopic site, if they lack the ability to generate signals to attract local stroma to support their growth. Experiments involving retrieval of solitary, marked, scattered tumor cells from organs where they were dormant showed that the cells were still viable in vitro and tumorigenic, if reunited with "permissive" stroma [33] by reinoculation in vivo. From these data it is difficult to escape the conclusion that the stroma plays a critical role in whether metastases are formed by disseminated cancer cells. Histopathological examination of many different early metastases obtained from autopsies on cancer patients [27], shows that the initial small colonies of tumor cells attract neighbouring host cells into their midst, to provide support for their growth [26, 27]. The similarities between these events in the early growth of metastases and those in primordial primary tumors are so close that it can be inferred that, unless tumor cells can activate mechanisms to recruit host mesenchymal cells, they cannot grow.

Many studies confirm that metastases reproducibly develop in some, but not all, organs seeded with cancer cells from the same primary tumor [33–37]. Hence, the fates of individual members of the same tumor cell population, distributed randomly in the body, after mixing in the maelstrom of the circulation, vary considerably in different locations. In some, they grow; in others they die; and in some they remain dormant. This, and other evidence cited above, provides powerful evidence that the stroma of the host organ is not passive in secondary tumor growth. Unless local stromal cells respond to signals from the tumor cells and provide them with resources to proliferate, the disseminated tumor cells do not thrive. Also, human tumor cells experimentally inoculated into some anatomical sites of animals, such as the *subcutaneous tissues*, generate tumors that do not metastasise but, when sister cells from the same tumor are inoculated *orthotopically*, they do so readily [38–42]. Co-inoculation with intercellular matrix components of the stroma also facilitates the local growth and metastasis of tumor cells [43], indicating that the non-cellular constituents of the stroma can influence tumor behaviour. Malignant and non-neoplastic components are therefore interactive and jointly produce the unique "emergent" properties [44] of each cancer, which can be exploited for novel advances in therapy discussed below.

Recent studies have provided evidence that some growing tumors recruit mesenchymal cells from remote locations, especially the bone marrow. For example, the work of Feng et al. [2] demonstrated that, in lethally irradiated mice reconstituted with green fluorescent protein (GFP) labelled marrow, implantation of tumor cells labelled with red fluorescent protein (RFP) was followed by the arrival of green mesenchymal cells in the growing tumors. The derivation and replenishment of stromal cells in primary and metastatic tumors can, therefore, occur from multiple local and distant sources. Such findings demonstrate the array of local and distant signalling resources that some tumors can mobilise and this poses a daunting challenge to successful therapy, but also an interesting possible opportunity (see section on potential live cell therapy below).

Small Cancers Can Regress: Cancer Screening Follow-Up Studies and Death Rates

Clinical observations on large cohorts of cancer patients confirm that tumors do not always and inevitably progress to accelerating growth, invasion and malignancy. Evidence from meta-analyses of several long-term follow-up studies on large scale cancer screening programs for a variety of cancer types have indicated that, whilst there is definitely an increase in detection of early stage cancers in the screened patients, the reduction of mortality from corresponding cancers is lower than was expected [32]. For example, while it is generally agreed that screening does prevent some cancer deaths, it is estimated that, for every breast cancer death averted 838 women must undergo mammographic screening for 6 years. This indicates that a significant proportion of tumors are indolent, do not advance or can regress. Other even more direct evidence that early stage carcinogenic processes can slow down, stop or reverse comes from follow up studies on histologically low-grade ductal carcinoma in situ of the breast in patients, who were treated by biopsy only or declined all treatment. These showed that this pre-malignant condition only progressed to invasive carcinoma of the breast in approximately 30% (11/29) of patients followed for 30 years [45]. Fully established metastatic cancer can also regress completely, as seen in patients with stage IV-S neuroblastoma, in whom tumor deposits in multiple organs shrink and eventually become undetectable [46]. These findings all agree well with experimental data on the reimplantation of naturally occurring lung metastases from breast cancers in mice which showed that, after transplantation of such secondary tumors into the mammary glands of fresh syngeneic mice, all formed local tumors but only 30% of these metastasized again in the new hosts [47]. Hence the available evidence shows that the balance of interactions between malignant and non-malignant cell populations within the nascent neoplasm sometimes changes, so that the tumor cells lose properties, such as the ability to grow in ectopic organs, which they had previously acquired.

Implications for Therapy

The aim of this book is to stimulate novel approaches to tumor therapy and to provide data indicating that the stroma of tumors, which is contributed by the host, is a valuable pathway by which to influence tumor cell behaviour and survival. This Chapter contributes a review of the pathological and pathophysiological properties of living tumors as a framework for considering the findings presented in other Chapters by investigators using this new approach to cancer therapy. A shift in emphasis from traditional oncotherapy should diminish toxic side effects on rapidly dividing non-malignant cell populations in other organs, such as the bone marrow, gut and skin, caused by current therapeutic efforts, which seek to directly attack the tumor cells themselves with systemic cytotoxic agents. For now, these agents together with radiation are the best therapeutic modalities available and rightly constitute the best standard of care, but the data discussed above indicate that refocusing research and therapeutic efforts on blocking the interactions at the tumor-host interface within the tumor microenvironment would be more effective at arresting malignant progression with less patient morbidity.

An example of a method using drugs that inhibit stromal mediators of tumor cell growth was successfully used by Sumida et al. [48] to impair the growth of xenografted human gastric carcinoma by blockade of PDGF-R signalling. These investigators demonstrated that carcinoma-associated stromal fibroblasts, pericytes and lymphatic endothelial cells expressed high levels of PDGF-R whereas carcinoma cells did not and treatment with high-dose imatinib and irinotecan in combination inhibited tumor growth and lymph node and peritoneal metastases by disrupting stroma-tumor signalling via this natural mediator.

The well documented genetic instability of tumor cells coupled with their rapid replication results in the intermittent emergence of resistant clones under the selection pressure of cytotoxic therapy. Such resistant clones can enable tumor recurrence and failure of treatment, making effective permanent eradication of malignant tumors by current cytotoxic agents and long-term (over 10 years) survival a difficult and infrequently achieved goal. In contrast, the non-neoplastic stromal cells supporting the tumor population do not display marked genetic instability, although mutations do occasionally occur in tumor stromal cells [49, 50]. The cellular and extra-cellular components of the tumor stroma might, therefore, constitute a better vehicle for anti-cancer therapy than standard approaches.

Turnover of Cells and Inert Materials: Tumor Dynamics

The turnover of tumor and host cells and of interstitial materials such as collagen, glycosaminoglycans and other molecules therefore varies according to the current behaviour of the lesion. This means that responses to therapy will differ between tumors and may even be different for the same tumor at different times. It also means that the response of the tumor is determined not only by the status of the tumor cells, but also by that of the non-neoplastic host cells and interstitial components of the stroma. As tumor cells have an absolute dependency upon the stroma for survival and growth, their fate can be affected by drug or cell therapies designed to focus on the tumor microenvironment. Concepts that need to be considered in using this new approach include:

- Penetration of intra-tumor space by therapeutic agents – diffusion through the extravascular, extracellular compartment. The rate of transit of molecules through the stroma of healthy or neoplastic tissue varies according to:
  - the viscosity/density of the amorphous gelatinous matrix in which cells and formed elements of the stroma (e.g. fibres) are suspended (e.g. cartilage or bone versus loose mesenchyme),
  - the compaction of stromal fibres (e.g. tendon versus loose mesenchyme)
  - the size and charge of the transiting molecules
- Retention of cytotoxic therapeutic agents – the design of agents to enable them to bind to stromal components, thereby hindering clearance and facilitating concentration within the tumor.

Potential Live Stromal Cell Cancer Therapy

The evidence that healthy host cells can and do contribute to regulating tumor cell kinetics, both positively and negatively, as described above, indicates that stromal cells have the potential to be used for tumor therapy. This is probably the most ambitious concept emerging from this study of the role of the stroma in tumor pathogenesis.

Options for cell-based therapy command serious attention when it is recalled that the human organism is a heterogeneous community of trillions of cells, of diverse specialized lineages, which are normally coordinated by powerful short range (inductive) and long range (endocrine) signals, established in embryo and maintained throughout life. It is, therefore, appropriate to consider how to harness specific components of these natural (non-immune) regulatory mechanisms in order to inhibit, control and subordinate malignant cell populations into harmonious behaviour. Early attempts to use mesenchymal stromal cells in cancer therapy have already begun and are reported to show promise [51–54]. To investigate this approach, non-cancerous living cells from the same organ can be sent into the lesions via intra-vascular or direct intra-tumor injection, as self-replicating vectors able to synthesise and deliver high concentrations of biologically-active inhibitor molecules on site, and actively participate in local tissue reorganization. Modern technology already exists to colour-code the effector cells with fluorescent tracker proteins and engineer them to produce molecules, already shown to be inhibitors of tumor growth and metastasis. It is predicted that the inoculated cells will multiply and be incorporated into each tumor under the influence of the same signals that the tumor uses to summon adjacent normal stromal cells to enter its corrupted community, multiply and supply its needs. Histopathological studies on tumor biopsies show that the rate of multiplication of stromal cells in tumors showing desmoplasia is astonishing. Being alive, the non-malignant cellular delivery vehicles can also co-adapt and evolve with their target population. The concept is to obstruct and undermine, from within, the support the cancer must necessarily derive from its stroma, in order to survive and grow, thereby instigating shrinkage of the lesion and the restoration of order. The resident stromal cells will already be under the influence of the tumor cell population to provide them with support, but the objective is for the engineered therapeutic stromal cells to oppose this.

Defining the Aims of Therapy: Eradication, Stasis, or Regression

Whilst the prognosis of various types of human tumors can be evaluated statistically in populations of patients, the fate of individual cancers is not predictable. It is not practically possible to observe all of the variables mentioned above continuously. However, knowledge of the dynamic evolution of the different components of tumors in general, indicates that frequent monitoring of the overall status of the tumor and of the patient is required for optimal therapeutic effects.

Knowledge of the life history of tumors discussed above indicates that it is essential to consider the aims of therapy for each patient. Patients seek help from their physicians at different stages of their illnesses and it is, therefore, often not possible to achieve complete eradication of the tumor, if they arrive in late stages of the illness. In such cases, physicians take the comfort and quality of life of the patient into account and may have to accept that some regression in tumor size or delay in growth is a more realistic aim than cure. The methods described in other Chapters of this book, based upon the structure and molecular composition of the tumor stroma, show that this new approach to treatment is a realistic option at any stage in cancer progression.

References

1. Tarin D (2013) Update on clinical and mechanistic aspects of paraneoplastic syndromes. Cancer Metastasis Rev 32(3–4):707–721. PubMed PMID: 24052434
2. Feng W, McCabe NP, Mahabeleshwar GH, Somanath PR, Phillips DR, Byzova TV (2008) The angiogenic response is dictated by beta3 integrin on bone marrow-derived cells. J Cell Biol 183(6):1145–1157. PubMed PMID: 19075116. Pubmed Central PMCID: 2600740
3. Wargotz ES, Silverberg SG (1988) Medullary carcinoma of the breast: a clinicopathologic study with appraisal of current diagnostic criteria. Hum Pathol 19(11):1340–1346. PubMed PMID: 2846422
4. Moffett BF, Baban D, Bao L, Tarin D (1992) Fate of clonal lineages during neoplasia and metastasis studied with an incorporated genetic marker. Cancer Res 52(7):1737–1743. PubMed PMID: 1551102
5. Baban D, Matsumura Y, Kocialkowski S, Tarin D (1993) Studies on relationships between metastatic and non-metastatic tumor cell populations using lineages labeled with dominant selectable genetic markers. Int J Dev Biol 37(1):237–243. PubMed PMID: 8507566
6. Yokoi K, Kojic M, Milosevic M, Tanei T, Ferrari M, Ziemys A (2014) Capillary-wall collagen as a biophysical marker of nanotherapeutic permeability into the tumor microenvironment. Cancer Res 74(16):4239–4246. PubMed PMID: 24853545. Pubmed Central PMCID: 4134692
7. Matsumura Y, Maeda H (1986) A new concept for macromolecular therapeutics in cancer chemotherapy: mechanism of tumoritropic accumulation of proteins and the antitumor agent smancs. Cancer Res 46(12 Pt 1):6387–6392. PubMed PMID: 2946403
8. Greish K (2010) Enhanced permeability and retention (EPR) effect for anticancer nanomedicine drug targeting. Methods Mol Biol 624:25–37. PubMed PMID: 20217587
9. Yasunaga M, Manabe S, Tarin D, Matsumura Y (2011) Cancer-stroma targeting therapy by cytotoxic immunoconjugate bound to the collagen 4 network in the tumor tissue. Bioconjug Chem 22(9):1776–1783. PubMed PMID: 21749132
10. Yasunaga M, Manabe S, Tarin D, Matsumura Y (2013) Tailored immunoconjugate therapy depending on a quantity of tumor stroma. Cancer Sci 104(2):231–237. PubMed PMID: 23121194
11. Yamagiwa K, Ichikawa K (1918) Experimental study of the pathogenesis of carcinoma. J Cancer Res 3:1–29
12. Orr JW (1938) The changes antecedent to tumor formation during the treatment of mouse skin with carcinogenic hydrocarbons. J Pathol Bacteriol 46:495–515
13. Billingham RE, Orr JW, Woodhouse DL (1951) Transplantation of skin components during chemical carcinogenesis with 20-methylcholanthrene. Br J Cancer 5:417–432
14. Marchant J, Orr JW (1953) Further attempts to analyse the role of epidermis and deeper tissues in experimental chemical carcinogenesis by transplantation and other method. Br J Cancer 7(3):329–341. PubMed PMID: 13106199. Pubmed Central PMCID: 2007903. Epub 1953/09/01. eng
15. Orr JW, Spencer AT (1972) Transplantation studies of the role of the stroma in epidermal carcinogenesis. In: Tarin D (ed) Tissue interactions in carcinogenesis. Academic, London, pp 291–303
16. Tarin D (1967) Sequential electron microscopical study of experimental mouse skin carcinogenesis. Int J Cancer 2(3):195–211. PubMed PMID: 6041987. Epub 1967/05/15. eng
17. Tarin D (1969) Fine structure of murine mammary tumors: the relationship between epithelium and connective tissue in neoplasms induced by various agents. Br J Cancer 23(2):417–425. PubMed PMID: 5788050. Pubmed Central PMCID: 2008267. Epub 1969/06/01. eng
18. Asbestos (1987) Available from: http://monographs.iarc.fr/ENG/Monographs/suppl7/Suppl7-20.pdf
19. Tarin D (1968) Further electron microscopic studies on the mechanism of carcinogenesis: the specificity of the changes in carcinogen-treated mouse skin. Int J Cancer 3(6):734–742. PubMed PMID: 5731980. Epub 1968/11/15. eng

20. Tarin D (1972) Morphological studies on the mechanism of carcinogenesis. In: Tarin D (ed) Tissue interactions in carcinogenesis. Academic, London, pp 227–289
21. Frithiof L (1972) Ultrastructural changes at the epithelial stromal junction in human oral pre-invasive and invasive carcinoma. In: Tarin D (ed) Tissue interactions in carcinogenesis. Academic, London, pp 161–189
22. Pinkus H (1972) Histological evidence of disturbed tissue interactions in human pre-neoplastic and neoplastic skin. In: Tarin D (ed) Tissue interactions in carcinogenesis. Academic, London, pp 95–126
23. Smith CJ (1972) The epithelial-connective tissue junction in the pathogenesis of human and experimental oral cancer. In: Tarin D (ed) Tissue interactions in carcinogenesis. Academic, London, pp 191–225
24. Sugar J (1972) Ultrastructural and histochemical changes during the development of cancer in various human organs. In: Tarin D (ed) Tissue interactions in carcinogenesis. Academic, London, pp 127–160
25. Matsumura Y, Sugiyama M, Matsumura S, Hayle AJ, Robinson P, Smith JC et al (1995) Unusual retention of introns in CD44 gene transcripts in bladder cancer provides new diagnostic and clinical oncological opportunities. J Pathol 177(1):11–20. PubMed PMID: 7472774. Epub 1995/09/01. eng
26. Tarin D (1976) Cellular interactions in neoplasia. In: Weiss L (ed) Fundamental aspects of metastasis. North Holland Publishing Co, Amsterdam, pp 151–187
27. Tarin D (2012) Clinical and biological implications of the tumor microenvironment. Cancer Microenviron 5: 95–112
28. Dawe C (1972) Epithelial-mesenchymal interactions in relation to the genesis of polyoma virus-induced tumors of mouse salivary gland. In: Tarin D (ed) Tissue interactions in carcinogenesis. Academic, London, pp 305–358
29. Tarin D (2012) Inappropriate gene expression in human cancer and its far-reaching biological and clinical significance. Cancer Metastasis Rev 31(1–2):21–39. PubMed PMID: 22138779
30. Pelosof LC, Gerber DE (2010) Paraneoplastic syndromes: an approach to diagnosis and treatment. Mayo Clin Proc 85(9):838–854. PubMed PMID: 20810794. Pubmed Central PMCID: 2931619. Epub 2010/09/03. eng
31. Darnell R, Posner J (2011) Paraneoplastic syndromes. Oxford University Press, Oxford, 496 p
32. Esserman L, Shieh Y, Thompson I (2009) Rethinking screening for breast cancer and prostate cancer. JAMA 302(15):1685–1692. PubMed PMID: 19843904. Epub 2009/10/22. eng
33. Suzuki M, Mose ES, Montel V, Tarin D (2006) Dormant cancer cells retrieved from metastasis-free organs regain tumorigenic and metastatic potency. Am J Pathol 169(2):673–681. PubMed PMID: 16877365. Pubmed Central PMCID: 1698784. Epub 2006/08/01. eng
34. Tarin D, Price JE, Kettlewell MG, Souter RG, Vass AC, Crossley B (1984) Mechanisms of human tumor metastasis studied in patients with peritoneovenous shunts. Cancer Res 44(8):3584–3592. PubMed PMID: 6744281
35. Paget S (1889) The distribution of secondary growths in cancer of the breast. Lancet 1:571–573
36. Hart IR, Fidler IJ (1980) Role of organ selectivity in the determination of metastatic patterns of B16 melanoma. Cancer Res 40(7):2281–2287. PubMed PMID: 7388794. Epub 1980/07/01. eng
37. Goodison S, Kawai K, Hihara J, Jiang P, Yang M, Urquidi V et al (2003) Prolonged dormancy and site-specific growth potential of cancer cells spontaneously disseminated from nonmetastatic breast tumors as revealed by labeling with green fluorescent protein. Clin Cancer Res 9:3808–3814. PubMed PMID: 14506175
38. Bresalier RS, Raper SE, Hujanen ES, Kim YS (1987) A new animal model for human colon cancer metastasis. Int J Cancer 39(5):625–630. PubMed PMID: 3032811
39. Morikawa K, Walker SM, Nakajima M, Pathak S, Jessup JM, Fidler IJ (1988) Influence of organ environment on the growth, selection, and metastasis of human colon carcinoma cells in nude mice. Cancer Res 48(23):6863–6871. PubMed PMID: 2846163

40. Naito S, von Eschenbach AC, Giavazzi R, Fidler IJ (1986) Growth and metastasis of tumor cells isolated from a human renal cell carcinoma implanted into different organs of nude mice. Cancer Res 46(8):4109–4115. PubMed PMID: 3731078
41. Stephenson RA, Dinney CP, Gohji K, Ordonez NG, Killion JJ, Fidler IJ (1992) Metastatic model for human prostate cancer using orthotopic implantation in nude mice. J Natl Cancer Inst 84(12):951–957. PubMed PMID: 1378502
42. Montel V, Mose ES, Tarin D (2006) Tumor-stromal interactions reciprocally modulate gene expression patterns during carcinogenesis and metastasis. Int J Cancer 119(2):251–263. PubMed PMID: 16482564. Epub 2006/02/17. eng
43. Bao L, Pigott R, Matsumura Y, Baban D, Tarin D (1993) Correlation of VLA-4 integrin expression with metastatic potential in various human tumor cell lines. Differentiation 52(3):239–246. PubMed PMID: 7683291. Epub 1993/03/01. eng
44. Tarin D (2011) Cell and tissue interactions in carcinogenesis and metastasis and their clinical significance. Semin Cancer Biol 21(2):72–82. PubMed PMID: 21147229. Epub 2010/12/15. eng
45. Sanders ME, Schuyler PA, Dupont WD, Page DL (2005) The natural history of low-grade ductal carcinoma in situ of the breast in women treated by biopsy only revealed over 30 years of long-term follow-up. Cancer 103(12):2481–2484. PubMed PMID: 15884091. Epub 2005/05/11. eng
46. Nickerson HJ, Matthay KK, Seeger RC, Brodeur GM, Shimada H, Perez C et al (2000) Favorable biology and outcome of stage IV-S neuroblastoma with supportive care or minimal therapy: a Children's Cancer Group study. J Clin Oncol 18(3):477–486. PubMed PMID: 10653863. Epub 2000/02/02. eng
47. Price JE, Carr D, Jones LD, Messer P, Tarin D (1982) Experimental analysis of factors affecting metastatic spread using naturally occurring tumors. Invasion Metastasis 2(2):77–112. PubMed PMID: 7188393. Epub 1982/01/01. eng
48. Sumida T, Kitadai Y, Shinagawa K, Tanaka M, Kodama M, Ohnishi M et al (2011) Anti-stromal therapy with imatinib inhibits growth and metastasis of gastric carcinoma in an orthotopic nude mouse model. Int J Cancer 128(9):2050–2062. PubMed PMID: 21387285. Epub 2011/03/10. eng
49. Patocs A, Zhang L, Xu Y, Weber F, Caldes T, Mutter GL et al (2007) Breast-cancer stromal cells with TP53 mutations and nodal metastases. N Engl J Med 357(25):2543–2551. PubMed PMID: 18094375. Epub 2007/12/21. eng
50. Akino T, Hida K, Hida Y, Tsuchiya K, Freedman D, Muraki C et al (2009) Cytogenetic abnormalities of tumor-associated endothelial cells in human malignant tumors. Am J Pathol 175(6):2657–2667. PubMed PMID: 19875502. Pubmed Central PMCID: 2789618. Epub 2009/10/31. eng
51. Bruna F, Plaza A, Arango M, Espinoza I, Conget P (2018) Systemically administered allogeneic mesenchymal stem cells do not aggravate the progression of precancerous lesions: a new biosafety insight. Stem Cell Res Ther 9(1):137. PubMed PMID: 29751770. Pubmed Central PMCID: 5948822
52. von Bahr L, Batsis I, Moll G, Hagg M, Szakos A, Sundberg B et al (2012) Analysis of tissues following mesenchymal stromal cell therapy in humans indicates limited long-term engraftment and no ectopic tissue formation. Stem Cells 30(7):1575–1578. PubMed PMID: 22553154
53. Le Blanc K, Davies LC (2018) MSCs-cells with many sides. Cytotherapy 20(3):273–278. PubMed PMID: 29434007
54. Prockop DJ, Brenner M, Fibbe WE, Horwitz E, Le Blanc K, Phinney DG et al (2010) Defining the risks of mesenchymal stromal cell therapy. Cytotherapy 12(5):576–578. PubMed PMID: 20735162

Cancer and Blood Coagulation

Yasuhiro Matsumura

Abstract We know that both intrinsic and extrinsic blood coagulation factors are involved in tumor vascular permeability as well as tumor-induced fibrin clot formation in the extra-vascular space. Inflow of fibroblasts and inflammatory cells into the tumor tissue occurs accompanying blood coagulation. Consequently, cancer induced blood coagulation generates insoluble fibrin rich tumor stroma. The stroma becomes a barrier preventing macromolecular DDS drugs from directly attacking cancer cells within the cancer tissue. To tackle this problem, we successfully developed a new strategy that uses a monoclonal antibody (mAb), generated in this laboratory that reacts only with human insoluble fibrin, and not with human fibrinogen or fibrin degradation product (FDP). Another advantage of this mAb is that it cross-reacts with mouse insoluble fibrin, but not with mouse fibrinogen or FDP. As a result of the unique properties of the mAb, it is not neutralized by soluble fibrinogen or soluble fibrin products in the body. Our anti-fibrin clot mAb recognizes an unexplored pocket that is only uncovered when a fibrin clot forms. The epitope in the pocket is a hydrophobic region on the Bβ-chain that interacts closely with a counterpart region on the γ-chain in a soluble state. Using the new mAb, we succeeded in constructing an antibody drug conjugate (ADC) that binds to polymerized fibrin and slowly releases a cytotoxic drug that, on account of its small molecular size, can diffuse through the dense stroma to kill cancer cells. Cancer and blood clotting research may lead to new therapeutic strategies as well as to the biological understanding of cancer.

Keywords Malignant cycle of blood coagulation · Cancer stroma · Insoluble fibrin · EPR effect · CAST therapy · Tumor vascular permeability · Intrinsic blood coagulation · Extrinsic blood coagulation

---

Y. Matsumura (✉)
Division of Developmental Therapeutics, Exploratory Oncology Research & Clinical Trial Center, National Cancer Center, Kashiwa, Japan
e-mail: yhmatsum@east.ncc.go.jp

Y. Matsumura, D. Tarin (eds.), *Cancer Drug Delivery Systems Based on the Tumor Microenvironment*, https://doi.org/10.1007/978-4-431-56880-3_2

Introduction

In the nineteenth century, the French physician Armand Trousseau described thrombophlebitis in patients with stomach cancer for the first time [1]. Today, a large body of clinical evidence indicates that abnormal coagulation occurs in patients with a variety of cancers, especially those of the pancreas, stomach and glioblastoma multiforme (GBM) [2]. Several prospective randomized clinical trials have demonstrated that low molecular weight heparins can be useful in the primary prevention of venous thromboembolic disease in cancer patients [3]. It should be noted that blood coagulation can also occur within the tissues in several pathological conditions including cancer. In studying this, we have established that both intrinsic and extrinsic coagulation factors may be involved in tumor vascular permeability as well as tumor-induced blood coagulation (Fig. 1) [4]. This knowledge can be used to advantage in cancer therapy, as described below.

Drug delivery system (DDS) formulations, including monoclonal antibody (mAb)-based drugs, are too large to pass through the normal vessel wall but can easily be extravasated from leaky tumor vessels and retained in solid tumor tissues for long periods of time because of the enhanced permeability and retention (EPR) effect [5]. Moreover, mAbs can target tumor cells actively. Therefore, to date, numerous mAbs recognizing molecules on tumor cell surfaces have been developed and conjugated with anticancer agents (ACAs), radioisotopes, or tox-

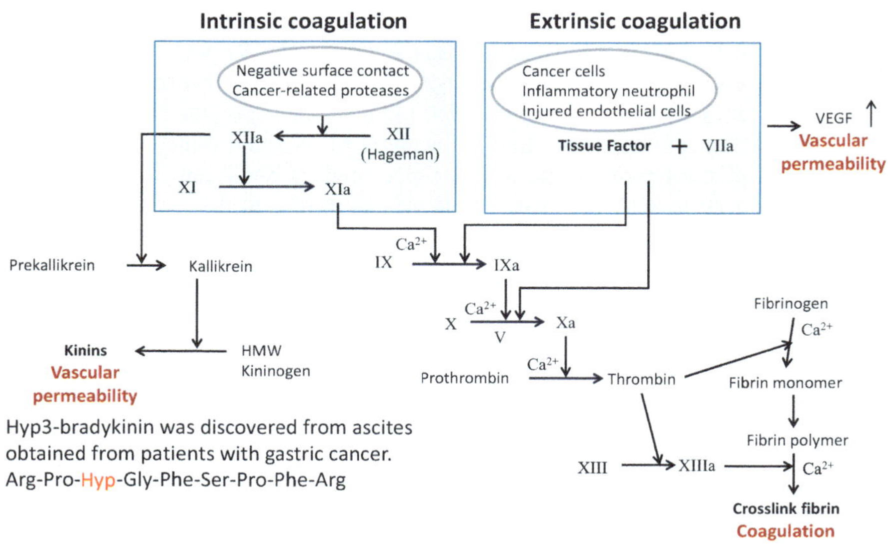

Fig. 1 Blood Coagulation Cascade in Cancer [4]. Both intrinsic and extrinsic coagulation factors are involved in cancer induced fibrin clot formation as well as tumor vascular permeability. The tissue factor is known to be the trigger protein of the extrinsic blood coagulation

ins. However, the use of high molecular weight agents presents a dilemma for cancer therapy, since the very properties that favor their high accumulation in a lesion also results in the low diffusion of these macromolecules within a tumor. Indeed, most human solid tumors possess abundant stroma that forms a barrier preventing macromolecular drugs from directly attacking cancer cells [4]. Tumor stroma is well known to comprise not only extra cellular matrix but also cellular components such as fibroblasts and white blood cells. Accordingly, crosstalk between the extra cellular matrix bound factors and the cells must be considered in order to understand tumor stromal biology. The leakiness and fragility of tumor blood vessels also allows fibrinogen to exude into the tissues where it can polymerize to form a fibrin coagulum, which is eventually replaced by collagen fibers. We have exploited this special aspect of tumor stromal biology to create a method by which both (a) binding of leaked macromolecular drug complexes to hold them in the tumor and (b) sustained release of small molecular cytotoxic agents from these complexes to allow them to diffuse through the dense stroma can be achieved simultaneously. This chapter, describes how collagenous cancer stroma can be initiated by cancer-induced blood coagulation and how these pathophysiological features of cancer, enable the formulation of a new onco-therapeutic modality.

Cancer Induced Blood Coagulation

Blood coagulation is a dynamic pathophysiological process and is classified into extrinsic and intrinsic pathways. Both extrinsic and intrinsic coagulation cascades are involved in tumor induced blood coagulation. In both systems, coagulation factor X is activated in a lesion and the activated factor X (Xa) converts prothrombin to thrombin. Thrombin cleaves the N-terminal region of fibrinogen to produce fibrinopeptides A and B allowing the resultant fibrin monomers to polymerize, and finally fibrin clots are formed.

Intrinsic Blood Coagulation

This only requires factors intrinsic to the blood. Coagulation factors such as factor XII that is also known as Hageman factor (HF), factor XI, prekallikrein (PK), and high molecular weight kininogen (HMK), are involved in the intrinsic blood coagulation system. This system initiates when blood contact with some negative surface in the body and it is, therefore, known as the "surface-mediated pathway or "contact system". Molecular events of the intrinsic coagulation system affect many important biochemical and cellular systems and play an important role in various pathophysiologic conditions including cancer [6, 7].

Extrinsic Blood Coagulation

This requires the participation of factors not found in normal blood. Tissue factor (TF), a 47-KDa,diffusely distributed, transmembrane glycoprotein that serves as an initiation factor for the extrinsic coagulation cascade, activates factor VIIa. The TF-VIIa complex activates factor X, and subsequently this protease cascade forms fibrin clots [8, 9]. TF is known to be highly expressed in various human cancers [10]. Cancer patients often suffer from a state of hypercoagulability and venous thrombosis, leading to patient morbidity and mortality [11, 12]. A study using a fibrinogen-deficient transgenic mouse model also indicated that fibrinogen appeared to be an important element affecting expression of the metastatic potential of circulating tumor cells [13]. Other studies have indicated that, TF plays an important role not only in blood coagulation but also in cell signaling in which the TF-VIIa complex phosphorylates extracellular-regulated kinase 1/2 (ERK1/2) via protease-activated receptor-2 (PAR-2) [14]. This complex promotes the expression of interleukin-8 (IL-8) and invasion in breast cancer cell lines [15].

Vascular Permeability Factors in Solid Tumors

Increased tumor vascular permeability is the most important event for the enhanced permeability and retention (EPR) effect seen in tumors [5–7]. The EPR effect can be augmented by at least two different vascular permeability factors. These factors affect normal as well as tumor vessels. Both factors are known to be produced in the tumor compartment. One of these factors is bradykinin and its derivatives, which we found as a side product in the intrinsic blood coagulation system in tumor tissue [5–7]. The other, which was reported by Dvorak et al., is called vascular permeability factor (VPF) [16]. Later, it was found that VPF was identical to vascular endothelial growth factor (VEGF) [17, 18] and also is now know to be produced when the extrinsic blood coagulation occurs within tumor tissue [19].

Kinin Production System

During investigation of the EPR effect, a bradykinin-producing cascade in solid tumor tissue was clarified (Fig. 1) [5–7]. Kinins are side-products of the intrinsic blood coagulation system in which HF is activated on a negatively charged surface of a pathological lesion including cancer. In solid cancer tissues or cancer ascites, any negatively charged substances derived from cancer cells can be an initiator of the intrinsic blood coagulation pathway. Kinins are generated from high molecular weight kininogen (KNG) by limited proteolysis with a serine protease (kallikrein), which is present in plasma as precursor, prekallikrein (PK).

Among kinins, bradykinin is the most potent permeability factor. The HF-PK-Kinin system can be activated easily by experimental artifacts, so that stringent precautions to prevent this activation were taken in our experiments. In that context, we adopted cancer ascites as a biological material for the experiment of kinin generation in tumors because it was impossible to examine the kinin generation cascade in solid tumors without causing artificial activation of the kinin generation system [5]. We showed that there was a highly elevated free bradykinin content in rodent and human ascitic fluid of gastric and ovarian cancers. It was also found in the pleural effusion of lung cancer patients [5]. The role of the bradykinin-generating system in the pathogenesis of cancer was explored by simultaneously measuring plasma PK, the precursor of kallikrein, which is the major enzyme responsible for bradykinin generation, and plasma KNG, which is precursor of bradykinin, in patients with various cancers [7]. It was found that plasma PK and KNG values were significantly lower in cancer patients compared with healthy volunteers. These data indicated that conversion of PK to kallikrein occurred with concomitant consumption of KNG by newly generated kallikrein for bradykinin generation. On the other hand, early stage cancer patients showed no significant difference from healthy volunteers [7]. During the investigation of the bradykinin generation system, we found bradykinin's derivative (^{3}hydroxyprolyl)-bradykinin (Hyp^3-bradykinin), in which the third amino acid, proline, in bradykinin is replaced by hydroxyproline [7, 20]. Ratios of bradykinin and Hyp^3-bradykinin were also examined in the kininogen of normal and cancer patients' sera. The results showed, however, that the content of Hyp^3-bradykinin in plasma KNG did not differ from healthy controls to a significant degree. This indicated that proline hydroxylation in bradykinin occurred in kininogen as a posttranslational modification [20, 21].

Vascular Permeability Factor (VPF)

Tissue factor (TF), the trigger protein for the extrinsic coagulation cascade, is known to induce production of VEGF in human tumor cells independent of its ability to activate the blood coagulation cascade [19]. VEGF was discovered by Dvorak et al. in the late 1970s as a factor inducing vascular permeability and it was called vascular permeability factor (VPF) at the time of the discovery [16, 22]. VPF (VEGF) has a molecular weight of approximately 40,000 and is produced by many types of tumor cells [23–25]. VPF (VEGF) is expressed constitutively at low levels in many normal tissues and is highly expressed by most human and animal cancers. VPF (VEGF) is essential for the development for the vascular system and for physiologic and pathologic angiogenesis including tumor angiogenesis [26]. Moreover, VPF (VEGF) upregulates the expression of various molecules including matrix metalloprotease, glucose transporter, nitric oxide synthase, mitogens, and anti-apoptotic factors.

The Malignant Cycle of Blood Coagulation

Apart from blood coagulation triggered by a molecule such as TF, cancer is a disease that can destroy the surrounding normal tissue including vasculature. Any malignant tumor can erode the surrounding normal tissue, and the more erosive types of cancer have more destructive actions. If these cancer clusters erode adjacent normal or tumor vessels, microscopic hemorrhage may occur at any place and at any time within or adjacent to cancer tissues, and fibrin clots immediately form in situ to stop the bleeding. The fibrin clots are subsequently replaced by collagenous tissue in a process similar to that in normal wound healing and other non-malignant diseases. Although there are many similarities between wound healing and the mechanisms underlying cancer-induced stroma, a fundamental difference between the two is that the pathophysiological generation of collagenous fibrous stroma in cancer can continue for as long as cancer cells survive in the body. Therefore, we call this the 'malignant cycle of blood coagulation' consisting of repetitive sequential iterations of cancerous invasion into vessels, hemorrhage, insoluble fibrin formation, and replacement with collagenous tissue (Fig. 2) [4]. Fibrin clot

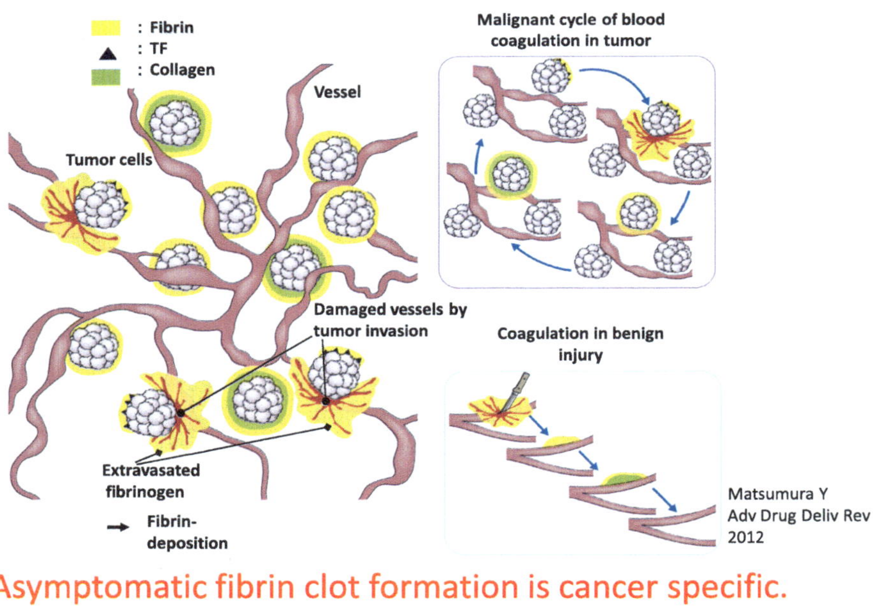

Fig. 2 Diagram of the 'malignant cycle of blood coagulation' in cancer tissue [4]. If cancer tissue erodes adjacent normal or tumor vessels, hemorrhage may occur, and fibrin clots should immediately form in situ to stop the bleeding. These clots are subsequently replaced by collagen, as occurs in the normal wound healing process. The pathophysiological condition in cancer lasts for as long as cancer cells survive in the body. The "malignant cycle of blood coagulation" generates versatile cancer stroma consisting of cancer invasion into vessels, hemorrhage, insoluble fibrin (IF) formation, and replacement with collagenous tissue

formation in non-malignant disorders such as cardiac infarction, brain infarction, injuries, and active rheumatoid arthritis usually occurs only at the onset or in the active state of the disease, is accompanied by some clinical symptoms. and subsequently disappears by plasmin digestion or replacement with collagen within a few weeks. On the other hand, fibrin clot formation in cancer occurs continuously, episodically and insidiously. In fact, tumor invasion and metastasis often progress without immediate symptoms (which is why imaging instruments are needed). When any symptoms accompanying cancer such as pain, intestinal obstruction, or macroscopic bleeding occur, the cancer is likely to involve sensory nerves and destruction of bones and larger blood vessels, and to occupy the whole lumen of a particular region of the intestine. Usually, patients with an advanced stage of cancer receive chemotherapy and it is worth noting that oncologists never treat such patients with such agents if they suffer from existing acute thrombotic complications, bleeding by injury, or active inflammation, because of the heightened risk of exacerbating hemorrhagic or thrombotic events. Therefore, we conclude that growth factors and tyrosine kinases never become tumor-specific molecules but that fibrin clots in cancer *tissues* of patients who may receive chemotherapy are actually tumor-specific [4], there being no clots in other tissues unless the patient concurrently has other diseases.

Anti-Insoluble Fibrin Monoclonal Antibody

Once blood coagulation occurs in the tumor tissue, insoluble fibrin (IF) is produced from fibrinogen in situ and is then immediately degraded through the fibrinolysis mechanism by plasmin produced only on the IF in the tumor tissue. The fibrin degradation products (FDP) are then dissolved in the circulation. Therefore IF exists only in pathological conditions, including cancer. Furthermore, it must be kept in mind that fibrin deposition in non-malignant diseases is characteristically accompanied by symptoms related to the particular illness. On the other hand, tumor-related fibrin deposition is not directly associated with any symptoms. Therefore, the development of a detection method involving fibrin clots is valuable from an oncological perspective (Fig. 3) [4] because it can assist the detection of occult malignant neoplasms.

Production of the Antibody

To develop anti-human fibrin mAb, we crushed a human fibrin clot into pieces and injected the suspension in saline into mice. The hybridoma producing anti-fibrinogen (mouse IgG clone K88-3) or anti-fibrin antibody (mouse IgM clone 102-10) was established using myeloma cells (P3U1) and lymph-node cells from the mouse with immunizing human fibrinogen or fibrin, the latter converted from

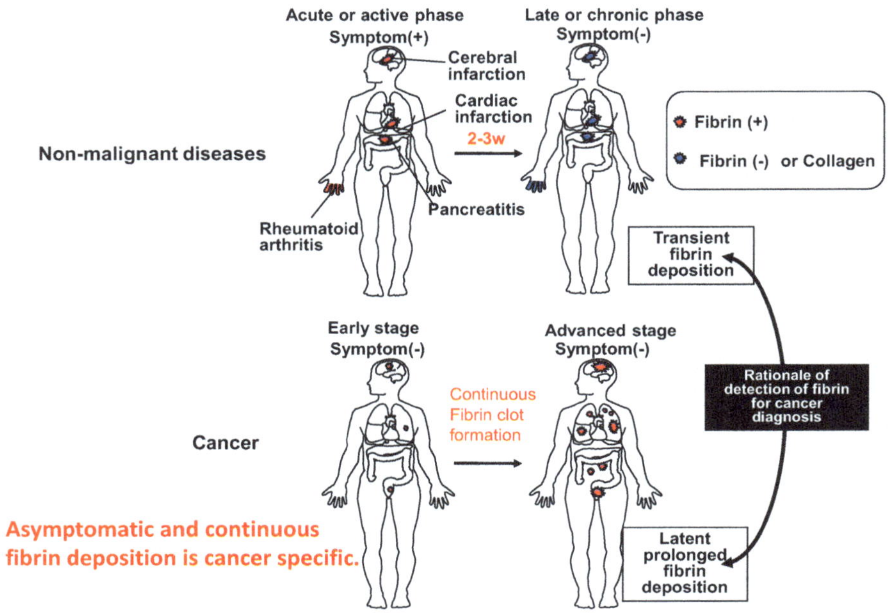

Fig. 3 Diagram of fibrin deposition in non-malignant and malignant diseases [4]. Fibrin deposition in non-malignant diseases such as infarction, arthritis, and trauma occurs only at their onset or during the active phase, and that this deposition subsequently disappears within a few weeks as a result of plasmin digestion and collagen replacement. Fibrin deposition in non-malignant diseases is inevitably accompanied by symptoms related to the particular condition. On the other hand, tumor-related fibrin deposition is not associated with any symptoms.

fibrinogen by thrombin cleavage. The heavy-chain variable and the kappa light-chain variable -region cDNAs were cloned into the vector for the human IgG1 expression. The vectors were transfected into the CHO cells. A stable clone (humanized IgG, clone 102-10) was isolated [29]. Consequently, we succeeded in obtaining a mAb (clone 102-10) that distinguished fibrin clots from fibrinogen, soluble fibrin (precursor of fibrin clot) [27], and D-dimer (degradation product of fibrin clot) [28], which are all soluble proteins (Figs. 4a and 5a).

Characterization of the Antibody

Although some anti-fibrin mAbs have been developed, none of them react exclusively with fibrin clots, but rather also react with fibrinogen, soluble fibrin, or D-dimer [30–36].

The specificity of the 102-10 mAb differed from existing anti-fibrin mAbs, NYB-T2G1 [37, 38] and MH-1 [35], as 102-10 mAb reacted only with fibrin clots (Fig. 5b). The most important point here is that the mAb is not neutralized by fibrin-

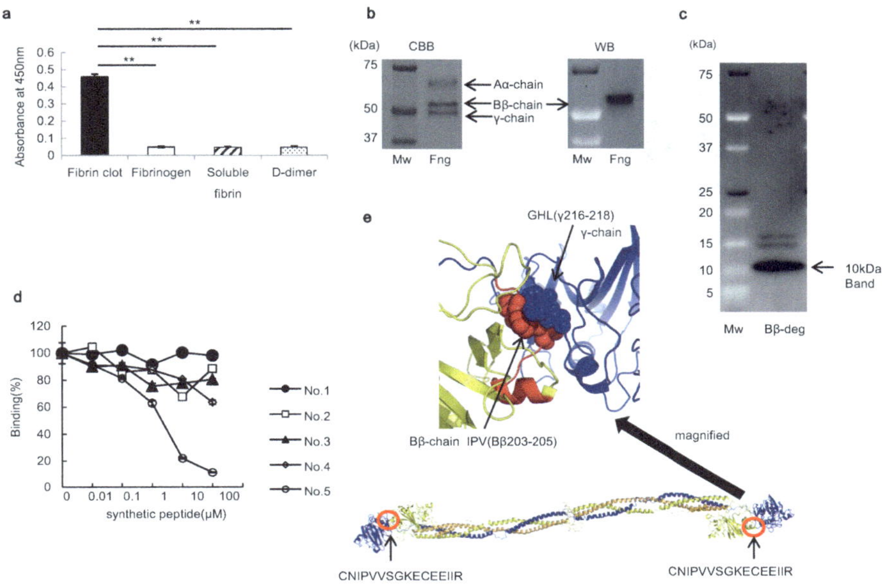

Fig. 4 Characterization of the anti-insoluble fibrin mAb 102-10 [39]. (**a**) 102-10 mAb was reactive to fibrin clots only (n = 6). The results are presented as the means ± s.d., **P < 0.01. (**b**) 102-10 mAb was reactive to the denatured Bβ-chain of fibrinogen (Fng). CBB (Coomassie Brilliant Blue) WB (western blotting) (**c**) Peptide of 10 kDa detected by 102-10 mAb following degradation of the Bβ-chain (Bβ-deg). (**d**) Only the No. 5 peptide inhibited the binding of 102-10 mAb to fibrin clots (n = 3; P < 0.01, No. 4 vs. No. 5). (**e**) The epitope of 102-10 (Bβ201–216; red) interacted with the γ-chain (blue), and Bβ203–205 (red ball) interacted with the region around γ216–218 (blue ball) in a hydrophobic manner. The whole structure of fibrinogen is shown (lower panel)

ogen and FDP in the blood despite the fact that fibrinogen, IF and FDP are transcribed from same gene and share the same amino acid sequences. Therefore, anti-IF mAb can easily reach the tumor tissue and bind to the IF in the tumor stroma [39]. In conducting epitope mapping for the 102-10 mAb, we found that it was reactive to fibrinogen under reducing and heat-denatured conditions in ELISAs and Western blots (WB) (Fig. 5c–e). This result indicates that the epitope was exposed under reducing conditions. Upon comparing Coomassie Brilliant Blue (CBB) staining with WB, only the Bβ-chain of fibrinogen reacted with 102-10 mAb (Fig. 4b). To confirm the location of the epitope of 102-10, the fibrinogen Bβ-chain was digested by lysyl endopeptidase, and a Bβ-chain-derived peptide of approximately 10 kDa was obtained, which was recognized by 102-10 mAb (Fig. 4c). This obtained peptide consisted of residues 149–234 of the Bβ-chain. To confirm the epitope's sequence, we prepared five synthetic peptides. Competitive inhibition tests revealed that only one of these peptides inhibited the binding of 102-10 mAb to fibrin clots, as measured by ELISA (Fig. 4d). The sequence of this synthetic peptide was CNIPVVSGKECEEIIR, which constitutes a hydrophobic region of residues 201–216 of the Bβ-chain. In view of the structure of fibrinogen, we can deduce that the

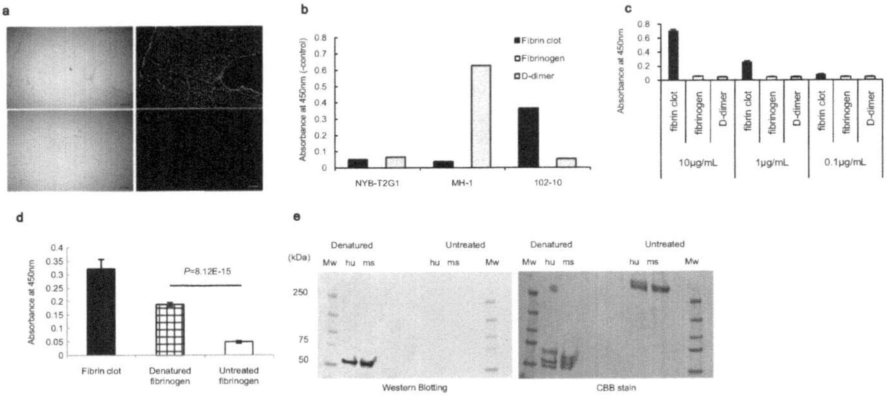

Fig. 5 Detailed studies of characterization of 102-10 [39]. (**a**) Immunohistochemistry of a fibrin clot with Alexa Fluor 647-conjugated 102-10 (upper) and control IgG (Cetuximab, lower). Only 102-10 mAb reacted with the fibrin clot. Scale bar, 100 μm. (**b**) Comparison of 102-10 mAb with previously studied anti-fibrin antibodies. (**c**) To clarify the specificity of the 102-10 mAB more clearly, we prepared fibrinogen plate, D-dimer plate, and fibrin clot plate. 0.1, 1.0, and 10.0 μg/mL peroxidase-conjugated mAb was incubated in each plates (n = 8). (**d**) Reactivity of 102-10 against fibrinogen under reducing, heat-shocked (denatured) conditions and untreated conditions (n = 7). The results are presented as the means ± SD. (**e**) 102-10 mAb was reactive to human (hu) and mouse (ms) fibrinogen under denatured conditions but not to fibrinogen under untreated conditions

epitope of the Bβ-chain of 102-10 mAb interacted with residues 206–220 of the γ-chain of fibrinogen (Fig. 4e: Protein Data Bank (PDB) code 3GHG24). Interestingly, these two amino acid sequences are completely conserved in mammals, birds, amphibians, and fish (Basic Local Alignment Search Tool (BLAST)), which suggests that these sites have major importance for blood coagulation in various species [39].

It is generally difficult to identify the epitope on a fibrin clot because X-ray analysis and NMR cannot be applied for insoluble proteins. However, using the 102-10 mAb against the fibrin clot, we identified the precise peptide sequence of the epitope and discovered a unique pit in which certain sequences on the Bβ-chain (containing the epitope of 102-10 mAb) and γ-chains of fibrinogen were hidden [39] (Fig. 6). To confirm this finding, we prepared a peptide antigen corresponding to the epitope region of 102-10 mAb and a peptide antigen corresponding to the counterpart region on the γ-chain. We then immunized mice with these two synthetic peptides, and we succeeded in establishing mAbs against the region on the Bβ-chain (anti-Bβ mAb) and the region on the γ-chain (anti-γ mAb). Both the anti-Bβ mAb and the anti-γ mAb recognized fibrin clots specifically [39] (Fig. 7). These findings imply that when fibrinogen develops into a fibrin clot, a conforma-

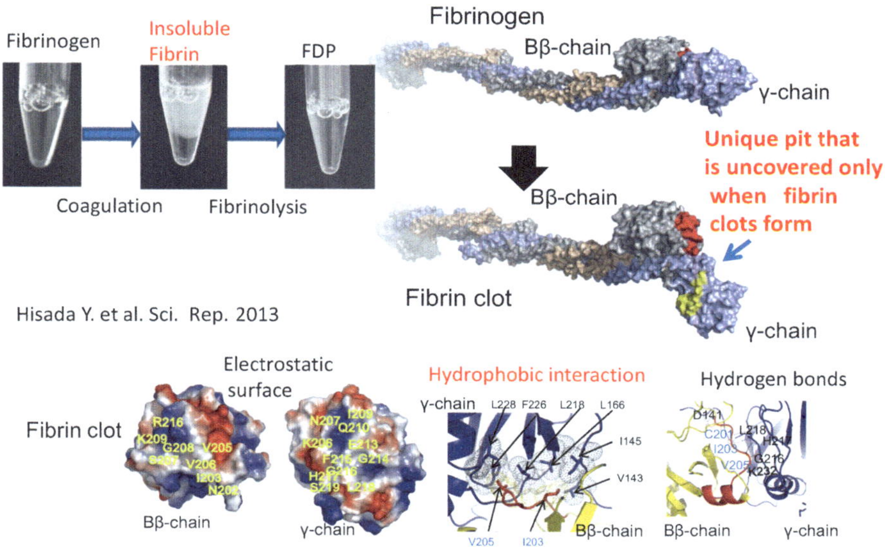

Fig. 6 Image of structural change from fibrinogen to fibrin clot and discovery of unique pits in fibrin clot [39]. Once blood coagulation occurs in the tumor tissue, insoluble fibrin (IF) is produced from fibrinogen in situ and is then immediately degraded through the fibrinolysis mechanism by plasmin produced only on the IF in the tumor tissue. The fibrin degradation products (FDP) are then dissolved in the circulation. IF exists only in pathological conditions, including cancer. 102-10 mAb reacts only with human IF, and not with human fibrinogen or FDP. 102-10mAb recognizes an epitope inside previously unidentified pits that are uncovered only when a fibrin clot forms. The epitope in the pits is a hydrophobic region on the Bβ-chain that interacts closely with a counterpart region on the γ-chain in a soluble state

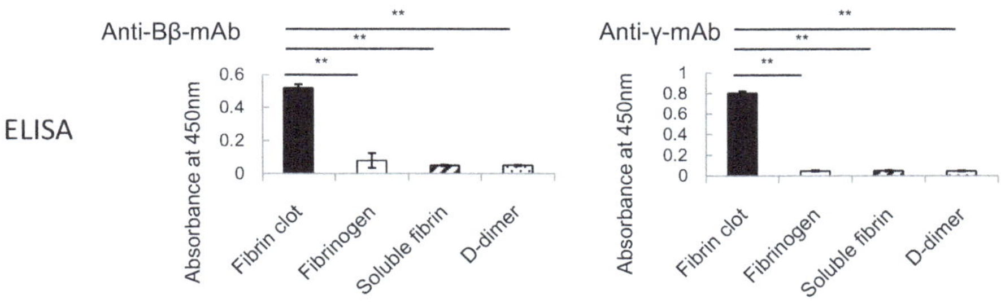

Fig. 7 Newly developed anti- Bβ mAb and anti-γ mAb [39]. Anti-Bβ mAb and anti-γ mAb were reactive only to fibrin clots (n = 8). The results are presented as the means ± s.d., **P < 0.01. The unique pit within the fibrin clot is proposed to be large enough for both mAbs to bind to the pit simultaneously

tional change occurs in the region that acts as an epitope for each mAb, which produces the unique pit (Fig. 6). To characterize this unique pit, competition ELISA was conducted using peroxidase-conjugated anti-Bβ mAb, peroxidase-conjugated anti-γ mAb, and each corresponding cold mAb. The binding of peroxidase-conjugated anti-Bβ mAb to the fibrin clot was blocked by the cold anti-Bβ mAb but not by the cold anti-γ mAb. Moreover, the binding of peroxidase-conjugated anti-γ mAb to the fibrin clot was blocked by the cold anti-γ mAb, but not by the cold anti-Bβ mAb. These findings indicate that the newly identified unique pit within fibrin clots was large enough for both anti-Bβ and anti-γ mAbs to bind simultaneously (Fig. 7).

The safety issue should also be considered for therapeutic use in clinical practice. We clarified that neither IF gel formation nor degradation was affected [40]. These previous data indicate that our anti-IF mAb may not cause thrombosis or bleeding as a side effect, although these should be clarified more precisely in terms of biochemical tests and pathological studies in future. As described earlier, the epitope sequence of the anti-IF mAb was highly conserved in mammals, and we confirmed that our anti-IF mAb cross-reacted with mouse and human IF clots [39]. Therefore, the results of this study may be reproduced in humans.

Immunohistochemistry for Various Diseases

To evaluate the reactivity of 102-10 mAb to intravital fibrin clots under various conditions in humans, normal tissue and tissue from non-malignant and malignant diseases were stained with 102-10 mAb. Strong fibrin deposition was observed in almost all cancer tissues, including brain, lung, pancreatic, and colorectal cancers, but not normal tissues (Fig. 8). In cases of non-malignant disease, fibrin clots were detected at the onset of cerebral infarction and cardiac infarction or in cases of serious inflammation such as acute pancreatitis. However, no clear fibrin clots were detected in the late or chronic phases of non-malignant conditions following the digestion of fibrin clots by plasmin and replacement by collagenous tissue (Fig. 8).

More invasive types of cancer are destructive of neighboring issues, resulting in higher fibrin clot deposition. In this context, we closely investigated fibrin deposition in surgically resected samples of glioma, one of the most erosive types of tumor. Immunohistochemistry was conducted for WHO-classified glioma tissue samples, and the level of fibrin deposition was graded into three subgroups: (−) fibrin clot not detectable; (+) faint but clear deposition; and (++) strong heterogeneous or diffuse deposition (Fig. 9). Fibrin deposition was observed in 100% (20/20) of grade 4 gliomas (glioblastoma multiforme), with strong fibrin deposition observed in 65% (13/20). In grade 3 gliomas, fibrin deposition was observed in 40% of cases (8/20), with strong fibrin deposition observed in 10% (2/20). In grade 1/2 gliomas, fibrin deposition was observed in 60% of cases (12/20), and strong fibrin deposition was observed in 10% (2/20) (Fig. 9).

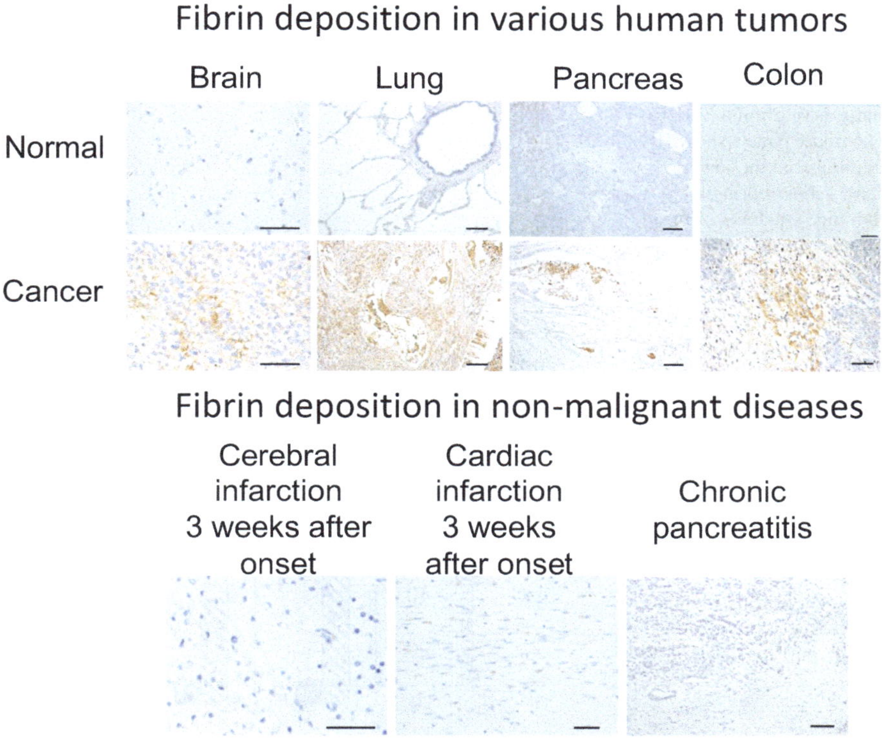

Fig. 8 Immunohistochemistry with 102-10 mAb in various human tissues [39]. Fibrin deposition was detected strongly in human cancer tissues, but not in normal human tissues or during the late or chronic phases of non-malignant diseases. Scale bar, 100 μm

These immunohistochemical studies strongly indicate that fibrin deposition in cancer tissue is positively correlated with the degree of malignancy of the cancer.

Cancer Diagnosis with the New Antibody

Because this unexplored pit is uncovered only when a fibrin clot forms and the amino acid sequence of the epitope of 102-10 mAb are completely conserved in mammals, the 102-10 mAb and other mAbs developed against human fibrin clots cross-reacted with mouse or rat fibrin clots. We also investigated the fate of fibrin clots in mice or rats using disease models such as cerebral infarction, incisional wounds, and arthritis (Fig. 10a). The results indicated that fibrin clot formation occurred only in the acute phase of non-malignant diseases, including thrombosis, inflammation, and trauma, and these clots virtually disappeared within 2 to 3 weeks

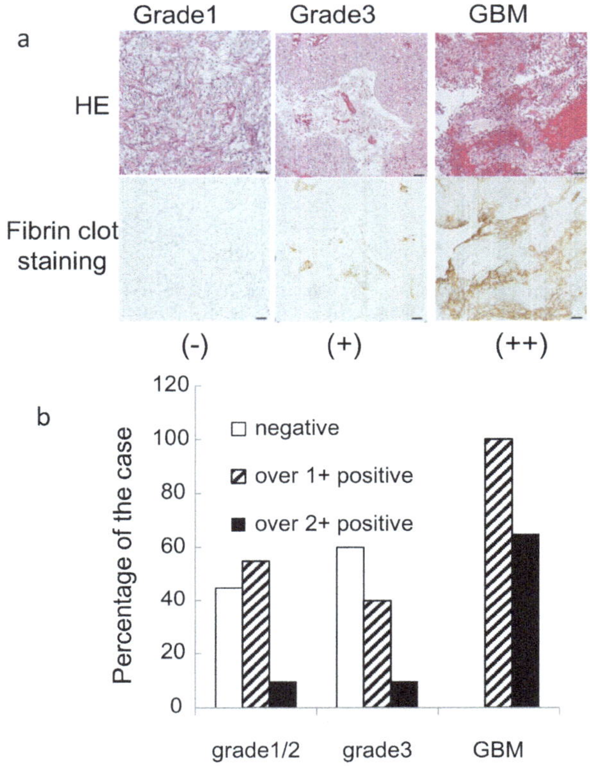

Fig. 9 Fibrin clot status according to WHO classified glioma grade. Immunohistochemical analysis of glioma [39]. (**a**) The upper panels show HE staining, and the lower panels show fibrin clot staining. The level of fibrin deposition was graded into three subgroups: (−) undetectable fibrin clot (upper panel, grade1); (+) faint but clear deposition (middle panel, grade3); and (++) strong heterogeneous or diffuse deposition (lower panel, GBM). (**b**) Percentage of fibrin deposition. The clinical stages of patients were classified as grade 1/2, grade 3, or grade 4 (GBM) according to the WHO grade classification. Each group consisted of 20 cases

and were substituted by collagen in the late phase. We next synthesized a PET probe of 89Zr-labelled 102-10 (Fig. 10b), which was injected into mice bearing chemically induced cutaneous tumors generated using 7,12-dimethylbenz[a]anthracene (DMBA) as an initiator and phorbol 12-myristate 13-acetate (PMA) as a promoter [41]. This autochthenous tumor type was selected as an appropriate experimental model for evaluating our immuno-probe because it exhibits remarkable fibrin deposition and abundant interstitial tissue, similar to clinical human cancer and unlike human tumor xenografts in mice [42]. By CT scan, the 102-10 mAb probe showed clear and more specific accumulation in tumors compared to control antibody (Fig. 10c). We next evaluated the specificity of the 102-10 mAb probe within the tumor tissue histologically. By hematoxylin and eosin (HE) staining, the tumor was shown to consist of areas with basophilic cancer cells and eosinophilic tumor stroma, and the accumulation of the 102-10 mAb probe was more pronounced in fibrin clot-positive tumor stroma (Fig.10d). Therefore, it is reasonable to assume that these PET/CT data from mouse experiments can be extrapolated to humans because 102-10 mAb can recognize both human and mouse fibrin clots and because humans and mice show common features of fibrin deposition.

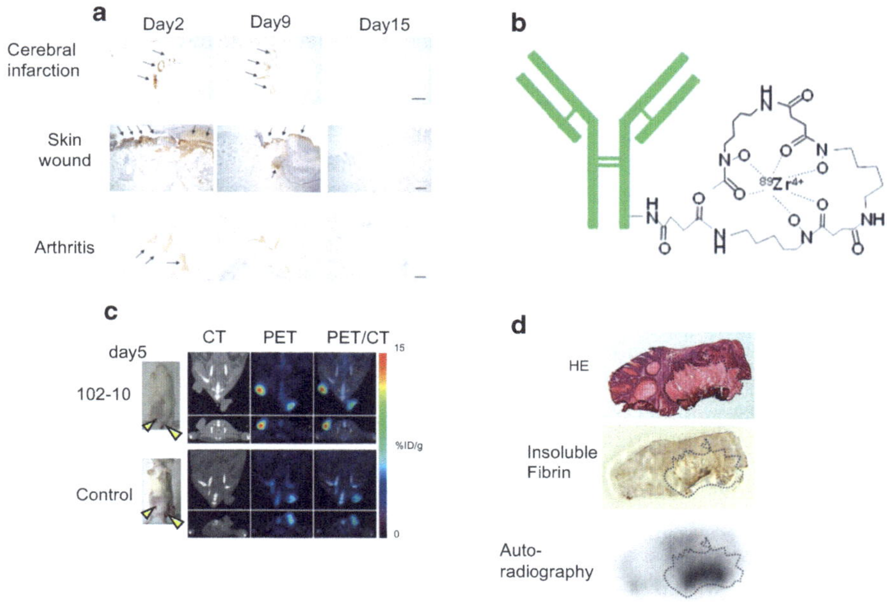

Fig. 10 The kinetics of fibrin clot deposition in several non-malignant disease models and PET/CT with 89Zr-labelled 102-10 in spontaneous tumor models [39]. (**a**) The kinetics of fibrin clot deposition in rat cerebral infarction (upper), mouse incisional wound (middle), and mouse arthritis (lower) were evaluated immunohistochemically. Fibrin clot deposition was observed at the acute phase (left and middle) but not at the late phase (right). Scale bar, 100 μm. (**b**) The PET probe was composed of the 102-10 mAb, a linker, and 89Zr. (**c**) The 89Zr-labeled 102-10 mAb probe showed clearer and more specific accumulation in tumors as compared to the control (cetuximab). (**d**) Accumulation of the 102-10 mAb probe

Cancer Stromal Targeting (CAST) Therapy

Low molecular weight (MW) anticancer drugs extravasate easily from normal vessels in the body causing drug adverse effects, but are cleared from the body rapidly. Conversely, high MW anti-tumor agents including antibodies against cancer cell antigens, accumulate selectively in tumors because of their leaky vasculature. However, most human solid tumors possess abundant intercellular connective tissue, hindering diffusion of such macromolecules. That is why immunoconjugate therapy for stroma rich common solid cancer has not yet proved successful in clinics. In order to overcome the above contradictory drawbacks, a small MW cytotoxic drug was conjugated with an antibody against particular components of tumor stroma. Stroma-targeting immunoconjugates bound to the stroma to create a scaffold, from which sustained release of cytotoxic agent occurred and subsequently diffused throughout the tumor tissue to damage both tumor cells and tumor vessels [4, 29, 43]. Primary tumors and their metastatic deposits both contain stroma which is morphologically similar but kinetic studies on human and animal tumor metastasis have shown that the stroma is functionally distinct in different organs. This causes metastasizing tumor cells from a given primary tumor to be able to colonize some organs but not others, although single tumor cells can be detected in all organs after dissemination by the systemic circulation [44, 45]. However, invasive primary tumors and their established metastatic deposits all retain their erosive destructive properties and abilities to mold the local stroma. The EPR effect occurs in all such lesions and local invasion can cause hemorrhage in the tissues. Therefore, repetitive cycles of IF production can occur in the metastatic lesions as it does in the primary lesion and this means that the strategy described above should constitute effective systemic anti-cancer treatment. Cancer-stroma targeting (CAST) therapy should be regarded as a new modality of oncological therapy, especially for refractory, stromal-rich cancers. Principal and concrete examples of CAST therapy are described in other chapters of this book.

References

1. Trousseau A (1865) Pegmasia alba dolens, vol 3. Balliere JB et Fils, Paris
2. Stein PD, Beemath A, Meyers FA, Skaf E, Sanchez J, Olson RE (2006) Incidence of venous thromboembolism in patients hospitalized with cancer. Am J Med 119:60–68
3. Petralia GA, Kakkar AK (2008) Anti-thrombotic therapy in cancer patients. In: Maragoudakis ME, Tsopanoglou NE (eds) Thrombin: physiology and disease. Springer, New York, pp 189–203
4. Matsumura Y (2012) Cancer stromal targeting (CAST) therapy. Adv Drug Deliv Rev 64:710–719
5. Matsumura and Maeda (1986) A new concept for macromolecular therapeutics in cancer chemotherapy: mechanism of tumoritropic accumulation of proteins and the antitumor agent smancs. Cancer Res 46:6387–6392
6. Matsumura Y, Kimura M, Yamamoto T, Maeda H (1988) Involvement of the kinin-generating cascade and enhanced vascular permeability in tumor tissue. Jpn J Cancer Res 79:327–1334
7. Matsumura Y, Maruo K, Kimura M, Yamamoto T, Konno T, Maeda H (1991) Kinin-generating cascade in advanced cancer patients a d in vitro study. Jpn J Cancer Res 82:732–741
8. Edgington TS, Mackman N, Brand K, Ruf W (1991) The structural biology of expression and function of tissue factor. Thromb Haemost 66:67–79
9. Ruf W, Edgington TS (1994) Structural biology of tissue factor, the initiator of thrombogenesis in vivo. FASEB J 8:385–390
10. van den Berg YW, Osanto S, Reitsma PH, Versteeg HH (2012) The relationship between tissue factor and cancer progression: insights from bench and bedside. Blood 119:924–932
11. Belting M, Ahamed J, Ruf W (2005) Signaling of the tissue factor coagulation pathway in angiogenesis and cancer. Arterioscler Thromb Vasc Biol 25:1545–1550
12. Sampson MT, Kakkar AK (2002) Coagulation proteases and human cancer. Biochem Soc Trans 30:201–207
13. Palumbo JS, Kombrinck KW, Drew AF, Drew AF, Grimes TS, Kiser JH, Degen JL, Bugge TH (2000) Fibrinogen is an important determinant of the metastatic potential of circulating tumor cells. Blood 96:3302–3309
14. Morris DR, Ding Y, Ricks TK, Gullapalli A, Wolfe BL, Trejo J (2006) Protease-activated receptor-2 is essential for factor VIIa and Xa-induced signaling, migration, and invasion of breast cancer cells. Cancer Res 66:307–314

15. Hjortoe GM, Petersen LC, Albrektsen T, Sorensen BB, Norby PL, Mandal SK, Pendurthi UR, Rao LV (2004) Tissue factor-factor VIIa-specific up-regulation of IL-8 expression in MDA-MB-231 cells is mediated by PAR-2 and results in increased cell migration. Blood 103:3029–3037
16. Senger DR, Galli SJ, Dvorak AM, Peruzzi CA, Harvey VS, Dvorak HF (1983) Tumor cells secret a vascular permeability factor that promotes accumulation of ascites fluid. Science 21:983–985
17. Ferrara N, Hillan KJ, Gerber HP, Novotny W (2004) Discovery and development of bevacizumab, an anti-VEGF antibody for treating cancer. Nat Rev Drug Discov 3:391–400
18. Dvorak HF, Rickles FR (2006) Malignancy and hemostasis. In: Colman RW, Marder VJ, Clowes AW, George JN, Goldharber SZ (eds) Hemostasis and thrombosis: basic principles and clinical practice fifth edition. Lippin Cott Williams $ Wilkins, Philadelphia, pp 851–873
19. Abe K, Shoji M, Chen J, Bierhause A, Danave I, Micko C, Casper K, Dillehay D, Nawroth PP, Rickles FR (1999) Regulation of vascular endothelial growth factor production and angiogenesis by the cytoplasmic tail of tissue factor. Proc Natl Acad Sci USA 96:8663–8668
20. Maeda H, Matsumura Y, Kato H (1988) Purification and identification of (Hydroxyprolyl3) bradykinin in ascitic fluid from a patient with gastric cancer. J Biol Chem 263:16051–16054
21. Kato H, Matsumura Y, Maeda H (1988) Isolation and identification of hydroxyproline analogues of bradykinin in human urine. FEBS Lett 232:252–254
22. Dvorak HF, Orenstein NS, Carvalho AC, Churchill WH, Dvorak AM, Galli SJ, Feder J, Bitzer AM, Rypysc J, Giovinco P (1979) Induction of a fibrin-gel investment: an early event in line 10 hepatocarcinoma growth mediated by tumor-secreted products. J Immunol 122:164–174
23. Dvorak HF, Orenstein NS, Carvalko AC, Churchill WH, Dvorak AM, Galli SJ, Feder J, Bitzer AM, Rypysc J, Giovinco P (1979) Induction of a fibrin-gel investment: an early event in line 10 hepatocarcinoma growth mediated by tumor secreted product. J Immunol 122:166–174
24. Dvorak HF, Senger DR, Dvorak AM, Harvey VS, McDonagh J (1985) Regulation of extravascular coagulation by microvascular permeability. Science 227:1059–1061
25. Senger DR, Perruzzi CA, Feder J, Dvorak HF (1986) A highly conserved vascular permeability factor secreted by a variety of human and rodent tumor cell lines. Cancer Res 46:5629–5632
26. Dvorak HF (2002) Vascular permeability factor/vascular endothelial growth factor: a critical cytokine in tumor angiogenesis and a potential target for diagnosis and therapy. J Clin Oncol 20:4368–4380
27. Soe G, Kohno I, Inuzuka K, Itoh Y, Matsuda M (1996) A monoclonal antibody that recognizes a neo-antigen exposed in the E domain of fibrin monomer complexed with fibrinogen or its derivatives: its application to the measurement of soluble fibrin in plasma. Blood 88:2109–2117
28. Wada H, Kobayashi T, Abe Y, Hatada T, Yamada N, Sudo A, Uchida A, Nobori T (2006) Elevated levels of soluble fibrin or D-dimer indicate high risk of thrombosis. J Thromb Haemost 4:1253–1258
29. Yasunaga M, Manabe S, Matsumura Y (2011) New concept of cytotoxic immunoconjugate therapy targeting cancer-induced fibrin clots. Cancer Sci 102:1396–1402
30. Pacella BL Jr, Hui KY, Haber E, Matsueda GR (1983) Induction of fibrin-specific antibodies by immunization with synthetic peptides that correspond to amino termini of thrombin cleavage sites. Mol Immunol 20:521–527
31. Laudano AP, Doolittle RF (1978) Synthetic peptide derivatives that bind to fibrinogen and prevent the polymerization of fibrin monomers. Proc Natl Acad Sci USA 75:3085–3089
32. Kudryk B, Rohoza A, Ahadi M, Chin J, Wiebe ME (1983) A monoclonal antibody with ability to distinguish between NH2-terminal fragments derived from fibrinogen and fibrin. Mol Immunol 20:1191–1200
33. Scheefers-Borchel U, Muller-Berghaus G, Fuhge P, Eberle R, Heimburger N (1985) Discrimination between fibrin and fibrinogen by a monoclonal antibody against a synthetic peptide. Proc Natl Acad Sci USA 82:7091–7095
34. Schielen WJ, Voskuilen M, Tesser GI, Nieuwenhuizen W (1989) The sequence A alpha-(148–160) in fibrin, but not in fibrinogen, is accessible to monoclonal antibodies. Proc Natl Acad Sci USA 86:8951–8954

35. Gargan PE, Graffney PJ, Pleasants JR, Ploplis VA (1993) A monoclonal antibody which recognises an epitopic region unique to the intact fibrin polymeric structure. Fibrinolysis 7:275–283
36. Hui KY, Haber E, Matsueda GR (1983) Monoclonal antibodies to a synthetic fibrin-like peptide bind to human fibrin but not fibrinogen. Science 222:1129–1132
37. Kudryk B, Rohoza A, Ahadi M, Chin J, Wiebe ME (1984) Specificity of a monoclonal antibody for the NH2-terminal region of fibrin. Mol Immunol 21:89–94
38. Rosebrough SF, Grossman ZD, McAfee JG, Kudryk BJ, Subramanian G, Ritter-Hrncirik CA, Witanowski LS, Tillapaugh-Fay G, Urrutia E (1987) Aged venous thrombi: radioimmunoimaging with fibrin-specific monoclonal antibody. Radiology 162:575–577
39. Hisada Y, Yasunaga M, Hanaoka S, Saijou S, Sugino T, Tsuji A, Saga T, Tsumoto K, Manabe S, Kuroda J, Kuratsu J, Matsumura Y (2013) Discovery of an uncovered region in fibrin clots and its clinical significance. Sci Rep 3:2604–2610
40. Obonai T, Fuchigami H, Furuya F, Kozuka N, Yasunaga M, Matsumura Y (2016) Tumor imaging by detecting fibrin clots in tumour stroma with anti-fibrin Fab fragment. Sci Rep 6:23613
41. Filler RB, Roberts SJ, Girardi M (2007) Cutaneous two-stage chemical carcinogenesis. CSH Protoc 2007:pdb prot4837
42. Hirakawa S, Kodama S, Kunstfeld R, Kajiya K, Brown LF, Detmar M (2005) VEGF-A induces tumor and sentinel lymph node lymphangiogenesis and promotes lymphatic metastasis. J Exp Med 201:1089–1099
43. Fuchigami H, Manabe S, Yasunaga M, Matsumura Y (2018) Chemotherapy payload of anti-insoluble fibrin antibody-drug conjugate is released specifically upon binding to fibrin. Sci Rep 8:14211–14219
44. Tarin D, Price JE, Kettlewell MG, Souter RG, Vass AC, Crossley B (1984) Mechanisms of human tumor metastasis studied in patients with peritoneovenous shunts. Cancer Res 44(8):3584–3592
45. Tarin D (2011) Cell and tissue interactions in carcinogenesis and metastasis and their clinical significance. Semin Cancer Biol 21(2):72–82

Tumor Blood Vessels as Targets for Cancer Therapy

Kyoko Hida, Nako Maishi, and Yasuhiro Hida

Abstract Tumor growth and metastasis are dependent on angiogenesis, which is the formation of new blood vessels. The newly formed blood vessels around the tumor supply oxygen and nutrients to the tumor, supporting its progression. Moreover, these blood vessels also serve as channels through which tumor cells metastasize to distant organs. Tumor blood vessels, and especially the endothelial cells lining tumor blood vessels [tumor endothelial cells (TECs)], are important targets in cancer therapy. Since newly formed tumor blood vessels originate from pre-existing normal vessels, tumor blood vessels and TECs traditionally have been considered the same as normal vasculature. Tumor blood vessels, however, have a distinctively abnormal phenotype, including morphological alterations. Recently, it has been revealed that TECs constitute a heterogeneous population, exhibiting characteristics induced largely by tumor microenvironmental factors. Furthermore, TECs induce cancer progression through metastasis passively but also actively. This chapter will review the mechanisms underlying tumor angiogenesis and discuss the biology of TECs, offering new perspectives on treatment strategies that can target tumor blood vessels from a personalized medicine perspective.

Keywords Tumor · Tumor endothelial cell · Angiogenesis · Metastasis · Drug delivery · Heterogeneity · Anti-angiogenic therapy · Microenvironment

---

K. Hida (✉) · N. Maishi
Vascular Biology and Molecular Pathology, Graduate School of Dental Medicine, Hokkaido University, Sapporo, Japan
e-mail: khida@den.hokudai.ac.jp

Y. Hida
Department of Cardiovascular and Thoracic Surgery, Faculty of Medicine, Hokkaido University, Sapporo, Japan

Y. Matsumura, D. Tarin (eds.), *Cancer Drug Delivery Systems Based on the Tumor Microenvironment*, https://doi.org/10.1007/978-4-431-56880-3_3

Tumor Angiogenesis

Blood vessels supply a tumor with oxygen and nutrients, in addition to removing waste products from tumor tissues. Thus, tumor blood vasculature has a central role in cancer metabolism. In addition, tumor vessels provide escape routes for metastatic tumor cells [1, 2] (Fig. 1). The angiogenic process that leads to the formation of new blood vessels is necessary for tumor initiation and progression. Cancer cells promote angiogenesis at an early stage of tumorigenesis by releasing angiogenic molecules to surrounding stromal tissue to stimulate the migration and proliferation of endothelial cells (ECs); this process is known as the angiogenic switch. Indeed, tumors cannot grow beyond 1–2 mm in size, and cannot metastasize unless this angiogenic switch is turned on. Tumors without angiogenesis typically go dormant and dormant tumors often have been observed during autopsies of individuals who have died of causes other than cancer [3]. Thus, regulating tumor blood vessel formation is an important strategy for cancer therapy (anti-angiogenic therapy).

Folkman J proposed such a concept in his 1971 landmark report, and since that time anti-angiogenic therapy has become an important strategy for cancer treatments. Thus, tumor ECs (TECs) that form the inner lining of tumor blood vessels, have been considered as important targets of cancer therapy. To date, numbers of anti-angiogenic drugs have been discovered and tested [4].

Traditionally, the advantages of this therapy have been considered as follows: (i) The survival of numerous cancer cells depends on a few TECs; therefore, targeting TECs may be more efficient than targeting tumor cells. (ii) Since TECs may exhibit common characteristics, regardless of different tumor types, a single, effective anti-angiogenic drug could be used to treat many types of cancers. (iii) Also, it is thought that TECs are genetically stable and, therefore, unlikely to acquire drug resistance, unlike tumor cells.

However, recent studies have revealed that the regulation of tumor angiogenesis is much more complicated.

Although vascular endothelial growth factor (VEGF) is the major angiogenic factor, other growth factors, such as the angiopoietins (Angs), basic fibroblast growth factor (bFGF), hepatocyte growth factor (HGF), chemokines, placental-derived growth factor (PDGF) [5], and ephrin families play important roles in tumor angiogenesis [6] (Table 1). For example, Ang-2 binds to the receptor tyrosine kinase Tie2, acting as an antagonist for Ang-1, which phosphorylates Tie2 and induces blood vessel maturation. On the other hand, Ang-2 inactivates Tie2, resulting in the initiation of angiogenesis. In addition, mural cells or fibroblasts secrete growth factors, such as PDGF or transforming growth factor beta families, which stimulate angiogenesis.

Furthermore, there are several endogenous angiogenesis inhibitors in the tumor microenvironment (Table 2). For instance, thrombospondin-1 (TSP-1) is regulated by tumor suppressor gene p53, such that its mutation causes downregulation of TSP-1 in cancers. Vasohibin [7], Down syndrome critical region-1 (DSCR-1) [8], and the Notch ligand Delta-like 4 (Dll4) are secreted by ECs and function as negative

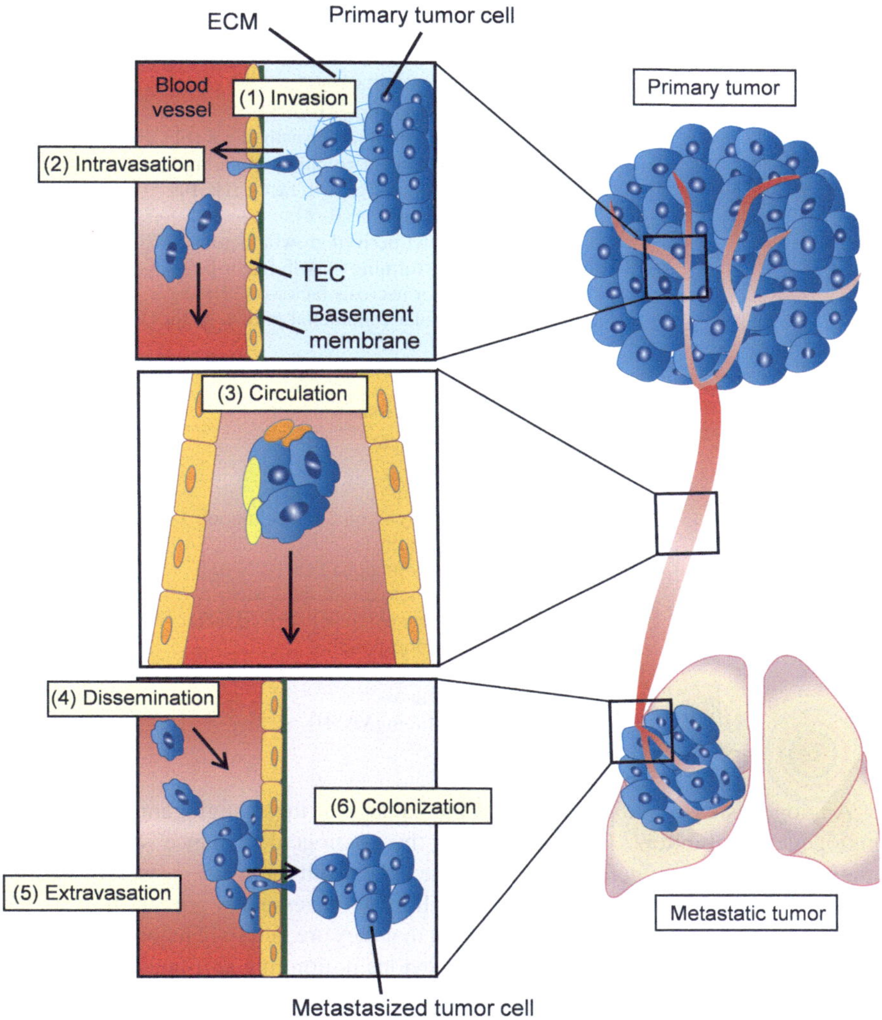

Fig. 1 Steps in the development of tumor metastasis. Tumor cells invade the surrounding stroma and ECM (*1*), intravasate into the blood stream (*2*), survive in the circulation with anoikis resistance (*4*), disseminate in the circulation and reach the distant organs (*5*), extravasate and invade into the parenchyma of distant tissues (*6*), adapt to the new microenvironment and proliferate to form metastatic colonization (*7*). Stromal cells have important roles in metastasis dissemination. (Reprinted from Maishi et al., Cancer Sci. (2017); in press with permission from Cancer Sci)

feedback mechanisms. Furthermore, there are endogenous angiogenesis inhibitors that are cleaved fragments of larger proteins, such as members of the coagulation system, or extracellular matrix family. Endostatin, for example, is a well-known endogenous angiogenesis inhibitor that is the cleaved C-terminal fragment of collagen XVIII [9].

Table 1 Angiogenic factors

Angiogenic factors
Acidic fibroblast growth factor (aFGF)
Angiogenin (Ang)
Angiopoietin (ANGPT)
Basic fibroblast growth factor (bFGF)
Granulocyte-colony-stimulating factor (G-CSF)
Hepatocyte growth factor (HGF)
Interleukin-8 (IL-8)
Platelet-derived growth factor (PDGF)
Transforming growth factor-β (TGF-β)
Tumor necrosis factor-α (TNF-α)
Vascular endothelial growth factor (VEGF)

Table 2 Angiogenesis inhibitors

Angiogenesis inhibitors
Angiostatin
Arrestin
Canstatin
Down syndrome critical region 1 (DSCR1)
Endostatin
Interferon alpha/beta (IFN-α/ β)
2-Methoxyestradiol (2-ME)
Pigment epithelial-derived factor (PEDF)
Thrombospondin 1
Tissue inhibitor of metalloproteinase 2 (TIMP-2)
Tumstatin
Vasohibin (VASH)

Angiogenesis is regulated based on the balance of these angiogenic factors and angiogenesis inhibitors. Once change in this angiogenic balance shifts into an angiogenic state, angiogenic factors activate ECs, resulting in new vessel formation.

And it has been reported that TECs that line blood vessels in primary tumor sites are different from normal ECs (NECs), and that they are heterogeneous among various tumor types. Moreover, despite the previous notion that anti-angiogenic drugs are less toxic than other cytotoxic drugs, it now has been reported that they sometimes have severe side effects, such as lethal hemoptysis [10, 11] and intestinal perforation [12, 13]. This is attributed to the fact that most of the current anti-angiogenic drugs inhibit the VEGF/VEGF receptor (VGEFR) pathway, which is important for even NEC survival.

It has become an important goal in cancer therapy to develop novel anti-angiogenic agents. Nevertheless, the achievement of this goal depends upon a thorough understanding of the biology of TECs.

Abnormality of Tumor Blood Vessels

It is well known that tumor blood vessels differ morphologically from normal blood vessels. Abnormalities in tumor blood vessels may be due to an imbalance of angiogenic factors and their inhibitors. Tumor vasculature demonstrates unorganized patterns, whereas normal vasculature shows an orderly, hierarchical branching pattern of arteries, veins, and capillaries [14].

Tumor blood vessels do not consist of regular monolayers of EC and do not have an normal barrier function [15]. Basement membranes vary in the thickness of type IV collagen layers. Unlike normal blood vessels, pericytes are associated loosely with the EC [16, 17]. These abnormalities result in tumor vessel leakiness, followed by an increase of interstitial fluid pressure in the tumor tissue, causing vessel collapse [18]. Thus, tumor blood vessels are often twisted in appearance, with uneven vessel diameters [18]. These abnormalities consequently cause blood flow in tumor vasculature to be generally chaotic. For example, some tumor vessels are not perfused with blood, while other vessels have a chaotic blood flow that may reverse in direction. This is one of the mechanisms of tumor hypoxia, regardless of the high vascularization, and is sometimes the cause of radiation resistance or insufficient delivery of anti-cancer drugs [19].

Abnormalities are not only observed in the architecture of tumor blood vessels but also there is a dysfunction in the TECs themselves. In normal blood vessels, ECs form a uniform and continuous monolayer, whereas TECs are irregular in shape and size. In addition, there are often gaps between adjacent TECs that cause the obstruction of the tumor cells proximal to these gaps [20]. Transcellular fenestrations also have been observed in TECs. Perhaps, these morphologically abnormalities in TECs cause hemorrhaging and plasma leakage, which has been described above. Furthermore, holes in the tumor blood vessels may be involved in a mechanism for tumor cell intravasation, an initial step of metastasis.

On the other hand, leaky blood vessels provide the advantage of nanodrug delivery accumulation in the tumor tissue. This is well known as the "enhanced permeability and retention (EPR) effect" [21, 22]. The EPR effects have been exploited for passive targeting of drugs into tumors.

It is well known that VEGF enhances the permeability of blood vessels [19], causing an increase of the interstitial fluid pressure (IFP) in tumor tissue, resulting in a collapse of the vessels. Thus, inhibition of VEGF signaling has been a strategy for vascular normalization, in which anti-angiogenic agents temporarily "normalize" both the structure and function of tumor vasculature, thereby, enabling more efficient delivery of oxygen and therapeutic drugs [19]. Vascular normalization by anti-VEGF drugs seems to be an attractive strategy regarding delivery of drugs to tumors, but it is difficult to predict the appropriate duration of use for optimal efficacy with these drugs. In some studies, long-term anti-VEGF therapy has caused undesirable ischemic changes; resulting in increased malignancy of cancer. As for mechanism, it has been suggested that epithelial mesenchymal transition (EMT) induced by hypoxia in the tumor cells, may be one of reason, although detail mechanism still remains [23].

Thus, the medical community has recognized the need for a proper window within which tumor vascular normalization treatments should be performed. However, predicting the correct window for normalization is not easy. A recent study reported that a decrease in plasma levels of apelin, a ligand of the APJ receptor, can predict blood vessel normalization [24]. However, it has not been fully elucidated yet how blood EC phenotypes are altered during or after anti-angiogenic therapy.

In addition, it has been reported that the tumor vasculature is heterogeneous in terms of structure and maturity with respect to pericyte coverage, tight junctions, and fenestration [25]. The EPR effect passively targets tumors and is dependent on these structural differences. On the other hand, active targeting of the tumor vasculature using anti-angiogenic therapy seems to be independent of these structural differences in blood vessels. This is because specific binding between a ligand and a surface protein on TECs plays a pivotal role in active-targeting strategy.

To develop a novel anti-angiogenic drug regimen with higher specificity to TECs, an understanding of the heterogeneity of tumor vasculature is important.

Characterization of Tumor Endothelial Cells

Recent evidence indicates that the cells of the vascular endothelium are heterogeneous and exhibit specialized phenotypes depending on their organs of origin and their functional states [26–28].

Until recently, most studies on tumor angiogenesis used NECs, human dermal microvascular ECs (HMVEC) and human umbilical vein ECs (HUVECs); however, TECs have different gene expression profiles compared to NECs [29, 30]. The first TEC isolation was reported by St. Croix et al. in 2000 [29], and since that time TECs have been isolated by cell sorting methods using antibodies against EC markers, such as intercellular adhesion molecule 2 (ICAM-2), CD144, CD105, and CD146.

There also have been several reports about the gene profile of TECs using global analysis [26–29, 31, 32].

We have isolated TECs and reported that they have several abnormalities [33, 34], including differences in responsiveness to epidermal growth factor (EGF) [35], adrenomedullin [36], and VEGF [37] when compared with NECs. These growth factors are important in conferring a pro-angiogenic phenotype in TEC. VEGF especially stimulates cell migration and enhances survival of TEC in an autocrine manner [33, 34]. A VEGF autocrine loop in TECs confers, at least in part, resistance to serum starvation, which is not observed in NECs [37, 38].

In addition, compared with NECs, TECs were more responsive to angiogenic growth factors, such as bFGF [37] and VEGF [39].

TEC also expresses stem cell markers [38, 40]. For example, the expression of embryonic renal transcription factor PAX2 in TECs induces apoptosis resistance and pro-angiogenic properties in those cells [41]. Other stem cell markers, such as P-gp/ABCB1, ALDH, and IL-6, are upregulated in TECs [42, 43].

Although there is a possibility that long-term cultured ECs have been selected for their activated proliferation and high survival rate during cell culture, TECs maintained their specific characteristics, at least in part, compared with that of NECs that were cultured under similar conditions. Thus, the use of cultured TECs has several potential benefits over isolated TECs that have not been cultured, reflecting their biology in the tumor microenvironment.

Drug Resistance in TECs

The TECs show several cytogenetic abnormalities, such as aneuploidy and abnormal centrosomes; these features have been demonstrated in mouse tumors [33] and human renal carcinomas [44]. These abnormalities are not an artifact of tumor cell contamination because the TECs used in these studies were purified using the diphtheria toxin (DT) method [33]. In addition, there are several reports that tumor cells transdifferentiate into TECs, which causes cytogenetic abnormalities in TECs cells [45, 46].

Because aneuploidy in TECs may cause genetic instability, it is not surprising that some TECs are drug resistant [38, 47]. For example, TECs isolated from a renal carcinoma were resistant to vincristine [38], and those from a hepatocellular carcinoma were resistant to both 5-fluorouracil and adriamycin [47, 48].

We have shown that TECs are resistant to several anti-cancer drugs, including paclitaxel, in tumors characterized by the upregulation of multidrug resistance gene-1 (MDR-1)/p-glycoprotein (P-gp), which is also known as the ABCB1 gene [48].

MDR-1 was upregulated by tumor-derived VEGF [48]. In this study, tumor-derived VEGF was implicated in drug resistance [48].

Heterogeneity of Tumor Endothelial Cells

It has been reported that ECs may be heterogeneous in terms of morphology, function, or gene expression. For example, leukocyte rolling velocity and arrest frequency at EC junctions are different from their central areas [49]. Inter-organ differences in ECs also have been reported [50]. In pre-existing blood vessels, stem cell-like ECs with a pro-angiogenic phenotype have been identified [51]. Additionally, in rat brain blood vessel ECs, the expressions of P-gp were heterogeneous, suggesting a lack of uniformity in the blood–brain barrier [52].

TECs are also heterogeneous [33, 38, 53]. We have demonstrated that some TECs show upregulated expressions of several stem cell markers, such as CD90, ABCB-1[48, 54] Sca-1and aldehyde dehydrogenase (ALDH) [42], which are often upregulated in cancer stem cells. Also, it has been reported that TEC in prostate cancer differentiate into osteoblast showing that TEC may have multipotency at least partially [53]. These stem-like TEC often show proangiogenic phenotype [54]. For example, ALDH^{high} TECs formed more tubes on Matrigel than ALDH^{low}, even under serum starvation, suggesting that they are more proangiogenic than ALDH^{low} TECs, although the specific role of ALDH in angiogenesis requires further study [42].

Within the tumor vasculature, the morphology and pericyte coverage of tumor blood vessels varies depending on tumor type and stage of progression [26, 50, 55]. Furthermore, we have reported that these heterogeneities [25] depend on tumor malignancy. The blood vessels of high metastatic tumors are more immature, with fewer pericytes than the blood vessels of low metastatic tumors [25]. These features contribute to the increased hypoxic nature of high metastatic tumors compared to low metastatic tumors. The TECs isolated from high metastatic tumors have higher proangiogenic phenotypes than TECs isolated from low metastatic tumors with higher expression of proangiogenic genes, such as VEGF, VEGFR2, MMPs and higher basal Akt activation [25, 56, 57].

As described above, TECs are resistant to several anti-cancer drugs. Higher levels of MDR1 mRNA are detected in high metastatic tumor-derived TECs compared with low metastatic tumor-derived TECs. Complex abnormal karyotypes and high aneuploidy rates are associated with cancer cells containing multidrug resistance genes [58]. However, it has been reported that even TECs are karyotypically aneuploid, whereas NECs are diploid not only in vitro and also in vivo [33, 44, 59]. In addition, a GTG-banding study demonstrated that high metastatic tumor-derived TECs have a more complex abnormal karyotype than low metastatic tumor-derived TECs [25].

Effects of Tumor Microenvironment on TECs

Interactions between tumor cells and stromal cells seem to be important for tumor progression, tumor malignancy, and metastasis. Indeed, ECs cocultured with highly metastatic tumor cell-conditioned medium showed upregulation of proangiogenic gene expressions and changes resembling some of the TEC phenotypes [25], suggesting that tumor-secreted factors influenced the associated ECs. Hypoxia may be another factor affecting the acquisition of TEC heterogeneity. As described previously, immature, leaky tumor blood vessels cause tumor hypoxia [60, 61]. The increased vascular permeability compromises blood flow in tumor blood vessels, which further decreases nutrient and oxygen delivery, thus causing physiological stress for the tumor, and sometimes even for the TECs. The persisting hypoxia, together with cytokine secretion, such as VEGF, induces the mobilization of bone marrow-derived endothelial progenitor cells to cancer [62]. Furthermore, it may cause genetic instability of EC. An excess of VEGF induces abnormal centrosome structures in NECs [63, 64]. Collectively, TECs show heterogeneous phenotypes, reflecting on the tumor microenvironment; these heterogeneities are required for developing novel strategies by better selection of anti-angiogenic therapies.

Acceleration of Cancer Progression by Tumor Endothelial Cells

Because of the immature structure with lack of smooth muscle cells and pericytes, tumor cells can intravasate easily through tumor blood vessels [20]. In addition to this passive entry route, TECs with an abnormal phenotype, as described above, may also provide signals that actively promote metastasis of tumor cells. Several reports, including ours, have shown that TECs release specific growth factors called angiocrine factors [65].

FGF4, secreted from B-cell lymphoma cells, activates FGFR1 in neighboring ECs, and upregulates the Notch ligand Jagged1 in ECs. In turn, Jagged1 on ECs reciprocally induces Notch2–Hey1 in lymphoma cells [66], which converts indolent tumor cells to more aggressive cells with properties of invasiveness and chemoresistance [67]. Other groups also have demonstrated that the Notch signals in ECs are involved in tumor progression and the stem cell phenotype [68, 69]. Notch activation in ECs promotes tumor cell migration and metastasis [70]. Such juxtacrine signals between ECs and tumor cells induce protumorigenic characteristics with increased expression of tumor-stimulating factors or reduced expression of tumor-suppressive factors. Thus, blocking such TEC-cancer cell interactions can be an effective method of targeting cancer stem cells.

Furthermore, we have reported the role of TECs in the initial steps of tumor metastasis [71]. Because TECs from high metastatic tumors express higher levels of growth factors and cytokines compared with TECs from low metastatic tumors, as described above, it was speculated that TECs may affect tumor cell behavior. *In vitro* data revealed that the ability of TECs to attract and adhere to tumor cells was greater in high metastatic tumors than in low metastatic tumors. Tumor cell transendothelial migration was enhanced on the high metastatic tumor -TEC monolayer. Biglycan, a small leucine-rich-repeat proteoglycan, was specifically upregulated in TECs, and TEC-biglycan facilitated the migration of Toll-like receptor (TLR)-expressing tumor cells via activation of NF-kB and extracellular signal-regulated kinase (ERK) signaling. Biglycan knockdown in TECs decreased the number of circulating tumor cells and lung metastases *in vivo*. In addition, biglycan levels in the plasma of cancer patients were higher than those in healthy volunteers. In particular, high levels of biglycan were detected in metastatic cases. These results suggested that TECs may provide a key molecule, such as biglycan, that allows tumor cells to break free resulting in hematogenous metastasis (Fig. 2). Collectively, altered TECs facilitate cancer progression and metastasis by orchestrating tumor cells and their microenvironment. Because ECs in primary tumors are in close contact with tumor cells during intravasation, it is not surprising that they have important roles in metastatic dissemination.

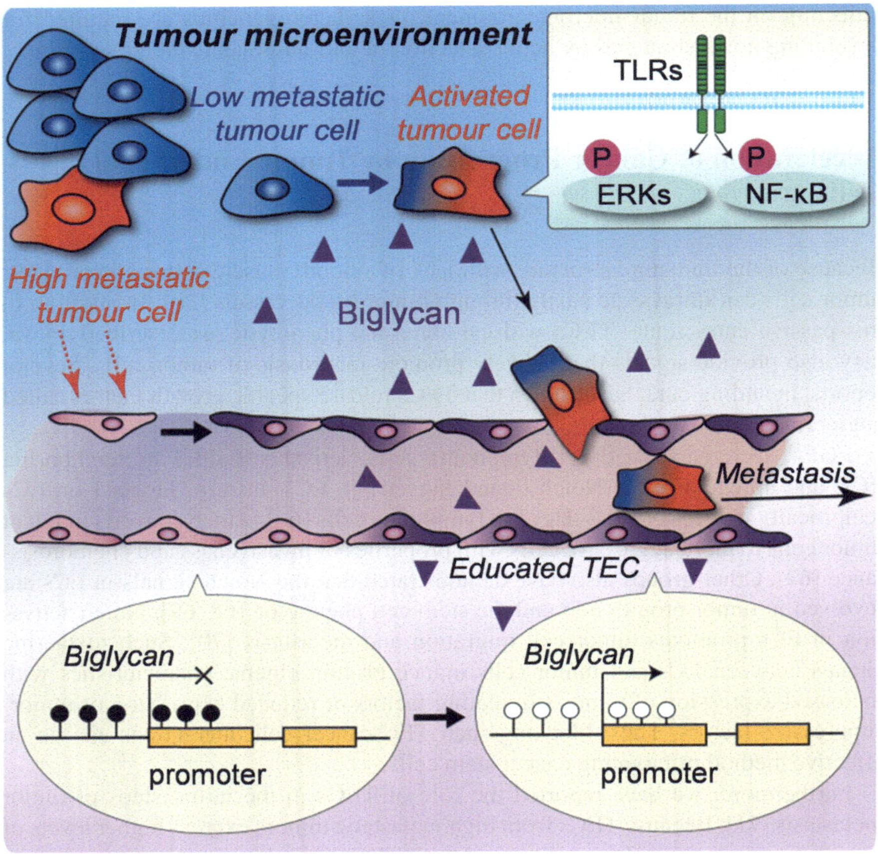

Fig. 2 Promotion of tumor metastasis by TECs through biglycan secretion. Tumor cells and microenvironment develop TECs. In turn, these educated TECs express high levels of biglycan with epigenetic modification, which stimulates tumor cells to metastasize via the activation of NF-κB and ERK signaling (P means phosphorylation). (Reprinted from Maishi et al., Sci Rep (2016) with permission from Sci Rep)

Active Targeting of Tumor Vasculature

The TECs may be "educated" by their parental tumors by tumor-derived soluble factors (e.g., growth factors, chemokines, and microRNA). Furthermore, the evidence cited above indicates that TECs may differ according to their tumor microenvironment and stage of progression. Thus, it may be possible to develop biomarkers from TEC that reflect heterogeneous tumor blood vasculature, allowing for better selection of anti-angiogenic therapies.

In this circumstance, active targeting of TECs is a promising strategy for developing personalized anti-angiogenic therapies for overcoming the limitations of current anti-angiogenic therapy. To date, TEC-specific markers have been reported [29, 30, 41, 72–82]. The types of TEC-specific markers vary, such as cell surface markers, secreted proteins, or cytoplasmic proteins. For drug delivery systems, active targeting of tumor blood vessels could be achieved via TEC cell surface markers. Several peptides that recognize TEC plasma membrane molecules have been reported. For example, a peptide ligand for CD13, the asparagine-glycine-arginine (NGR) motif peptide, is overexpressed in TECs [83]. NGR motif peptides have been used successfully to deliver drugs as well as liposomes to the tumor vasculature [84]. In addition, the arginine-glycine-aspartic acid (RGD) peptide has been used for drug delivery because it can recognize integrin $\alpha V\beta 3$, and is expressed in TECs and in some cancer cells [85]. The multi-functional envelope nanodevice (MEND) [86, 87] is a liposomal small interfering RNA (siRNA) drug delivery system that was developed in recent years. Furthermore, a MEND composed of a pH-sensitive cationic lipid, YSK05 (YSK-MEND), has achieved an increased efficacy of gene reduction [88]. Harashima et al. applied the YSK system as an active targeting system to knockdown angiogenesis-related genes in the tumor vasculature using cyclic RGD as a ligand; this was designated as RGD-MEND [89]. It is possible that siRNA may be a useful drug for this purpose because of its ability to inhibit the expression of specific genes [90]. Regarding the targeting of heterogeneous TECs, TEC marker silencing could be an attractive strategy. Recently, it was reported that tumor growth was successfully inhibited via a new anti-angiogenic therapy that delivered VEGFR2 siRNA to TECs in the RGD-MEND [89]. RGD-MEND caused significant gene silencing in TECs without causing severe toxicity to NECs [89]. In addition, RNA interference was confirmed in TECs by the 5' RACE-PCR system for rapid amplification of cDNA ends, indicating that siRNA delivery had occurred specifically to tumor blood vessels. This system is a promising method for actively targeting TECs and may be applied for personalized anti-cancer therapies.

Conclusion

This chapter summarized the significance of heterogeneity in TECs. Although the mechanism for TEC heterogeneity is still unclear, TEC phenotypes raise several important issues regarding cancer therapeutics. One of the most critical points is the observation that even stromal cells can be abnormal in the tumor microenvironment. A novel drug delivery system to actively target heterogeneous tumor vasculature is a promising anti-cancer strategy. Further studies on TEC biology are necessary for selecting suitable drug delivery systems.

Acknowledgments We thank Editors from Enago, for editing a draft of this manuscript.

We also thank members of the Department of Vascular Biology, Institute for Genetic Medicine, Hokkaido University.

Conflict of Interest None declared.

References

1. Folkman J (1971) Tumor angiogenesis: therapeutic implications. N Engl J Med 285(21):1182–1186
2. Folkman J, Kerbel R (2002) Role of angiogenesis in tumor growth and metastasis. Clinical translation of angiogenesis inhibitors. Semin Oncol 29(6 Suppl 16):15–18
3. Nielsen M, Thomsen JL, Primdahl S, Dyreborg U, Andersen JA (1987) Breast cancer and atypia among young and middle-aged women: a study of 110 medicolegal autopsies. Br J Cancer 56(6):814–819
4. Folkman J (2007) Angiogenesis: an organizing principle for drug discovery? Nat Rev Drug Discov 6(4):273–286
5. Bergers G, Benjamin LE (2003) Tumorigenesis and the angiogenic switch. Nat Rev Cancer 3(6):401–410
6. Goh PP, Sze DM, Roufogalis BD (2007) Molecular and cellular regulators of cancer angiogenesis. Curr Cancer Drug Targets 7(8):743–758
7. Watanabe K, Hasegawa Y, Yamashita H, Shimizu K, Ding Y, Abe M et al (2004) Vasohibin as an endothelium-derived negative feedback regulator of angiogenesis. J Clin Invest 114(7):898–907
8. Minami T, Yano K, Miura M, Kobayashi M, Suehiro J, Reid PC et al (2009) The down syndrome critical region gene 1 short variant promoters direct vascular bed-specific gene expression during inflammation in mice. J Clin Invest 119(8):2257–2270
9. Folkman J (2006) Antiangiogenesis in cancer therapy–endostatin and its mechanisms of action. Exp Cell Res 312(5):594–607
10. Johnson DH, Fehrenbacher L, Novotny WF, Herbst RS, Nemunaitis JJ, Jablons DM et al (2004) Randomized phase II trial comparing bevacizumab plus carboplatin and paclitaxel with carboplatin and paclitaxel alone in previously untreated locally advanced or metastatic non-small-cell lung cancer. J Clin Oncol 22(11):2184–2191
11. Keedy VL, Sandler AB (2007) Inhibition of angiogenesis in the treatment of non-small cell lung cancer. Cancer Sci 98(12):1825–1830
12. Kindler HL, Friberg G, Singh DA, Locker G, Nattam S, Kozloff M et al (2005) Phase II trial of bevacizumab plus gemcitabine in patients with advanced pancreatic cancer. J Clin Oncol 23(31):8033–8040
13. Saif MW, Elfiky A, Salem RR (2007) Gastrointestinal perforation due to bevacizumab in colorectal cancer. Ann Surg Oncol 14(6):1860–1869
14. McDonald DM, Choyke PL (2003) Imaging of angiogenesis: from microscope to clinic. Nat Med 9(6):713–725
15. Hashizume H, Baluk P, Morikawa S, McLean JW, Thurston G, Roberge S et al (2000) Openings between defective endothelial cells explain tumor vessel leakiness. Am J Pathol 156(4):1363–1380
16. Kalluri R (2003) Basement membranes: structure, assembly and role in tumour angiogenesis. Nat Rev Cancer 3(6):422–433
17. Baluk P, Hashizume H, McDonald DM (2005) Cellular abnormalities of blood vessels as targets in cancer. Curr Opin Genet Dev 15(1):102–111
18. McDonald DM, Baluk P (2002) Significance of blood vessel leakiness in cancer. Cancer Res 62(18):5381–5385
19. Jain RK (2005) Normalization of tumor vasculature: an emerging concept in antiangiogenic therapy. Science 307(5706):58–62
20. Chang YS, di Tomaso E, McDonald DM, Jones R, Jain RK, Munn LL (2000) Mosaic blood vessels in tumors: frequency of cancer cells in contact with flowing blood. Proc Natl Acad Sci U S A 97(26):14608–14613
21. Matsumura Y, Maeda H (1986) A new concept for macromolecular therapeutics in cancer chemotherapy: mechanism of tumoritropic accumulation of proteins and the antitumor agent smancs. Cancer Res 46(12 Pt 1):6387–6392

22. Fang J, Nakamura H, Maeda H (2011) The EPR effect: unique features of tumor blood vessels for drug delivery, factors involved, and limitations and augmentation of the effect. Adv Drug Deliv Rev 63(3):136–151
23. Paez-Ribes M, Allen E, Hudock J, Takeda T, Okuyama H, Vinals F et al (2009) Antiangiogenic therapy elicits malignant progression of tumors to increased local invasion and distant metastasis. Cancer Cell 15(3):220–231
24. Zhang L, Takara K, Yamakawa D, Kidoya H, Takakura N (2015) Apelin as a marker for monitoring the tumor vessel normalization window during antiangiogenic therapy. Cancer Sci 107(1):36–44
25. Ohga N, Ishikawa S, Maishi N, Akiyama K, Hida Y, Kawamoto T et al (2012) Heterogeneity of tumor endothelial cells: comparison between tumor endothelial cells isolated from high- and low-metastatic tumors. Am J Pathol 180(3):1294–1307
26. Langenkamp E, Molema G (2009) Microvascular endothelial cell heterogeneity: general concepts and pharmacological consequences for anti-angiogenic therapy of cancer. Cell Tissue Res 335(1):205–222
27. Arap W, Kolonin MG, Trepel M, Lahdenranta J, Cardo-Vila M, Giordano RJ et al (2002) Steps toward mapping the human vasculature by phage display. Nat Med 8(2):121–127
28. Trepel M, Arap W, Pasqualini R (2002) In vivo phage display and vascular heterogeneity: implications for targeted medicine. Curr Opin Chem Biol 6(3):399–404
29. St Croix B, Rago C, Velculescu V, Traverso G, Romans KE, Montgomery E et al (2000) Genes expressed in human tumor endothelium. Science 289(5482):1197–1202
30. Seaman S, Stevens J, Yang MY, Logsdon D, Graff-Cherry C, St Croix B (2007) Genes that distinguish physiological and pathological angiogenesis. Cancer Cell 11(6):539–554
31. Carson-Walter EB, Watkins DN, Nanda A, Vogelstein B, Kinzler KW, St Croix B (2001) Cell surface tumor endothelial markers are conserved in mice and humans. Cancer Res 61(18):6649–6655
32. Nanda A, St Croix B (2004) Tumor endothelial markers: new targets for cancer therapy. Curr Opin Oncol 16(1):44–49
33. Hida K, Hida Y, Amin DN, Flint AF, Panigrahy D, Morton CC et al (2004) Tumor-associated endothelial cells with cytogenetic abnormalities. Cancer Res 64(22):8249–8255
34. Hida K, Klagsbrun M (2005) A new perspective on tumor endothelial cells: unexpected chromosome and centrosome abnormalities. Cancer Res 65(7):2507–2510
35. Amin DN, Hida K, Bielenberg DR, Klagsbrun M (2006) Tumor endothelial cells express epidermal growth factor receptor (EGFR) but not ErbB3 and are responsive to EGF and to EGFR kinase inhibitors. Cancer Res 66(4):2173–2180
36. Tsuchiya K, Hida K, Hida Y, Muraki C, Ohga N, Akino T et al (2010) Adrenomedullin antagonist suppresses tumor formation in renal cell carcinoma through inhibitory effects on tumor endothelial cells and endothelial progenitor mobilization. Int J Oncol 36(6):1379–1386
37. Matsuda K, Ohga N, Hida Y, Muraki C, Tsuchiya K, Kurosu T et al (2010) Isolated tumor endothelial cells maintain specific character during long-term culture. Biochem Biophys Res Commun 394(4):947–954
38. Bussolati B, Deambrosis I, Russo S, Deregibus MC, Camussi G (2003) Altered angiogenesis and survival in human tumor-derived endothelial cells. FASEB J 17(9):1159–1161
39. Matsuda K, Ohga N, Hida Y, Muraki C, Tsuchiya K, Kurosu T et al (2010) Isolated tumor endothelial cells maintain specific character during long-term culture. Biochem Biophys Res Commun 394:947–954
40. Bussolati B, Grange C, Bruno S, Buttiglieri S, Deregibus MC, Tei L et al (2006) Neural-cell adhesion molecule (NCAM) expression by immature and tumor-derived endothelial cells favors cell organization into capillary-like structures. Exp Cell Res 312(6):913–924
41. Fonsato V, Buttiglieri S, Deregibus MC, Puntorieri V, Bussolati B, Camussi G (2006) Expression of Pax2 in human renal tumor-derived endothelial cells sustains apoptosis resistance and angiogenesis. Am J Pathol 168(2):706–713

42. Ohmura-Kakutani H, Akiyama K, Maishi N, Ohga N, Hida Y, Kawamoto T et al (2014) Identification of tumor endothelial cells with high aldehyde dehydrogenase activity and a highly angiogenic phenotype. PLoS One 9(12):e113910
43. Hida K, Maishi N, Akiyama K, Ohmura-Kakutani H, Torii C, Ohga N et al (2017) Tumor endothelial cells with high aldehyde dehydrogenase activity show drug resistance. Cancer Sci 108:2195–2203
44. Akino T, Hida K, Hida Y, Tsuchiya K, Freedman D, Muraki C et al (2009) Cytogenetic abnormalities of tumor-associated endothelial cells in human malignant tumors. Am J Pathol 175(6):2657–2667
45. Ricci-Vitiani L, Pallini R, Biffoni M, Todaro M, Invernici G, Cenci T et al (2010) Tumour vascularization via endothelial differentiation of glioblastoma stem-like cells. Nature 468(7325):824–828
46. Wang R, Chadalavada K, Wilshire J, Kowalik U, Hovinga KE, Geber A et al (2010) Glioblastoma stem-like cells give rise to tumour endothelium. Nature 468(7325):829–833
47. Xiong YQ, Sun HC, Zhang W, Zhu XD, Zhuang PY, Zhang JB et al (2009) Human hepatocellular carcinoma tumor-derived endothelial cells manifest increased angiogenesis capability and drug resistance compared with normal endothelial cells. Clin Cancer Res 15(15):4838–4846
48. Akiyama K, Ohga N, Hida Y, Kawamoto T, Sadamoto Y, Ishikawa S et al (2012) Tumor endothelial cells acquire drug resistance by MDR1 up-regulation via VEGF signaling in tumor microenvironment. Am J Pathol 180(3):1283–1293
49. Mundhekar AN, Bullard DC, Kucik DF (2006) Intracellular heterogeneity in adhesiveness of endothelium affects early steps in leukocyte adhesion. Am J Physiol Cell Physiol 291(1):C130–C137
50. Molema G (2010) Heterogeneity in endothelial responsiveness to cytokines, molecular causes, and pharmacological consequences. Semin Thromb Hemost 36(3):246–264
51. Naito H, Kidoya H, Sakimoto S, Wakabayashi T, Takakura N (2011) Identification and characterization of a resident vascular stem/progenitor cell population in preexisting blood vessels. EMBO J 31(4):842–855
52. Saubamea B, Cochois-Guegan V, Cisternino S, Scherrmann JM (2012) Heterogeneity in the rat brain vasculature revealed by quantitative confocal analysis of endothelial barrier antigen and P-glycoprotein expression. J Cereb Blood Flow Metab 32(1):81–92
53. Dudley AC, Khan ZA, Shih SC, Kang SY, Zwaans BM, Bischoff J, Klagsbrun M (2008) Calcification of multipotent prostate tumor endothelium. Cancer cell 14(3):201–211
54. Naito H, Wakabayashi T, Kidoya H, Muramatsu F, Takara K, Eino D, Yamane K, Iba T, Takakura N (2016) Endothelial side population cells contribute to tumor angiogenesis and anti-angiogenic drug resistance. Cancer Res 76(11):3200–3210
55. Nagy JA, Dvorak HF (2012) Heterogeneity of the tumor vasculature: the need for new tumor blood vessel type-specific targets. Clin Exp Metastasis 29(7):657–662
56. Bussolati B, Assenzio B, Deregibus MC, Camussi G (2006) The proangiogenic phenotype of human tumor-derived endothelial cells depends on thrombospondin-1 downregulation via phosphatidylinositol 3-kinase/Akt pathway. J Mol Med (Berl) 84(10):852–863
57. Adya R, Tan BK, Punn A, Chen J, Randeva HS (2008) Visfatin induces human endothelial VEGF and MMP-2/9 production via MAPK and PI3K/Akt signalling pathways: novel insights into visfatin-induced angiogenesis. Cardiovasc Res 78(2):356–365
58. Duensing S, Munger K (2002) Human papillomaviruses and centrosome duplication errors: modeling the origins of genomic instability. Oncogene 21(40):6241–6248
59. Hida K, Hida Y, Shindoh M (2008) Understanding tumor endothelial cell abnormalities to develop ideal anti-angiogenic therapies. Cancer Sci 99(3):459–466
60. Sato Y (2011) Persistent vascular normalization as an alternative goal of anti-angiogenic cancer therapy. Cancer Sci 102(7):1253–1256
61. Helfrich I, Scheffrahn I, Bartling S, Weis J, von Felbert V, Middleton M et al (2010) Resistance to antiangiogenic therapy is directed by vascular phenotype, vessel stabilization, and maturation in malignant melanoma. J Exp Med 207(3):491–503

62. Gao D, Nolan D, McDonnell K, Vahdat L, Benezra R, Altorki N et al (2009) Bone marrow-derived endothelial progenitor cells contribute to the angiogenic switch in tumor growth and metastatic progression. Biochim Biophys Acta 1796(1):33–40
63. Taylor SM, Nevis KR, Park HL, Rogers GC, Rogers SL, Cook JG et al (2010) Angiogenic factor signaling regulates centrosome duplication in endothelial cells of developing blood vessels. Blood 116(16):3108–3117
64. Kondoh M, Ohga N, Akiyama K, Hida Y, Maishi N, Towfik AM et al (2013) Hypoxia-induced reactive oxygen species cause chromosomal abnormalities in endothelial cells in the tumor microenvironment. PLoS One 8(11):e80349
65. Butler JM, Kobayashi H, Rafii S (2010) Instructive role of the vascular niche in promoting tumour growth and tissue repair by angiocrine factors. Nat Rev Cancer 10(2):138–146
66. Cao Z, Ding BS, Guo P, Lee SB, Butler JM, Casey SC et al (2014) Angiocrine factors deployed by tumor vascular niche induce B cell lymphoma invasiveness and chemoresistance. Cancer Cell 25(3):350–365
67. Cao Z, Scandura JM, Inghirami GG, Shido K, Ding BS, Rafii S (2017) Molecular checkpoint decisions made by subverted vascular niche transform indolent tumor cells into Chemoresistant Cancer stem cells. Cancer Cell 31(1):110–126
68. Zhu TS, Costello MA, Talsma CE, Flack CG, Crowley JG, Hamm LL et al (2011) Endothelial cells create a stem cell niche in glioblastoma by providing NOTCH ligands that nurture self-renewal of cancer stem-like cells. Cancer Res 71(18):6061–6072
69. Pedrosa AR, Trindade A, Carvalho C, Graca J, Carvalho S, Peleteiro MC et al (2015) Endothelial Jagged1 promotes solid tumor growth through both pro-angiogenic and angiocrine functions. Oncotarget 6(27):24404–24423
70. Wieland E, Rodriguez-Vita J, Liebler SS, Mogler C, Moll I, Herberich SE et al (2017) Endothelial Notch1 activity facilitates metastasis. Cancer Cell 31(3):355–367
71. Maishi N, Ohba Y, Akiyama K, Ohga N, Hamada J, Nagao-Kitamoto H et al (2016) Tumour endothelial cells in high metastatic tumours promote metastasis via epigenetic dysregulation of biglycan. Sci Rep 6:28039
72. van Beijnum JR, Dings RP, van der Linden E, Zwaans BM, Ramaekers FC, Mayo KH et al (2006) Gene expression of tumor angiogenesis dissected: specific targeting of colon cancer angiogenic vasculature. Blood 108(7):2339–2348
73. Lu C, Bonome T, Li Y, Kamat AA, Han LY, Schmandt R et al (2007) Gene alterations identified by expression profiling in tumor-associated endothelial cells from invasive ovarian carcinoma. Cancer Res 67(4):1757–1768
74. Buckanovich RJ, Sasaroli D, O'Brien-Jenkins A, Botbyl J, Hammond R, Katsaros D et al (2007) Tumor vascular proteins as biomarkers in ovarian cancer. J Clin Oncol 25(7):852–861
75. Muraki C, Ohga N, Hida Y, Nishihara H, Kato Y, Tsuchiya K et al (2012) Cyclooxygenase-2 inhibition causes antiangiogenic effects on tumor endothelial and vascular progenitor cells. Int J Cancer 130(1):59–70
76. Yamamoto K, Ohga N, Hida Y, Maishi N, Kawamoto T, Kitayama K et al (2012) Biglycan is a specific marker and an autocrine angiogenic factor of tumour endothelial cells. Br J Cancer 106(6):1214–1223
77. Maishi N, Ohga N, Hida Y, Akiyama K, Kitayama K, Osawa T et al (2012) CXCR7: a novel tumor endothelial marker in renal cell carcinoma. Pathol Int 62(5):309–317
78. Osawa T, Ohga N, Akiyama K, Hida Y, Kitayama K, Kawamoto T et al (2013) Lysyl oxidase secreted by tumour endothelial cells promotes angiogenesis and metastasis. Br J Cancer 109(8):2237–2247
79. Osawa T, Ohga N, Hida Y, Kitayama K, Akiyama K, Onodera Y et al (2012) Prostacyclin receptor in tumor endothelial cells promotes angiogenesis in an autocrine manner. Cancer Sci 103(6):1038–1044
80. Alam MT, Nagao-Kitamoto H, Ohga N, Akiyama K, Maishi N, Kawamoto T et al (2014) Suprabasin as a novel tumor endothelial cell marker. Cancer Sci 105(12):1533–1540

81. Otsubo T, Hida Y, Ohga N, Sato H, Kai T, Matsuki Y et al (2014) Identification of novel targets for antiangiogenic therapy by comparing the gene expressions of tumor and normal endothelial cells. Cancer Sci 105(5):560–567
82. Zhuang X, Herbert JM, Lodhia P, Bradford J, Turner AM, Newby PM et al (2015) Identification of novel vascular targets in lung cancer. Br J Cancer 112(3):485–494
83. Pasqualini R, Koivunen E, Kain R, Lahdenranta J, Sakamoto M, Stryhn A et al (2000) Aminopeptidase N is a receptor for tumor-homing peptides and a target for inhibiting angiogenesis. Cancer Res 60(3):722–727
84. Pastorino F, Brignole C, Marimpietri D, Cilli M, Gambini C, Ribatti D et al (2003) Vascular damage and anti-angiogenic effects of tumor vessel-targeted liposomal chemotherapy. Cancer Res 63(21):7400–7409
85. Arap W, Pasqualini R, Ruoslahti E (1998) Cancer treatment by targeted drug delivery to tumor vasculature in a mouse model. Science 279(5349):377–380
86. Kogure K, Akita H, Yamada Y, Harashima H (2008) Multifunctional envelope-type nano device (MEND) as a non-viral gene delivery system. Adv Drug Deliv Rev 60(4–5):559–571
87. Hatakeyama H, Akita H, Harashima H (2011) A multifunctional envelope type nano device (MEND) for gene delivery to tumours based on the EPR effect: a strategy for overcoming the PEG dilemma. Adv Drug Deliv Rev 63(3):152–160
88. Sato Y, Hatakeyama H, Sakurai Y, Hyodo M, Akita H, Harashima H (2012) A pH-sensitive cationic lipid facilitates the delivery of liposomal siRNA and gene silencing activity in vitro and in vivo. J Control Release 163(3):267–276
89. Sakurai Y, Hatakeyama H, Sato Y, Hyodo M, Akita H, Ohga N et al (2014) RNAi-mediated gene knockdown and anti-angiogenic therapy of RCCs using a cyclic RGD-modified liposomal-siRNA system. J Control Release 173:110–118
90. Whitehead KA, Langer R, Anderson DG (2009) Knocking down barriers: advances in siRNA delivery. Nat Rev Drug Discov 8(2):129–138

Stromal Barriers Within the Tumor Microenvironment and Obstacles to Nanomedicine

Hiroyoshi Y. Tanaka and Mitsunobu R. Kano

Abstract Nanomedicine has been heralded as the elusive "magic bullet" of cancer chemotherapeutics ever since the description of the Enhanced Permeability and Retention (EPR) effect in 1986. However, decades of research have often shown a great discrepancy between the salient effects seen in preclinical models and the suboptimal effects seen in clinical trials. This indicates that various obstacles exist within human tumors that impede the delivery and penetration of nanomedicine and that the EPR effect itself may be necessary but not sufficient for an efficacious nanomedicine formulation. Furthermore, these obstacles may be absent or much weaker in often used preclinical models, pointing at the importance of developing novel, clinically relevant preclinical models for testing the efficacy of nanomedicine. It is becoming increasingly clear that the various cellular and extracellular matrix components of the tumor stroma that together consist the tumor microenvironment (TME) play an important role in determining the efficiency of nanomedicine penetration into the tumor. We refer to the impediments that these stromal components of the TME pose to nanomedicine as "stromal barriers". In this chapter, we review the factors affecting nanomedicine delivery with a particular emphasis on the stromal barriers within the TME. We also review the preclinical models available for testing the efficacy of nanomedicine, and how novel models might be developed to further our understanding of the principles governing nanomedicine delivery and penetration into tumors.

Keywords Enhanced permeability and retention (EPR) effect · Nanomedicine delivery · Stromal barriers · Tumor microenvironment

---

H. Y. Tanaka
Department of Pharmaceutical Biomedicine, Okayama University Graduate School of Medicine, Dentistry and Pharmaceutical Sciences, Okayama, Japan

M. R. Kano (✉)
Department of Pharmaceutical Biomedicine, Okayama University Graduate School of Medicine, Dentistry and Pharmaceutical Sciences, Okayama, Japan

Department of Pharmaceutical Biomedicine, Okayama University Graduate School of Interdisciplinary Science and Engineering in Health Systems, Okayama, Japan
e-mail: mikano-tky@umin.net

Y. Matsumura, D. Tarin (eds.), *Cancer Drug Delivery Systems Based on the Tumor Microenvironment*, https://doi.org/10.1007/978-4-431-56880-3_4

Abbreviations

ABC	accelerated blood clearance
ADC	antibody-drug conjugate
CAF	cancer-associated fibroblast
CAST	cancer stromal targeting
CDX	cell-line-derived xenograft
CPP	cell-penetrating peptide
ECM	extracellular matrix
EMT	epithelial-to-mesenchymal transition
EndMT	endothelial-to-mesenchymal transition
EPR	enhanced permeability and retention
GEMM	genetically engineered mouse model
HA	hyaluronic acid
IFP	interstitial fluid pressure
LOX	lysyl oxidase
MSC	mesenchymal stem cell
PDX	patient-derived xenograft
PEG	polyethylene glycol
PFT	pericyte-to-fibroblast transition
RES	reticuloendothelial system
TAM	tumor-associated macrophage
TCGA	The Cancer Genome Atlas
TEC	tumor endothelial cell
TEM	tumor endothelial marker
TGF-β	transforming growth factor-beta
TME	tumor microenvironment
VEGF	vascular endothelial growth factor

Introduction

Decades of intensive cancer research resulted in the discovery and successful development of various chemotherapeutic agents, the epitome of which is the existence of numerous molecularly targeted drugs that have revolutionized cancer chemotherapy [1–3]. However, despite the remarkable progress being made, complete eradication of tumor cells is often not attainable and eventually relapse ensues. Side effects also remain a significant issue. A major reason for this shortcoming is that drug molecules diffuse and freely distribute throughout the body with little specificity, thus failing to accumulate *en masse* at the disease site. This results in both a failure to demonstrate efficacy due to an inability to achieve local concentrations high enough to exert anti-tumor effects and deleterious side effects due to the non-specific distribution of drugs in vulnerable healthy tissue.

The Enhanced Permeability and Retention (EPR) effect theory, first described by Matsumura and Maeda in 1986 [4], posits that the tumor neovasculature is immature—demonstrating large fenestrations [5]—thus resulting in extravasation of macromolecules that in normal vessels of healthy tissues would remain within the bloodstream (*enhanced permeability*). In addition, the extravasated macromolecules accumulate within the tumor tissue because intratumoral lymphatics are underdeveloped and the tumor is poorly drained (*enhanced retention*). Therefore, nanomedicine delivery via the EPR effect has emerged as the "magic bullet" (Paul Ehrlich) of cancer chemotherapeutics. This is because, theoretically, the EPR effect should enable the selective and efficient accumulation of nanomedicine—macromolecules, supramolecules, or solid particles designed and/or used for a therapeutic/diagnostic purpose with a size usually ranging from 10 to 200 nm [6, 7]—in tumors. The EPR effect thus has served as the theoretical underpinning of the formulation and development of various nanomedicines over recent years (for a comprehensive review on the EPR effect, the reader is referred to [8, 9]).

Indeed, the liposomal formulation of doxorubicin (Doxil) and albumin-bound paclitaxel (Abraxane), both of which are clinically approved nanomedicine formulations of traditionally used small molecular weight chemotherapeutics, demonstrate improved safety and tolerability profiles compared to the respective conventional formulations of each [10, 11]. However, these are the very few—albeit significant—success stories. Despite the fact that nanomedicine works miraculously well in pre-clinical tumor models, very few formulations have actually made it to the bedside. Moreover, even those formulations which have been approved for clinical use often offer marginal, or at best modest, improvements in efficacy over conventional therapy [12, 13].

There is, therefore, a clear discrepancy between the salient efficacy of nanomedicine formulations demonstrated pre-clinically and the sub-optimal efficacy observed in the clinic. Such a discrepancy is typified by MCC-465, a formulation in which doxorubicin is encapsulated in immunoliposomes. MCC-465 demonstrated clear efficacy in pre-clinical models [14] but failed to show any anti-tumor effect in a phase I clinical trial [15]. Taken at face value, this gap between the bench and bedside suggests that there are therapeutic obstacles in human disease that are apparently much weaker or absent in the experimental models traditionally used to validate novel nanomedicine formulations. Accordingly, a fuller understanding of the mechanisms governing nanomedicine efficacy is clearly needed so that we may resolve this gap. This chapter thus aims to provide an overview of the factors affecting nanomedicine efficacy. We particularly focus on the various obstacles that the tumor microenvironment (TME) poses to nanomedicine in terms of efficient delivery.

Obstacles to Successful Nanomedicine Delivery

As elegantly conceptualized by Blanco et al. among others [16–21], it is important to keep in mind the path an intravenously administered nanomedicine must pass to successfully locate a tumor target. Before a nanomedicine can exert its antitumor effects, it must efficiently clear each of the following steps:

Step 1, evasion of clearance: evade clearance in the kidneys or by the reticuloendothelial system (RES)
Step 2, selective delivery: be preferentially delivered to the tumor neovasculature while avoiding non-specific distribution
Step 3, margination: demonstrate proper margination in the bloodstream
Step 4, extravasation: exit from the tumor blood vessel into the tumor tissue
Step 5, stromal penetration: penetrate the tumor stroma/interstitium and locate tumor cells
Step 6, internalization: be internalized by tumor cells and successfully reach its target-of-action while avoiding drug detoxification and/or efflux

Challenges exist at each and every step. The failure of administered nanomedicines to deliver more than minimal therapeutic effects is due to the inability of the nanomedicine to successfully surmount one or more of these obstacles. These impediments to successful nanomedicine delivery can be divided largely into the following, not necessarily mutually exclusive, factors: obstacles to tumor localization (constituted mainly by **steps 1–3**), intratumoral accumulation (mainly **steps 3** and **4**), intratumoral penetration (mainly **step 5**), and internalization (mainly **step 6**) (Fig. 1). Obstacles to tumor localization and intratumoral accumulation together result in an attenuation of the EPR effect to a level insufficient to attain a therapeutic response. Furthermore, obstacles to intratumoral penetration and internalization suggest that the EPR effect, though necessary, is by itself insufficient to realize a truly efficacious nanomedicine. That is, in desmoplastic, stroma-dense tumors such as pancreatic cancer, strategies to augment the EPR effect may lead to improvement of tumor localization and intratumoral accumulation, but the nanomedicine may still be unable to reach the tumor cells due to hindrance by the fibrotic tissue [22].

Consequently, a successful nanomedicine strategy must, ideally, incorporate and integrate the following strategies. Each strategy corresponds to the impediments described above:

Strategy 1: improve tumor localization
Strategy 2: increase intratumoral accumulation
Strategy 3: enhance stromal penetration
Strategy 4: optimize internalization and avoid drug detoxification/efflux

This can, at least theoretically, be achieved either by the generation of a multifunctional nanomedicine formulation addressing all of the abovementioned obstacles. Perhaps more realistically, a combination of nanomedicine(s) and/or drug(s) each targeting a subset of the obstacles might be used. Examples of the latter strategy

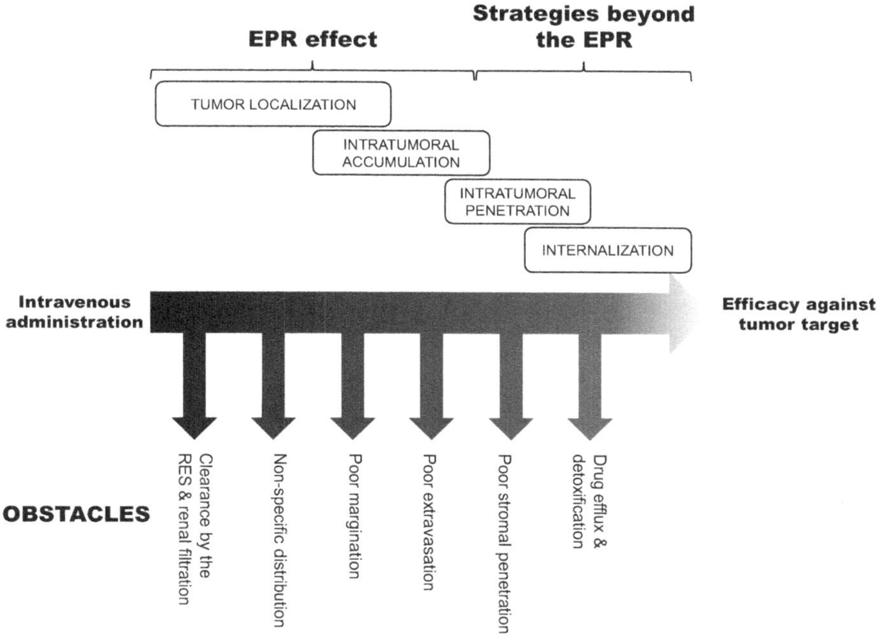

Fig. 1 The various obstacles to nanomedicine delivery and penetration into tumors. After intravenous administration, a nanomedicine formulation must avoid various obstacles before locating a tumor target and demonstrating therapeutic efficacy. These obstacles include clearance by the reticuloendothelial system (RES) and renal filtration, non-specific distribution, poor margination, poor extravasation, poor stromal penetration, and drug efflux and/or detoxification. These obstacles can further be organized into four categories: obstacles to 1) tumor localization, 2) intratumoral accumulation, 3) intratumoral penetration, and 4) internalization. The Enhanced Permeability and Retention (EPR) effect may be leveraged to increase tumor localization and intratumoral accumulation, but strategies beyond the EPR effect are necessary to increase intratumoral penetration and internalization

are albumin-bound paclitaxel together with gemcitabine [23] and nanoliposomal irinotecan with fluorouracil and folinic acid [24], both of which have been shown to extend survival in metastatic pancreatic cancer compared to monotherapy. In any case, a comprehensive understanding of the obstacles which limit nanomedicine delivery and efficacy is necessary to successfully overcome the obstacles, one by one.

Factors which Affect Nanomedicine Delivery

Numerous factors are known to affect nanomedicine delivery. It is worth noting, however, that there are many discordant findings to date. This perhaps points at material-specific characteristics (*e.g.*, what holds true for nano-sized micelles may not be true for metallic nanoparticles, *etc.*), cell type, cell line, disease type, and/or species-specific effects (*e.g.*, epithelial cells *vs.* macrophages, pancreatic cancer cell lines MiaPaCa-2 *vs.* BxPC-3, colon cancer *vs.* pancreatic cancer, mice *vs.* human, *etc.*), or simply issues in reproducibility of nanomedicine formulations especially across research groups. We have thus yet to arrive at a practical and universal guiding design principle of nanomedicine formulations. Therefore, this section will simply aim to give a general overview of factors which have been reported to affect nanomedicine delivery, and avoid delving into the details of individual papers.

Though, as mentioned above, a myriad of factors are known to affect nanomedicine delivery, they can largely be classified into the two following categories:

Factor 1: Properties of the nanoparticle
Factor 2: Properties of the tumor tissue

By "property", we refer to anything biological, chemical, and/or physical. Though the above dichotomy is most surely over-simplistic in certain aspects, it conceptually corresponds to the schematic of obstacles mentioned above. That is, each barrier requires optimal tuning of different aspects of the above two factors, with an overall larger contribution of **factor 2** in the later steps of the delivery process. For example, **step 1** and **step 2** in nanomedicine delivery is before the nanoparticle even reaches the tumor tissue; therefore, it is probably more rewarding to tune **factor 1** rather than to try interfering with **factor 2**.

In contrast, for **step 3**, **factor 1** (such as nanoparticle size and shape which have been fairly well studied) has a clear impact on margination dynamics [25–31], but there is also an indirect effect of **factor 2**. As solid tumors have an immature, leaky neovasculature [5], and also demonstrate increased solid stress and interstitial fluid pressure (IFP) which could lead to vessel deformation and/or collapse [32–34], the hemorheological parameters of tumor neovessels are likely much different from those of healthy, physiological vessels [35, 36]. Moreover, for **step 4** and **step 5** which are likely dependent on the properties of the nanoparticle (such as size, as described later in this chapter) relative to that of the tumor blood vessel (such as fenestration size of the blood vessel for **step 4**, and the amount and density of fibrous tissue for **step 5**), one could adjust **factor 1** [37, 38] to optimally correspond to the given tumor blood vessel. Alternatively, one could also attempt to favorably alter the tumor blood vessel or target the fibrous tissue itself [36, 39, 40] (**factor 2**). Likewise, in **step 6**, the importance of nanoparticle size, shape, and charge (**factor 1**) have been well studied and an increasing number of **factor 1** parameters are now appreciated to be tunable properties in improving cellular uptake. In addition, because cells have intrinsic detoxification [41, 42] and drug efflux mechanisms [43–45] which attenuate therapeutic efficacy even after successful internalization of the nanomedicine (**factor 2**), pharmacological inhibition of these processes is likely an important supplementary strategy.

Throughout the process of nanomedicine delivery, successful distribution of the nanomedicine to the tumor depends on a favorable interaction between the nanoparticle and its surrounding biological (not necessarily limited to the tumor) microenvironment. In this regard, tuning of **factor 1** and optimization of **factor 2** are complementary strategies in attaining this goal—although the difficulty of each strategy may vary according to the interaction in question. Thus, both a thorough understanding of the biophysical microenvironment that the nanomedicine encounters once administered *in vivo* and also the ability to finely tune the biophysicochemical properties of the nanoparticle are necessary to promote an optimally efficacious and safe interaction. The next sub-section will provide an overview of general considerations regarding **factor 1**, and the ensuing sections will focus on **factor 2**.

Properties of the Nanoparticle

The properties of a given nanomedicine are multifactorial. It can be described in physical (*e.g.*, size and morphology), chemical (*e.g.*, chemical composition), and/or biological (*e.g.*, biocompatibility) terms. Because of the interdependent nature of many of these properties, it is important to keep in mind that whilst a researcher may intend to change only a particular property *X* of a nanoparticle (as an example, let *X* be size), there may be other properties which may be altered together with *X*. This could, for example, be charge—which may be controlled if one uses a neutral nanoparticle so as to make charge independent of size, or surface area—which is virtually impossible to control as it necessarily changes with size if morphology is kept similar.

Another possible caveat is that the diverse mechanisms leading to nanomedicine accumulation within tumor tissues are not completely understood. Therefore, even when a given parameter appears not to have any effect on nanomedicine accumulation within tumors, it may be that distinct (and perhaps unknown) mechanisms are being mobilized with the *net* effect being zero. For example, nanoparticles with certain chemical compositions have been reported to engage a differential set of internalization mechanisms depending on charge (reviewed in [46]). Thus, even if two nanomedicine formulations with different charges *a* and *b* demonstrate a comparable level of tumor cell uptake, it may be that nanoparticle with charge *a* preferentially engages clathrin-dependent endocytosis, while nanoparticle with charge *b* preferentially engages clathrin-independent endocytosis. However, both nanoparticles may each engage its respective endocytotic mechanism to a similar extent such that the amount endocytosed does not demonstrate an appreciable difference. It would clearly be wrong to conclude from these results that "nanoparticle charge does not affect cellular uptake and is thus not a property that must be tuned". The use of multiple formulations and/or cross-validation across multiple experimental models is necessary to avoid such a mistake.

A similar problem arises when only a single step is put into consideration. For example, let us consider the case in which one optimized the nanoparticle surface chemistry and charge to direct efficient cellular uptake (**step 6**). While optimized for cellular uptake, differences in nanoparticle surface chemistry and charge have been shown across diverse nanoparticle formulations to affect the composition of protein corona formation within the bloodstream [47–51]. This, in turn, will likely affect the distribution and/or clearance of the nanoparticle [52, 53]—a change which may or may not be synergistic with the optimization of **step 6**. The *net* effect *in vivo*, therefore, remains unclear from a simple optimization of a single step in the multi-step process culminating in a nanomedicine formulation successfully exerting its therapeutic effect.

The choice of experimental models is especially crucial here. Both the conspicuous and tacit assumptions and limitations of the model used must carefully be assessed. Continuing with the example of "optimizing the nanoparticle for efficient tumor cell uptake" used above, it is common practice to use cultured tumor cells on 2D plastic culture-ware when assessing nanoparticle uptake. However, it has been known for some time, that cells cultured on plastic—orders of magnitude stiffer than biological tissues [54]—in most cases demonstrate phenotypes largely different from those under physiological conditions. Indeed, substrate stiffness regulates the uptake efficiency of nanoparticles [55]: thus, findings *in vitro* may not necessarily directly translate into nanoparticle uptake efficiency *in vivo*.

Some general principles of experimental design and interpretation can be derived from the above discussion. First, results of experiments analyzing the effect of a particular property of a nanoparticle should be interpreted with caution as both known and unknown confounding factors may exist. Second, to increase the generalizability of the findings, multiple experimental models should be utilized for cross-validation when possible. Third, nanoparticles should ideally be comprehensively designed to direct optimal behavior at each of the steps leading to intratumoral accumulation.

That being said, it is now fairly clear that physicochemical properties such as nanoparticle size, charge and charge density, shape, stiffness/elasticity, hydrophobicity, surface area (roughness and porosity), and surface chemistry have a clear impact on the behavior of nanoparticles both *in vitro* and *in vivo* [37, 46, 56–61]. Furthermore, the presence of internal disorder has been demonstrated within single nanoparticles, thus suggesting that nanoparticles cannot be merely comprehended as nanometer-sized pieces of bulk material [62]. Whether such internal disorder affects nanomedicine delivery has not yet been directly investigated, but it may offer new insights into the unique pharmacokinetic behavior that nanoparticles demonstrate compared to conventional small-molecules.

Numerous studies on various nanoparticle formulations have been performed to assess how the properties of the nanoparticle affect nanomedicine penetration and how these properties may be tuned for an optimal penetration of tumors. The optimization of nanoparticle properties is of paramount importance to overcome the barrier at **step 1**: opsonization/phagocytosis by the RES and exclusion from the kidneys [19, 20]. It has now become common practice to modify nanoparticles by polymers such as polyethylene glycol (PEG) to evade elimination by the RES, and to design nanoparticles larger than 10 nm to decrease renal filtration. However, decorating the surface of nanomedicine with PEG tends to also hinder uptake by target cells (known as the "PEG dilemma") [7, 63], and further advances are clearly warranted. Unfortunately, a detailed inspection of these studies, though valuable and interesting, is beyond the scope of this chapter, and the reader is referred to reviews published elsewhere [46, 57–61].

Properties of the Tumor Tissue

As aforementioned, successful intratumoral accumulation of nanomedicine and subsequent exertion of its therapeutic effects rely on an optimal relationship between the properties intrinsic to the nanoparticle (such as those described in the previous subsection) and to the tumor, as we describe in this and following sections.

Biological Properties of the Tumor Tissue

Properties of the tumor tissue can be largely divided, though not mutually exclusively, into biological and physical factors. "Biological properties" encompass a wide range of aspects, including the histotype of the tumor (*e.g.*, adenocarcinoma *vs*. squamous cell carcinoma), the molecular subtype of the tumor, and histopathological characteristics such as vascularity, vessel architecture, and stromal content and phenotype. Our current knowledge of how histopathological characteristics alter nanomedicine efficacy, especially in **steps 3–5**, will be discussed in detail in section "Stromal Barriers to Nanomedicine Penetration Within the TME".

Especially with the advent of next-generation sequencing technologies, which have accelerated the comprehensive characterization of the molecular profile of tumors (the most notable of which has been carried out by The Cancer Genome Atlas [TCGA] project such as [64–68] among many others), it is now widely appreciated that even tumors with the same histotype originating from the same organ are highly heterogeneous. Importantly, this heterogeneity exists not only between patients but also in a spatiotemporal manner in individual patients [69, 70], which could lead to heterogeneity in the EPR effect [9]. There is now an increasing emphasis on moving forward from correlation and classification of tumors to understanding the importance of heterogeneity—now known to exist even at the single-cell level [71–73]—in pathogenesis and clinical outcome [74–76].

In regards to nanomedicine, our increased knowledge of the molecular subtypes of various cancers has largely not yet been systematically correlated with nanomedicine efficacy. However, a particularly notable advance brought forth by the data surge has been the facilitation of discovering genes specifically up-regulated in particular tumors which in turn has increasingly been informing the development of actively targeted nanomedicine formulations (discussed later in further detail in the section "Strategies to Overcome Stromal Barriers for Enhanced Nanomedicine Penetration"). Active targeting relies on a molecular target specifically expressed in the TME and a targeting moiety within the nanoparticle formulation which interacts with the intended molecular target [77–80]. Needless to say, the comprehensive molecular profile data of a given tumor will also offer guidance regarding effective drug choices [81, 82]. In the precision medicine era [83], we can envisage a strategy whereby a given patient's tumor is comprehensively characterized molecularly and assessed for the optimal targeting moiety. This moiety will then be linked to drugs and/or drug carriers with the most likely chance of demonstrating efficacy.

Physical Properties of the Tumor Tissue

By "physical properties", we refer to the various abnormal mechanical forces that characterize tumors. These include high IFP [84], elevated solid stress [85], increased extracellular matrix (ECM) stiffness [86], as well as the stagnant and/or disturbed flow within tumor neovessels resulting from its aberrant branching and tortuosity [87, 88].

The physical properties of the tumors—the characteristic, abnormal mechanical forces at play within tumors—presumably arise as a result of various biological processes associated with tumorigenesis. For example, high IFP results from increased vascular permeability, poor lymphatic drainage, intratumoral fibrosis, and/or ECM contraction [84]. Furthermore, increased solid stress within tumors comprises of externally applied stress, swelling stress, and residual stress as conceptualized in a recent, excellent review by Stylianopoulos [85]. Externally applied stress occurs since tumor cells proliferate abnormally, and the tumor volume expands at the expense of the surrounding, normal host tissue [33, 89, 90]. Swelling stress in tumors has been shown to result from increased hyaluronic acid (HA) production within tumors, and/or the abnormal tonicity of the tumor cells which perhaps is a response to the compressive solid stresses exerted onto tumor cells by its surrounding environment [85, 91, 92]. Residual stress is any solid stress that remains after the tumor has been excised and results, at least in part, from tumor growth itself [33].

These mechanical forces have a clear impact on nanomedicine delivery. Because distribution into tumors rely either on diffusion (determined by the concentration gradient between the blood vessel and the tumor) or convection (determined by hydrostatic and oncotic pressure gradients between the blood vessel and the tumor) [58, 93], an elevated IFP and/or solid stress will result in decreased intratumoral penetration of nanomedicine (effect on **step 4** and **5**). Furthermore, when IFP and solid stress elevate beyond a certain threshold, it may result in vascular deformation or collapse resulting in poor and/or intermittent perfusion of tumors (effect on **steps 2–4**)—even when tumors histopathologically appear to be highly vascular.

While abnormal physical properties of the tumor may arise as a consequence of biological processes, physical properties merit consideration individually as a factor which must be addressed for successful nanomedicine and not merely as a subordinate to biological properties. This is first of all because, from an experimental perspective, methods to analyze biological properties and physical properties are largely different. Therefore, without specialized methodologies and equipment to measure the mechanical forces at play, much of our understanding of the biomechanics of tumors remain speculative or theoretical (driven by mathematical/

physical models) at best. Indeed, many facets of the biomechanics of tumors remain largely understudied to date [85]. Furthermore, it is increasingly becoming clear, thanks to the rapidly evolving field of mechanobiology and mechanotransduction [94, 95], that mechanical forces exert a strong effect on and is an essential determinant of the biological behavior of tumor cells and its surrounding stromal cells [96–105]. For example, increased ECM stiffness is sufficient to induce activation of fibroblasts, resulting in additional deposition of ECM components and contraction of the ECM network which further increases ECM stiffness [96, 106]—forming a vicious loop. This suggests that the physical properties of the tumor are in itself a therapeutic target [105]. Indeed, targeting ECM components or mechanotrasducing pathways drastically remodel the TME and result in altered therapeutic responses [86, 96, 106–109], as we will discuss further later.

As with the biological properties of tumors, it is becoming increasingly clear that the TME is also highly heterogeneous from a physical perspective [110–112]. This will likely lead to the non-uniform distribution of nanomedicine. However, the mechanisms, both biological and physical, leading to this heterogeneity is incompletely understood. This is partly due to the general lack of feasible experimental models and techniques to monitor mechanical forces at high spatiotemporal resolution. Advances in this direction should greatly inform nanoparticle design and strategies of therapeutic interventions to the TME and are much awaited.

Stromal Barriers to Nanomedicine Penetration Within the TME

Analyzing the histopathological characteristics of the tumor to be targeted by nanomedicine is very important as it gives us an idea of what structures lie in between the blood vessel and where tumor cells exist—the "stromal barriers" within the TME that nanomedicine must overcome to exert its therapeutic effect [18].

Let us begin by considering a simple, but a particularly illuminating example. The difference in vascular structure is a histopathological characteristic that is an important determinant of intratumoral nanomedicine accumulation. Tumors with relatively high permeability to nanoparticles are generally characterized by low pericyte coverage, while those which are poorly permeable are characterized by abundant pericyte coverage [35, 113]. For example, pancreatic cancer is characterized by blood vessels with abundant pericyte coverage, whereas coverage is less prominent in colon cancer. Diminishing pericyte coverage of tumor neovessels in pancreatic cancer with an inhibitor of Transforming Growth Factor-β (TGF-β) signaling results in an enhancement of the EPR effect [37, 39], whereas the benefit of TGF-β inhibition in colon cancer is negligible [40]. Conversely, experimentally increasing pericyte coverage of neovessels in colon cancers results in the enhanced vascular integrity of these tumors [114]. These findings extend the obligatory role of pericytes in the maintenance of vessel stability [115] to the hindrance of nanoparticle extravasation.

It is also important in what environment tumor neovessels are situated. According to a simple conceptual framework proposed by Smith et al., tumors can be divided into the tumor-vessel type and stroma-vessel type [116]. Tumor-vessel type tumors are characterized by tumor neovessels which exist in close proximity to tumor cells with very little in between. In contrast, stroma-vessel type tumors are characterized by tumor neovessels that are embedded in thick, stromal tissue. This difference in histopathology is crucial: it affects the response of these tumors to various inhibitors, suggesting that blood vessels of each tumor have different functional consequences [40, 116].

We now, therefore, give an overview of the main cellular constituents of the TME (the reader is referred to [117] for a concise but instructive graphic demonstration) which are known to affect the intratumoral penetration of nanomedicine. This will give us an idea of what factors we must incorporate into preclinical models (discussed in the section "Development of relevant pre-clinical models of the TME") and what factors may be targeted to enhance nanomedicine efficacy (discussed in the section "Strategies to Overcome Stromal Barriers for Enhanced Nanomedicine Penetration").

Endothelial Cells

Endothelial cells show great plasticity [118, 119] and tissue-specific molecular and morphological characteristics [120] (*e.g.*, large fenestrations in liver sinusoids *vs.* blood-brain barrier formation in blood vessels of the brain). The great variety that exists in endothelial phenotype likely affects to what extent endothelial cells affect nanoparticle extravasation in a given tissue under various physiological and/or pathological conditions (**step 4**). Indeed, endothelial cells found within tumors have been shown to demonstrate tumor-specific characteristics, giving rise to the term "tumor endothelial cells (TECs)" [121, 122]. TEC-specific expression of certain "marker" proteins, referred to as tumor endothelial markers (TEMs) [123, 124], will enable the specific targeting of nanomedicine to tumors via active targeting [125] (**steps 1 and 2**). How the TEC phenotype affects vascular bursts (**step 4**)—dynamic vents in the tumor vasculature which contribute to the intratumoral accumulation of nanoparticles [38]—has not yet been assessed but is an interesting question. Endothelial cells have also been reported to contribute to intratumoral fibrosis (**step 5**) via the process of endothelial-to-mesenchymal transition (EndMT) [126].

Pericytes

Pericytes, which are the perivascular stromal cells covering the endothelial lining of the blood microvessels, are important regulators of vascular integrity as described above [115]. As such, pericytes exert important effects on the extravasation of nanoparticles (**step 4**). Pericytes, like endothelial cells, have been shown to also contribute to intratumoral fibrosis (**step 5**) via the process of pericyte-to-fibroblast transition (PFT) [127].

Fibroblasts

Fibroblasts within tumors are referred to as cancer-associated fibroblasts (CAFs), a highly heterogeneous population originating from various cell types [128] including resident fibroblasts, mesenchymal stem cells (MSCs) [129], cancer cells via the process of epithelial-to-mesenchymal transition (EMT) [130–132], endothelial cells [126], and pericytes [127]. It should be noted, however, that the specific contribution of EMT and EndMT to intratumoral fibrosis is a matter of significant debate [133, 134].

Fibroblasts secrete abundant ECM which poses a direct physical barrier to nanomedicine passage through the stroma (**step 5**). The ECM also gives rise to physical forces (*e.g.*, IFP, solid stress, substrate stiffness) which affect nanomedicine penetration through vascular compression/collapse (**steps 1–4**). Furthermore, not only is the ECM abundantly deposited but is also abnormally structured [135–137]. This altered ECM architecture directs phenotypic change of cells of the TME, such as myofibroblastic activation [135, 138, 139] and perhaps also induction of EndMT [140] and EMT [100] via substrate stiffening. The processes further enforce the fibrotic response within tumors (**step 5**). The role of CAFs in the pathogenesis of tumors is complex, with both pro- and anti-tumor functions reported [128]. We will discuss this point further in the section "Strategies to Overcome Stromal Barriers for Enhanced Nanomedicine Penetration".

Immune Cells

Various types of immune cells exist in the tumors including the tumor-associated macrophages (TAMs), B and T lymphocytes, natural killer cells, and dendritic cells, each with tumor-promoting and tumor-suppressive subgroups [141]. The immune cell composition within the TME is likely dependent on not only cancer type but differs between patients [142] and within a patient in a spatiotemporal manner [143, 144]. The intratumoral immune response significantly contributes to "non-resolving inflammation" [145] which gives rise to multiple hallmarks of cancer [146]. Notably, in regards to nanomedicine, inflammation affects vascular function and permeability [119] (**steps 1–4**), promotes the fibrotic response [119, 128, 130] (**step 5**), and mediates drug resistance for example through the regulation of the expression of multiple drug efflux transporters [147] (**step 6**). Also, the presence of immune cells, especially macrophages, will inevitably result in off-target phagocytosis of nanoparticles [17].

Development of Relevant Pre-Clinical Models of the TME

There is now an increased appreciation of cancer as a complex disease, with a greater emphasis on the active role that the TME plays (a comparison of [148] and [146], which have been written by the same authors 11 years apart, is particularly illuminating). However, development of experimental models that recapitulate the complexity of the disease is somewhat lagging. The presence of stromal barriers within the TME of various tumors that impede penetration of nanomedicine underscores the necessity of experimental models that sufficiently integrate the various essential components of the TME. In this section, we provide an overview of the present status and future directions in the development of clinically relevant pre-clinical experimental models of the TME.

in vivo *Models of the TME*

The current *de facto* standard of demonstrating the therapeutic efficacy of a particular nanomedicine formulation is *in vivo*, usually with murine tumor models. Murine tumor models can generally be divided into cell-line-derived xenograft (CDX) models, genetically engineered mouse models (GEMM), and patient-derived xenograft (PDX) models [149]. Especially in regards to the TME, model selection is of paramount importance. For example, most commonly used cell-line-based xenografts of pancreatic cancer fail to show appreciable levels of fibrosis [150] with the notable exception of the BxPC-3 cell-line which demonstrates prominent fibrosis occupying up to 60% of the tumor area [151]. GEMMs and PDX models have been shown to demonstrate fibrosis and more closely mirror human histopathology.

CDX models are generated by inoculation of established cancer cell lines and can further be classified by the site to which the tumor cells are inoculated: ectopic (usually subcutaneous) or orthotopic. Though ectopically inoculated cancer cells do recruit host murine stromal cells to a certain extent, it is usually insufficient to adequately recapitulate human histopathology. Orthotopic CDX models are somewhat better in this regard but are usually more technically demanding than subcutaneous CDX models. Though CDX models are widely used for its technical ease and short experimental duration, it is not without its drawbacks. First, as abovementioned, simple inoculation of tumor cells fails to recapitulate histopathological characteristics of the original tumor. Second, most published studies use a restricted number of cell lines [150], which perhaps is advantageous for reproducibility, but fails to represent the interpatient heterogeneity that exists. Third, molecular profiles of cell lines may be different from those of clinical tumors [152, 153] in certain cases because cancer cells with certain mutations cannot be propagated by standard culturing [154]. Fourth, human tumor cell lines must be inoculated into immunodeficient mice, which precludes the accurate assessment of immunological processes. Finally, the predictive value of the efficacy of therapeutic agents by CDX models is dependent on tumor type, showing utility in certain tumors but not in others [155–157].

GEMMs have been developed to overcome some of the shortcomings of the CDX model [158]. GEMMs are usually generated by the recapitulation of genetic aberrations present in human cancer. For example, GEMMs of human pancreatic cancer have been developed by recapitulating clinically relevant mutations in *Kras*, in combination with *Tp53*, *Cdkn2a*, *Smad4*, and *Tgfbr2* [159–162]. The tumors formed in these GEMMs more closely mirror the histopathology of human pancreatic cancer than CDX models. Together with the advent of various genome-editing technologies [163–165], GEMMs hold great potential in disease modeling and analyses [166]. However, GEMM tumors generally tend to require more time to develop and to be more heterogeneous in terms of latency and growth; though this is in a sense reminiscent of human disease, it makes experiments more challenging.

To generate the PDX model, tumor specimens obtained from patients are implanted into immunodeficient mice. Though the methodology itself has been known for some time [167], it was only with the recognition of the shortcomings of the CDX model that the PDX model gained momentum [168]. PDX models retain/recapitulate many of the features of human histopathology, although human stromal cells gradually get replaced by host murine stromal cells upon subsequent serial passages [168, 169]. A salient feature of the PDX model is that it can guide treatment choices on a personal basis [170, 171]. Taking this a step further, a virtual "clinical trial" can be conducted—which may best be expressed as the "avatar model" [169]. Indeed, Gao et al. have tested this idea by generating a large library of PDX tumors with a diverse set of driver mutations and evaluating the therapeutic response to 62 treatments across six indications [172]. Finally, we can envisage a future in which "co-clinical trials" whereby clinical trials are performed in parallel in paired PDX models become normal practice. Co-clinical trials, a concept which was initially proposed using GEMMs [173, 174], are expected to expedite identification of biomarkers predictive of therapeutic response and assessment of novel treatment combinations [169, 173]. There are some drawbacks to the PDX model, however. First, PDX models take months to develop consuming considerable resources; this may pose a limitation on the use of PDX models in precision medicine especially for rapidly progressing tumors. Second, engraftment efficiency is not uniform and has been associated with adverse clinicopathological features in the patient of origin [175, 176]. As in the case of CDX models, this may result in a failure to represent the full spectrum of human disease. Finally, as with human tumor CDX models, PDX models are also maintained in immunodeficient mice. The generation of mice with a humanized immune system and engraftment of PDX models in such mice have been proposed as a solution and warrants further study [177].

in vitro *Models of the TME*

Given the intricate web of interactions between various types of cells in the TME, *in vivo* studies should be complemented with *in vitro* studies to fully shed light on the mechanisms governing the generation and maintenance of the TME and how it may affect nanomedicine efficacy. Therefore, the development of clinically relevant *in vitro* models of the TME is crucial, especially when rigorous analyses on the molecular scale are required.

While conventional *in vitro* studies were performed in 2D (*i.e.*, cells cultured on a flat, plastic surface), there is accumulating evidence that the topological arrangement of cells is an important determinant of cellular phenotype [178–180] and that cells actively respond to flow conditions [181–183]. Therefore, while 2D assays remain an important part of our experimental repertoire, it is becoming ever important to think in 3D [184]. Recently, remarkable progress has been made in 3D culture techniques and microfluidic technologies. These experimental techniques are greatly increasing our ability to model more and more complex phenomena. For example, there is now a wide range of techniques [185–188] which enable culturing of cells in 3D such as the hydrogel matrices which provide scaffolds with tunable properties [189], spheroids [190], organoids/organotypic cultures [191–193], and layer-by-layer techniques [194, 195]. Together with advances in microfluidics [196], these two technologies combined enable the generation of organs-on-chips [197] or cancer-on-chips [198–200] which are expected to integrate most, if not all, relevant components of the TME. Though each of the above technologies merits detailed inspection, it is beyond the scope of this chapter and the reader is referred to reviews dedicated to each technology published elsewhere.

Let us then consider what aspects of the TME might best be modeled *in vitro* and the engineering considerations to be made in each case. We will follow the steps that an intravenously administered nanomedicine takes before locating a tumor target and exerting its therapeutic effects.

Step 1, Evasion of Clearance The RES has traditionally been modeled *in vitro* by the standard culture of phagocytic cells (often macrophage cell lines) in 2D. However, it is now known that the mechanical and physical environment alters macrophage phenotype [201]. Indeed, as aforementioned, substrate stiffness alters nanoparticle uptake [55]. Therefore, to better mimic *in vivo* conditions, biomaterials with clinically relevant stiffness may be used instead of plain plastic to culture phagocytic cells. Furthermore, macrophages are cells with remarkable plasticity [202] and become educated by the TME to assume a TAM phenotype [203]. Addressing this factor will require co-culturing of multiple cell types other than phagocytic cells to mimic the heterotypic cellular interactions governing the phenotype of phagocytic cells and the consequent opsonization of nanoparticles. 3D culture techniques offer great promise in realizing the complex crosstalk, but imaging techniques for visualization and analytical tools are also in need of development.

Regarding renal clearance, it has proven difficult to model this step *in vitro*, but organs-on-chips technology may eventually solve this issue. Establishment of relevant cell lines retaining functional integrity for use in the organs-on-chips is a major hurdle.

Step 2, Selective Delivery Modeling this step in its entirety *in vitro* is a huge challenge, if not impossible, as it involves the whole body. However, we may in the future be able to assess to what extent nanoparticles associate (non-)specifically with vessels of various tissues/organs by creating a microfluidic platform recapitulating the characteristics of vessels in each tissue/organ [120]. Although most experiments to date have used generic endothelial cells (usually derived from the umbilical vein or the dermal microvasculature), use of TECs which demonstrate strikingly different features from their normal counterpart [121, 122] may offer better insights into the mechanisms underlying nanoparticle delivery and distribution.

Step 3, Margination Microfluidic systems, together with hydrodynamic simulations, may enable the *in vitro* assessment of margination dynamics. A challenge in this direction is to be able to measure and/or estimate clinically relevant parameters of flow and microfluidic design (*e.g.*, flow rate, pulsatile *vs.* continuous flow, morphological parameters such as branching and cross-sectional shape, fluid viscosity, *etc.*).

Step 4, Extravasation Conventional assays were based on the transwell permeability assay. There are a few shortcomings of this assay. First, hydrostatic pressure is the only physical force present in the model that drives nanoparticles through the endothelial cell monolayer. Therefore, the role of the often disturbed flow seen in the tumor neovasculature cannot be assessed. Second, generic endothelial cells have often been used with little regard to the heterogeneity and plasticity of endothelial cells [118, 120, 121]. These setbacks can be overcome by the use of microfluidic systems and/or organs-on-chips. However, acquiring and maintaining endothelial cells with clinically relevant phenotypes may be difficult. Furthermore, as is the case for modeling **step 3**, measuring and/or estimating clinically relevant parameters of flow and microfluidic design may be a challenge.

Step 5, Stromal Penetration The generation of 3D models, ideally with perfused vessels, of clinically relevant thickness is necessary to attain this goal. Fibrotic tissue consisting of fibroblasts with or without co-cultured tumor cells have already been reported and used to analyze drug penetration [204–208] and ECM architecture [137]. Vascular structures have also been successfully introduced [209]. Future challenges in this direction are to be able to increase ECM content (such as in [210]), to precisely regulate ECM architecture, and to manipulate physical factors (*e.g.*, IFP, solid stress, *etc.*). In the end, a comprehensive analysis regarding the predictive value of these models must be performed; until then, it remains an open question whether these models are successful *in vitro* models of the TME.

Step 6, Internalization Considerations to be made for modeling this step is similar to that of **step 1**.

Strategies to Overcome Stromal Barriers for Enhanced Nanomedicine Penetration

As outlined in the section "Obstacles to Successful Nanomedicine Delivery", a successful nanomedicine must incorporate the following strategies:

Strategy 1: improve tumor localization
Strategy 2: increase intratumoral accumulation
Strategy 3: enhance stromal penetration
Strategy 4: optimize internalization and avoid drug detoxification/efflux

We provide a general overview of the present status and future considerations to be made for the realization of each strategy.

Strategy 1: Improving Tumor Localization

This strategy aims to mainly overcome barriers presented by **steps 1–3**. As mentioned earlier, nanoparticle design, most notably nanoparticle size, charge, and surface chemistry is important to evade opsonization by the RES and to avoid renal clearance. Engagement of specific biological measures, for example the installation of "don't eat me" signals such as soluble CD47 to inhibit phagocytosis by macrophages [211, 212], may allow specific inhibition of uptake in only macrophages while leaving uptake by target tumor cells intact—thus circumventing the PEG dilemma.

A complementary strategy is active targeting which is accomplished by a targeting moiety on the nanoparticle that exploits a characteristic specific to the tumor. This is in contrast to passive targeting which simply relies on the EPR effect for tumor targeting [79, 80]. Targeting moieties may be antibodies or its fragments, proteins (*e.g.*, transferrin), peptides (*e.g.*, cRGD peptide), aptamers, or small molecules (*e.g.*, folic acid and sugars). However, the formation of protein coronas on the surface of nanoparticles once intravenously administered may negatively affect the ability of the nanoparticle to be specifically targeted [52, 53] while it may also enable the avoidance of clearance by the RES [213]. With an increased understanding of the rules governing protein corona formation and its biological consequences [214], we may in the future be able to engineer nanoparticles in such a way that precisely specifies the nature of the protein corona formed. We may then be able to minimize unintended interactions while leaving active targeting moieties intact.

Application of external cues such as magnetic force [215] has also been demonstrated to enhance tumor localization.

For actively targeted nanoparticles to recognize cognate target molecules, they must first marginate to blood vessel walls. Margination which enables the interaction of the nanoparticle with the vascular wall is affected by nanoparticle parameters such as shape, size, and density [25–31]. Active targeting moieties may further enhance interaction of nanoparticles with the vascular wall, but the density of the moiety must also be tuned for optimal results [216, 217]. A systematic analysis of the effect of altering hemorheological parameters on margination dynamics is currently lacking but may offer novel complementary strategies to enhance tumor localization of nanomedicine.

Strategy 2: Increasing Intratumoral Accumulation

This strategy aims to mainly overcome barriers presented by **steps 3** and **4**. **Step 3** has been discussed in the previous sub-section. An additional layer of complexity is that a successful nanoparticle-vessel wall interaction does not necessarily entail a successful uptake [57].

Overcoming the obstacles posed by **step 4** has been studied extensively and can broadly be understood as constituting a continuum from EPR augmentation to vascular normalization [18, 35]. For highly permeable, dysfunctional vessels, vascular normalization via the controlled use of anti-angiogenics improves vascular perfusion and enhances delivery of nanomedicine [36]. On the other hand, for tumors with poorly permeable vessels, anti-angiogenics are less beneficial [40]. Instead, vasoactive agents such as bradykinin and nitrogen oxide (NO) which increase endothelial permeability lead to enhanced EPR effect [9]. Targeting pericyte coverage-rich vessels by TGF-β inhibition has also proven to be promising [39, 218].

Optimizing nanoparticle parameters to direct optimal extravasation is also important. Perhaps the clearest example is size modulation: intratumoral accumulation of nanoparticles in poorly permeable tumors depend on size [37]. Interestingly, the predominant mechanism by which nanoparticles extravasate seems to differ depending on size. It has been reported that 30 nm micelles extravasate from static pores within the tumor neovessels while 70 nm particles depend on localized, intermittent surges in vascular permeability—a phenomenon referred to as "vascular bursts" or "nano-eruptions" [38]. The vascular burst phenomenon urges us to renew the image of leaky vessels with static pores that the EPR effect theory conjures, and invites us to more carefully analyze the dynamic nature of nanoparticle extravasation. The spatiotemporally dynamic regulation of vascular permeability is also backed by observations that nanoparticles of the same size when administered temporally apart do not co-localize [219]. Because the precise mechanisms underlying vascular bursts is hitherto unknown, it cannot yet be systematically exploited for therapy. However, it offers a unique opportunity of spatiotemporally controlled extravasation and intratumoral entrapment of nanomedicine.

Strategy 3: Enhancing Stromal Penetration

This strategy aims to mainly overcome the barriers posed by **step 5**. Fibrosis poses a formidable obstacle to the interstitial transport of nanoparticles via several mechanisms. First, fibroblasts and the ECM components that they actively secrete, physically obstruct nanoparticle passage. Second, ECM components such as collagen I become increasingly crosslinked via enzymes such as lysyl oxidase (LOX), and this results in increased substrate stiffness [220]. This, in turn, promotes a myofibroblastic phenotype which results in ECM contraction—further diminishing nanoparticle passage. Third, excessive deposition of HA in the extracellular milieu induces tumor swelling. This increases the solid stress of tumors and leads to vascular deformation and/or collapse.

Strategies to overcome this can be divided into the following, not necessarily mutually exclusive, methods: stromal ablation, stromal reprogramming, and ECM targeting. An objective common to these strategies is stress alleviation—recovering vascular perfusion and interstitial passage of nanomedicine by attenuating the physical forces impeding them [85]. Stromal ablation refers to any strategy which aims to quantitatively diminish stromal content. However, while short-term inhibition of pro-fibrotic signaling (such as inhibition of Sonic Hedgehog [SHH]) resulted in increased nanomedicine delivery [221], ablation of CAFs in GEMMs of pancreatic cancer resulted in disease progression [222–224]. Thus, especially in light of increasing recognition that fibrotic elements constitute a heterogeneous population [128, 225, 226], a more sophisticated strategy of reprogramming pro-tumor CAFs into tumor-suppressive CAFs has been proposed [227, 228]. Targeting the mechanobiological apparatus of stromal elements may be a possible mechanism to systematically induce stromal reprogramming. Accordingly, numerous agents are currently under trial [105]. The ECM targeting strategy aims to destroy the abnormal ECM architecture underlying the physical barriers that the fibrotic tissue poses. This can be done by inhibition of matrix crosslinking enzymes such as LOX [229] or directly destructing ECM by delivery of degrading enzymes [91, 230].

Notably, a strategy alternative to stromal ablation has been proposed by some groups in which the ECM abundantly present in the tumor stroma instead serves as a scaffold from which anti-tumor agents are released [22, 231, 232]. To this end, antibodies are raised against ECM components to which anti-tumor agents are conjugated. These antibody-drug conjugates (ADCs) recognize their cognate antigens in the tumor stroma from which conjugated drug payloads are released via cleavage. This is best known as the cancer stromal targeting (CAST) therapy [22] and is described in more detail in other chapters of this book.

Strategy 4: Optimizing Internalization and Avoiding Drug Detoxification/Efflux

This strategy aims to mainly overcome barriers posed by **step 6**. As mentioned above, nanoparticle parameters to a great extent direct cellular uptake and thus must be optimized. After internalization, drugs must be safeguarded from prematurely being detoxified or removed from the cell via drug efflux pumps. Concurrent use of detoxification enzyme (such as glutathione S-transferase [41] or UDP-glucuronosyltransferase [42]) inhibitors and/or drug efflux pump (such as those of the ATP-binding cassette family of transporters [43–45]) inhibitors may prove useful.

A complementary strategy is the subcellular targeting of nanomedicine: nanoparticles designed to release its payload at a designated intracellular structure [233]. Most nanoparticles are internalized and delivered to endosomes, where they eventually become targeted for degradation unless the drugs succeed in escaping into the cytosol [234–236]. General strategies for subcellular targeting can be divided into engagement of endogenous mechanisms of localization to subcellular structures (*i.e.*, conjugation of folate results in localization to endosomes), use of peptides such as cell-penetrating peptides (CPPs) which direct cytosolic localization, or the use of stimuli-responsive nanocarriers [237, 238] which enable endosomal escape among various other tunable features.

Other Considerations

An effective nanomedicine must be both efficacious and safe. The safety of long-term administration of nanoparticles is a question which remains incompletely understood. This is especially a pertinent concern because cancer is a chronic disease which usually requires repeated administrations.

The immunogenicity of PEG moieties, as observed in the accelerated blood clearance (ABC) phenomenon, should serve as a cautionary example. The ABC phenomenon occurs when repeated doses of PEGylated nanoparticles result in the generation of anti-PEG IgM antibodies. The antibodies cause a more rapid clearance of the PEGylated nanoparticles after repeated exposures [239–241]. The occurrence of the ABC phenomenon has been correlated with therapeutic consequences clinically [242]. Numerous other modifications as a substitute to PEGylation have been proposed, and the reader is referred to reviews elsewhere [242–244]. More generally, the immunogenicity of nanoparticles—the determinants of immunogenicity and the mechanisms by which it develops—is incompletely understood and warrants future study [245].

Furthermore, pioneering work by Case and colleagues has demonstrated that nanoparticles can propagate DNA damage across cellular barriers [246, 247]. More specifically, they have shown that maternal exposure to cobalt-chromium nanoparticles, which do not cross the placental barrier, results in neuronal damage in the fetus [248]. Though the generalizability of these findings to nanoparticles of different chemical composition is currently unclear, it does demonstrate how much we still do not know about the cellular response to nanoparticles and the importance of systems-level assessment of nanoparticle toxicity.

Concluding Remarks

With numerous trials, errors, and disappointments, we have learned that there exist formidable obstacles to a truly efficacious nanomedicine. With increased understanding of the TME, it is now clear that stromal barriers within the TME impede nanoparticle penetration. Various complementary strategies, which can be realized only through concerted efforts of numerous disciplines, must be implemented to comprehensively target the stromal barriers and to achieve the full potential of nanomedicine.

References

1. DeVita VT, Chu E (2008) A history of Cancer chemotherapy. Cancer Res 68:8643–8653
2. Ma WW, Adjei AA (2009) Novel agents on the horizon for Cancer therapy. CA Cancer J Clin 59:111–137
3. Gotwals P, Cameron S, Cipolletta D et al (2017) Prospects for combining targeted and conventional cancer therapy with immunotherapy. Nat Rev Cancer 17:286–301
4. Matsumura Y, Maeda H (1986) A new concept for macromolecular therapeutics in cancer chemotherapy: mechanism of tumoritropic accumulation of proteins and the antitumor agent smancs. Cancer Res 46:6387–6392
5. Hashizume H, Baluk P, Morikawa S, McLean JW, Thurston G, Roberge S, Jain RK, McDonald DM (2000) Openings between defective endothelial cells explain tumor vessel leakiness. Am J Pathol 156:1363–1380
6. Markman JL, Rekechenetskiy A, Holler E, Ljubimova JY (2013) Nanomedicine therapeutic approaches to overcome cancer drug resistance. Adv Drug Deliv Rev 65:1866–1879
7. Cabral H, Miyata K, Osada K, Kataoka K (2018) Block copolymer micelles in Nanomedicine applications. Chem Rev 118:6844–6892
8. Fang J, Nakamura H, Maeda H (2011) The EPR effect: unique features of tumor blood vessels for drug delivery, factors involved, and limitations and augmentation of the effect. Adv Drug Deliv Rev 63:136–151
9. Maeda H (2015) Toward a full understanding of the EPR effect in primary and metastatic tumors as well as issues related to its heterogeneity. Adv Drug Deliv Rev 91:3–6
10. Barenholz Y (Chezy) (2012) Doxil® — the first FDA-approved nano-drug: lessons learned. J Control Release 160:117–134
11. Hawkins MJ, Soon-Shiong P, Desai N (2008) Protein nanoparticles as drug carriers in clinical medicine. Adv Drug Deliv Rev 60:876–885
12. O'Brien MER, Wigler N, Inbar M et al (2004) Reduced cardiotoxicity and comparable efficacy in a phase III trial of pegylated liposomal doxorubicin HCl (CAELYX/Doxil) versus conventional doxorubicin for first-line treatment of metastatic breast cancer. Ann Oncol Off J Eur Soc Med Oncol 15:440–449

13. Gradishar WJ, Tjulandin S, Davidson N, Shaw H, Desai N, Bhar P, Hawkins M, O'Shaughnessy J (2005) Phase III trial of nanoparticle albumin-bound paclitaxel compared with Polyethylated Castor oil–based paclitaxel in women with breast Cancer. J Clin Oncol 23:7794–7803
14. Hosokawa S, Tagawa T, Niki H, Hirakawa Y, Nohga K, Nagaike K (2003) Efficacy of immunoliposomes on cancer models in a cell-surface-antigen-density-dependent manner. Br J Cancer 89:1545–1551
15. Matsumura Y, Gotoh M, Muro K et al (2004) Phase I and pharmacokinetic study of MCC-465, a doxorubicin (DXR) encapsulated in PEG immunoliposome, in patients with metastatic stomach cancer. Ann Oncol Off J Eur Soc Med Oncol 15:517–525
16. Blanco E, Shen H, Ferrari M (2015) Principles of nanoparticle design for overcoming biological barriers to drug delivery. Nat Biotechnol 33:941–951
17. Miao L, Lin CM, Huang L (2015) Stromal barriers and strategies for the delivery of nanomedicine to desmoplastic tumors. J Control Release 219:192–204
18. Tanaka HY, Kano MR (2018) Stromal barriers to nanomedicine penetration in the pancreatic tumor microenvironment. Cancer Sci 109:2085–2092
19. Nie S (2010) Understanding and overcoming major barriers in cancer nanomedicine. Nanomedicine (Lond) 5:523–528
20. Ranganathan R, Madanmohan S, Kesavan A, Baskar G, Krishnamoorthy YR, Santosham R, Ponraju D, Rayala SK, Venkatraman G (2012) Nanomedicine: towards development of patient-friendly drug-delivery systems for oncological applications. Int J Nanomedicine 7:1043–1060
21. Nichols JW, Bae YH (2012) Odyssey of a cancer nanoparticle: from injection site to site of action. Nano Today 7:606–618
22. Matsumura Y (2012) Cancer stromal targeting (CAST) therapy. Adv Drug Deliv Rev 64:710–719
23. Von Hoff DD, Ervin T, Arena FP et al (2013) Increased survival in pancreatic cancer with nab-paclitaxel plus gemcitabine. N Engl J Med 369:1691–1703
24. Wang-Gillam A, Li C-P, Bodoky G et al (2016) Nanoliposomal irinotecan with fluorouracil and folinic acid in metastatic pancreatic cancer after previous gemcitabine-based therapy (NAPOLI-1): a global, randomised, open-label, phase 3 trial. Lancet 387:545–557
25. Gentile F, Chiappini C, Fine D, Bhavane RC, Peluccio MS, Cheng MM-C, Liu X, Ferrari M, Decuzzi P (2008) The effect of shape on the margination dynamics of non-neutrally buoyant particles in two-dimensional shear flows. J Biomech 41:2312–2318
26. Doshi N, Prabhakarpandian B, Rea-Ramsey A, Pant K, Sundaram S, Mitragotri S (2010) Flow and adhesion of drug carriers in blood vessels depend on their shape: a study using model synthetic microvascular networks. J Control Release 146:196–200
27. Müller K, Fedosov DA, Gompper G (2015) Margination of micro- and nano-particles in blood flow and its effect on drug delivery. Sci Rep 4:4871
28. Lee S-Y, Ferrari M, Decuzzi P (2009) Design of bio-mimetic particles with enhanced vascular interaction. J Biomech 42:1885–1890
29. Thompson AJ, Mastria EM, Eniola-Adefeso O (2013) The margination propensity of ellipsoidal micro/nanoparticles to the endothelium in human blood flow. Biomaterials 34:5863–5871
30. Vahidkhah K, Bagchi P (2015) Microparticle shape effects on margination, near-wall dynamics and adhesion in a three-dimensional simulation of red blood cell suspension. Soft Matter 11:2097–2109
31. Toy R, Hayden E, Shoup C, Baskaran H, Karathanasis E (2011) The effects of particle size, density and shape on margination of nanoparticles in microcirculation. Nanotechnology 22:115101
32. Roose T, Netti PA, Munn LL, Boucher Y, Jain RK (2003) Solid stress generated by spheroid growth estimated using a linear poroelasticity model☆. Microvasc Res 66:204–212

33. Stylianopoulos T, Martin JD, Chauhan VP et al (2012) Causes, consequences, and remedies for growth-induced solid stress in murine and human tumors. Proc Natl Acad Sci 109:15101–15108
34. Helmlinger G, Netti PA, Lichtenbeld HC, Melder RJ, Jain RK (1997) Solid stress inhibits the growth of multicellular tumor spheroids. Nat Biotechnol 15:778–783
35. Kano MR (2014) Nanotechnology and tumor microcirculation. Adv Drug Deliv Rev 74:2–11
36. Jain RK (2005) Normalization of tumor vasculature: an emerging concept in antiangiogenic therapy. Science 307:58–62
37. Cabral H, Matsumoto Y, Mizuno K et al (2011) Accumulation of sub-100 nm polymeric micelles in poorly permeable tumours depends on size. Nat Nanotechnol 6:815–823
38. Matsumoto Y, Nichols JW, Toh K et al (2016) Vascular bursts enhance permeability of tumour blood vessels and improve nanoparticle delivery. Nat Nanotechnol 11:533–538
39. Kano MR, Bae Y, Iwata C et al (2007) Improvement of cancer-targeting therapy, using nanocarriers for intractable solid tumors by inhibition of TGF-beta signaling. Proc Natl Acad Sci 104:3460–3465
40. Kano MR, Komuta Y, Iwata C, Oka M, Shirai Y, Morishita Y, Ouchi Y, Kataoka K, Miyazono K (2009) Comparison of the effects of the kinase inhibitors imatinib, sorafenib, and transforming growth factor-β receptor inhibitor on extravasation of nanoparticles from neovasculature. Cancer Sci 100:173–180
41. Townsend DM, Tew KD (2003) The role of glutathione-S-transferase in anti-cancer drug resistance. Oncogene 22:7369–7375
42. Tukey RH, Strassburg CP (2000) Human UDP-Glucuronosyltransferases: metabolism, expression, and disease. Annu Rev Pharmacol Toxicol 40:581–616
43. International Transporter Consortium, Giacomini KM, Huang S-M, et al (2010) Membrane transporters in drug development. Nat Rev Drug Discov 9:215–236
44. Gottesman MM, Fojo T, Bates SE (2002) Multidrug resistance in cancer: role of ATP–dependent transporters. Nat Rev Cancer 2:48–58
45. Schinkel AH, Jonker JW (2003) Mammalian drug efflux transporters of the ATP binding cassette (ABC) family: an overview. Adv Drug Deliv Rev 55:3–29
46. Fröhlich E (2012) The role of surface charge in cellular uptake and cytotoxicity of medical nanoparticles. Int J Nanomedicine 7:5577–5591
47. Caracciolo G, Pozzi D, Candeloro De Sanctis S, Laura Capriotti A, Caruso G, Samperi R, Laganà A (2011) Effect of membrane charge density on the protein corona of cationic liposomes: interplay between cationic charge and surface area. Appl Phys Lett 99:033702
48. Bewersdorff T, Vonnemann J, Kanik A, Haag R, Haase A (2017) The influence of surface charge on serum protein interaction and cellular uptake: studies with dendritic polyglycerols and dendritic polyglycerol-coated gold nanoparticles. Int J Nanomedicine Volume 12:2001–2019
49. Sakulkhu U, Mahmoudi M, Maurizi L, Salaklang J, Hofmann H (2015) Protein Corona composition of Superparamagnetic Iron oxide nanoparticles with various Physico-chemical properties and coatings. Sci Rep 4:5020
50. Chandran P, Riviere JE, Monteiro-Riviere NA (2017) Surface chemistry of gold nanoparticles determines the biocorona composition impacting cellular uptake, toxicity and gene expression profiles in human endothelial cells. Nanotoxicology 11:507–519
51. Lundqvist M, Stigler J, Elia G, Lynch I, Cedervall T, Dawson KA (2008) Nanoparticle size and surface properties determine the protein corona with possible implications for biological impacts. Proc Natl Acad Sci 105:14265–14270
52. Tenzer S, Docter D, Kuharev J et al (2013) Rapid formation of plasma protein corona critically affects nanoparticle pathophysiology. Nat Nanotechnol 8:772–781
53. Salvati A, Pitek AS, Monopoli MP, Prapainop K, Bombelli FB, Hristov DR, Kelly PM, Åberg C, Mahon E, Dawson KA (2013) Transferrin-functionalized nanoparticles lose their targeting capabilities when a biomolecule corona adsorbs on the surface. Nat Nanotechnol 8:137–143

54. Discher DE (2005) Tissue cells feel and respond to the stiffness of their substrate. Science 310:1139–1143
55. Huang C, Butler PJ, Tong S, Muddana HS, Bao G, Zhang S (2013) Substrate stiffness regulates cellular uptake of nanoparticles. Nano Lett 13:1611–1615
56. Guo P, Liu D, Subramanyam K, Wang B, Yang J, Huang J, Auguste DT, Moses MA (2018) Nanoparticle elasticity directs tumor uptake. Nat Commun 9:130
57. Duan X, Li Y (2013) Physicochemical characteristics of nanoparticles affect circulation, biodistribution, cellular internalization, and trafficking. Small 9:1521–1532
58. Durymanov MO, Rosenkranz AA, Sobolev AS (2015) Current approaches for improving Intratumoral accumulation and distribution of Nanomedicines. Theranostics 5:1007–1020
59. Sun Q, Ojha T, Kiessling F, Lammers T, Shi Y (2017) Enhancing tumor penetration of Nanomedicines. Biomacromolecules 18:1449–1459
60. Albanese A, Tang PS, Chan WCW (2012) The effect of nanoparticle size, shape, and surface chemistry on biological systems. Annu Rev Biomed Eng 14:1–16
61. Zhao F, Zhao Y, Liu Y, Chang X, Chen C, Zhao Y (2011) Cellular uptake, intracellular trafficking, and cytotoxicity of Nanomaterials. Small 7:1322–1337
62. Gilbert B, Huang F, Zhang H, Waychunas GA, Banfield JF (2004) Nanoparticles: strained and stiff. Science 305:651–654
63. Hatakeyama H, Akita H, Harashima H (2013) The polyethyleneglycol dilemma: advantage and disadvantage of PEGylation of liposomes for systemic genes and nucleic acids delivery to tumors. Biol Pharm Bull 36:892–899
64. Bell D, Berchuck A, Birrer M et al (2011) Integrated genomic analyses of ovarian carcinoma. Nature 474:609–615
65. Verhaak RGW, Hoadley KA, Purdom E et al (2010) Integrated genomic analysis identifies clinically relevant subtypes of Glioblastoma characterized by abnormalities in PDGFRA, IDH1, EGFR, and NF1. Cancer Cell 17:98–110
66. Network CGA (2012) Comprehensive molecular characterization of human colon and rectal cancer. Nature 487:330–337
67. Cancer Genome Atlas Research Network (2012) Comprehensive genomic characterization of squamous cell lung cancers. Nature 489:519–525
68. Network CGA (2012) Comprehensive molecular portraits of human breast tumours. Nature 490:61–70
69. Swanton C (2012) Intratumor heterogeneity: evolution through space and time. Cancer Res 72:4875–4882
70. Navin N, Kendall J, Troge J et al (2011) Tumour evolution inferred by single-cell sequencing. Nature 472:90–94
71. Lambrechts D, Wauters E, Boeckx B et al (2018) Phenotype molding of stromal cells in the lung tumor microenvironment. Nat Med. https://doi.org/10.1038/s41591-018-0096-5
72. Patel AP, Tirosh I, Trombetta JJ et al (2014) Single-cell RNA-seq highlights intratumoral heterogeneity in primary glioblastoma. Science 344:1396–1401
73. Khoo BL, Chaudhuri PK, Ramalingam N, Tan DSW, Lim CT, Warkiani ME (2016) Single-cell profiling approaches to probing tumor heterogeneity. Int J Cancer 139:243–255
74. Liotta L, Petricoin E (2000) Molecular profiling of human cancer. Nat Rev Genet 1:48–56
75. Collisson EA, Sadanandam A, Olson P et al (2011) Subtypes of pancreatic ductal adenocarcinoma and their differing responses to therapy. Nat Med 17:500–503
76. Moffitt RA, Marayati R, Flate EL et al (2015) Virtual microdissection identifies distinct tumor- and stroma-specific subtypes of pancreatic ductal adenocarcinoma. Nat Genet 47:1168–1178
77. Bazak R, Houri M, El Achy S, Kamel S, Refaat T (2015) Cancer active targeting by nanoparticles: a comprehensive review of literature. J Cancer Res Clin Oncol 141:769–784
78. Byrne JD, Betancourt T, Brannon-Peppas L (2008) Active targeting schemes for nanoparticle systems in cancer therapeutics. Adv Drug Deliv Rev 60:1615–1626

79. Bertrand N, Wu J, Xu X, Kamaly N, Farokhzad OC (2014) Cancer nanotechnology: the impact of passive and active targeting in the era of modern cancer biology. Adv Drug Deliv Rev 66:2–25
80. Danhier F, Feron O, Préat V (2010) To exploit the tumor microenvironment: passive and active tumor targeting of nanocarriers for anti-cancer drug delivery. J Control Release 148:135–146
81. Dancey JE, Bedard PL, Onetto N, Hudson TJ (2012) The genetic basis for Cancer treatment decisions. Cell 148:409–420
82. Hyman DM, Taylor BS, Baselga J (2017) Implementing genome-driven oncology. Cell 168:584–599
83. Collins FS, Varmus H (2015) A new initiative on precision medicine. N Engl J Med 372:793–795
84. Heldin C-H, Rubin K, Pietras K, Östman A (2004) High interstitial fluid pressure — an obstacle in cancer therapy. Nat Rev Cancer 4:806–813
85. Stylianopoulos T (2017) The solid mechanics of Cancer and strategies for improved therapy. J Biomech Eng 139:021004
86. Lampi MC, Reinhart-King CA (2018) Targeting extracellular matrix stiffness to attenuate disease: from molecular mechanisms to clinical trials. Sci Transl Med 10:eaao0475
87. Jain RK (1988) Determinants of tumor blood flow: a review. Cancer Res 48:2641–2658
88. Ruoslahti E (2002) Specialization of tumour vasculature. Nat Rev Cancer 2:83–90
89. Nia HT, Liu H, Seano G et al (2016) Solid stress and elastic energy as measures of tumour mechanopathology. Nat Biomed Eng 1:0004
90. Stylianopoulos T, Martin JD, Snuderl M, Mpekris F, Jain SR, Jain RK (2013) Coevolution of solid stress and interstitial fluid pressure in tumors during progression: implications for vascular collapse. Cancer Res 73:3833–3841
91. Voutouri C, Polydorou C, Papageorgis P, Gkretsi V, Stylianopoulos T (2016) Hyaluronan-derived swelling of solid tumors, the contribution of collagen and Cancer cells, and implications for Cancer therapy. Neoplasia 18:732–741
92. McGrail DJ, McAndrews KM, Brandenburg CP, Ravikumar N, Kieu QMN, Dawson MR (2015) Osmotic regulation is required for Cancer cell survival under solid stress. Biophys J 109:1334–1337
93. Tredan O, Galmarini CM, Patel K, Tannock IF (2007) Drug resistance and the solid tumor microenvironment. JNCI J Natl Cancer Inst 99:1441–1454
94. Ingber DE (1997) Tensegrity: the architectural basis of cellular Mechanotransduction. Annu Rev Physiol 59:575–599
95. DuFort CC, Paszek MJ, Weaver VM (2011) Balancing forces: architectural control of mechanotransduction. Nat Rev Mol Cell Biol 12:308–319
96. Calvo F, Ege N, Grande-Garcia A et al (2013) Mechanotransduction and YAP-dependent matrix remodelling is required for the generation and maintenance of cancer-associated fibroblasts. Nat Cell Biol 15:637–646
97. Goetz JG, Minguet S, Navarro-Lérida I et al (2011) Biomechanical remodeling of the microenvironment by stromal Caveolin-1 favors tumor invasion and metastasis. Cell 146:148–163
98. Zanconato F, Cordenonsi M, Piccolo S (2016) YAP/TAZ at the roots of Cancer. Cancer Cell 29:783–803
99. Das T, Safferling K, Rausch S, Grabe N, Boehm H, Spatz JP (2015) A molecular mechanotransduction pathway regulates collective migration of epithelial cells. Nat Cell Biol 17:276–287
100. Wei SC, Fattet L, Tsai JH et al (2015) Matrix stiffness drives epithelial–mesenchymal transition and tumour metastasis through a TWIST1–G3BP2 mechanotransduction pathway. Nat Cell Biol 17:678–688
101. Parekh A, Weaver AM (2009) Regulation of cancer invasiveness by the physical extracellular matrix environment. Cell Adhes Migr 3:288–292

102. Lu P, Weaver VM, Werb Z (2012) The extracellular matrix: a dynamic niche in cancer progression. J Cell Biol 196:395–406
103. Shieh AC (2011) Biomechanical forces shape the tumor microenvironment. Ann Biomed Eng 39:1379–1389
104. Paszek MJ, Zahir N, Johnson KR et al (2005) Tensional homeostasis and the malignant phenotype. Cancer Cell 8:241–254
105. Mohammadi H, Sahai E (2018) Mechanisms and impact of altered tumour mechanics. Nat Cell Biol 20:766–774
106. Laklai H, Miroshnikova YA, Pickup MW et al (2016) Genotype tunes pancreatic ductal adenocarcinoma tissue tension to induce matricellular fibrosis and tumor progression. Nat Med 22:497–505
107. Levental KR, Yu H, Kass L et al (2009) Matrix crosslinking forces tumor progression by enhancing integrin signaling. Cell 139:891–906
108. Provenzano PP, Cuevas C, Chang AE, Goel VK, Von Hoff DD, Hingorani SR (2012) Enzymatic targeting of the Stroma ablates physical barriers to treatment of pancreatic ductal adenocarcinoma. Cancer Cell 21:418–429
109. Dolor A, Szoka FC (2018) Digesting a path forward: the utility of collagenase tumor treatment for improved drug delivery. Mol Pharm 15:2069–2083
110. Malandrino A, Mak M, Kamm RD, Moeendarbary E (2018) Complex mechanics of the heterogeneous extracellular matrix in cancer. Extrem Mech Lett 21:25–34
111. Streitberger K-J, Reiss-Zimmermann M, Freimann FB, Bayerl S, Guo J, Arlt F, Wuerfel J, Braun J, Hoffmann K-T, Sack I (2014) High-resolution mechanical imaging of Glioblastoma by multifrequency magnetic resonance Elastography. PLoS One 9:e110588
112. Liu T, Babaniyi OA, Hall TJ, Barbone PE, Oberai AA (2015) Noninvasive in-vivo quantification of mechanical heterogeneity of invasive breast carcinomas. PLoS One 10:e0130258
113. Zhang L, Nishihara H, Kano MR (2012) Pericyte-coverage of human tumor vasculature and nanoparticle permeability. Biol Pharm Bull 35:761–766
114. Kamei R, Tanaka HY, Kawano T, Morii C, Tanaka S, Nishihara H, Iwata C, Kano MR (2017) Regulation of endothelial Fas expression as a mechanism of promotion of vascular integrity by mural cells in tumors. Cancer Sci 108:1080–1088
115. Armulik A, Genové G, Betsholtz C (2011) Pericytes: developmental, physiological, and pathological perspectives, problems, and promises. Dev Cell 21:193–215
116. Smith NR, Baker D, Farren M et al (2013) Tumor stromal architecture can define the intrinsic tumor response to VEGF-targeted therapy. Clin Cancer Res 19:6943–6956
117. Balkwill FR, Capasso M, Hagemann T (2012) The tumor microenvironment at a glance. J Cell Sci 125:5591–5596
118. Aird WC (2012) Endothelial cell heterogeneity. Cold Spring Harb Perspect Med 2:a006429–a006429
119. Dejana E, Hirschi KK, Simons M (2017) The molecular basis of endothelial cell plasticity. Nat Commun 8:14361
120. Nolan DJ, Ginsberg M, Israely E et al (2013) Molecular signatures of tissue-specific microvascular endothelial cell heterogeneity in organ maintenance and regeneration. Dev Cell 26:204–219
121. Hida K, Ohga N, Akiyama K, Maishi N, Hida Y (2013) Heterogeneity of tumor endothelial cells. Cancer Sci 104:1391–1395
122. Dudley AC (2012) Tumor endothelial cells. Cold Spring Harb Perspect Med 2:a006536–a006536
123. Seaman S, Stevens J, Yang MY, Logsdon D, Graff-Cherry C, St. Croix B (2007) Genes that distinguish physiological and pathological angiogenesis. Cancer Cell 11:539–554
124. St Croix B, Rago C, Velculescu V et al (2000) Genes expressed in human tumor endothelium. Science 289:1197–1202

125. Yoshikawa M, Mukai Y, Okada Y et al (2013) Robo4 is an effective tumor endothelial marker for antibody-drug conjugates based on the rapid isolation of the anti-Robo4 cell-internalizing antibody. Blood 121:2804–2813
126. Zeisberg EM, Potenta S, Xie L, Zeisberg M, Kalluri R (2007) Discovery of endothelial to Mesenchymal transition as a source for carcinoma-associated fibroblasts. Cancer Res 67:10123–10128
127. Hosaka K, Yang Y, Seki T et al (2016) Pericyte–fibroblast transition promotes tumor growth and metastasis. Proc Natl Acad Sci 113:E5618–E5627
128. LeBleu VS, Kalluri R (2018) A peek into cancer-associated fibroblasts: origins, functions and translational impact. Dis Model Mech 11:dmm029447
129. Quante M, Tu SP, Tomita H et al (2011) Bone marrow-derived Myofibroblasts contribute to the Mesenchymal stem cell niche and promote tumor growth. Cancer Cell 19:257–272
130. Kalluri R, Weinberg RA (2009) The basics of epithelial-mesenchymal transition. J Clin Invest 119:1420–1428
131. Lamouille S, Xu J, Derynck R (2014) Molecular mechanisms of epithelial–mesenchymal transition. Nat Rev Mol Cell Biol 15:178–196
132. Nieto MA, Huang RY-J, Jackson RA, Thiery JP (2016) EMT: 2016. Cell 166:21–45
133. Tarin D (2013) Role of the host stroma in cancer and its therapeutic significance. Cancer Metastasis Rev 32:553–566
134. Tarin D, Thompson EW, Newgreen DF (2005) The fallacy of epithelial Mesenchymal transition in Neoplasia. Cancer Res 65:5996–6001
135. Amatangelo MD, Bassi DE, Klein-Szanto AJP, Cukierman E (2005) Stroma-derived three-dimensional matrices are necessary and sufficient to promote Desmoplastic differentiation of Normal fibroblasts. Am J Pathol 167:475–488
136. Lee H-O, Mullins SR, Franco-Barraza J, Valianou M, Cukierman E, Cheng JD (2011) FAP-overexpressing fibroblasts produce an extracellular matrix that enhances invasive velocity and directionality of pancreatic cancer cells. BMC Cancer 11:245
137. Tanaka HY, Kitahara K, Sasaki N et al (2019) Pancreatic stellate cells derived from human pancreatic cancer demonstrate aberrant SPARC-dependent ECM remodeling in 3D engineered fibrotic tissue of clinically relevant thickness. Biomaterials 192:355–367
138. Tomasek JJ, Gabbiani G, Hinz B, Chaponnier C, Brown RA (2002) Myofibroblasts and mechano-regulation of connective tissue remodelling. Nat Rev Mol Cell Biol 3:349–363
139. Karsdal MA, Nielsen SH, Leeming DJ et al (2017) The good and the bad collagens of fibrosis – their role in signaling and organ function. Adv Drug Deliv Rev 121:43–56
140. Zhang H, Chang H, Wang L, Ren K, Martins MCL, Barbosa MA, Ji J (2015) Effect of polyelectrolyte film stiffness on endothelial cells during endothelial-to-Mesenchymal transition. Biomacromolecules 16:3584–3593
141. Pitt JM, Marabelle A, Eggermont A, Soria J-C, Kroemer G, Zitvogel L (2016) Targeting the tumor microenvironment: removing obstruction to anticancer immune responses and immunotherapy. Ann Oncol 27:1482–1492
142. Wellenstein MD, de Visser KE (2018) Cancer-cell-intrinsic mechanisms shaping the tumor immune landscape. Immunity 48:399–416
143. Jiménez-Sánchez A, Memon D, Pourpe S et al (2017) Heterogeneous tumor-immune microenvironments among differentially growing metastases in an ovarian Cancer patient. Cell 170:927–938. e20
144. Lee SS-Y, Bindokas VP, Kron SJ (2017) Multiplex three-dimensional optical mapping of tumor immune microenvironment. Sci Rep 7:17031
145. Nathan C, Ding A (2010) Nonresolving inflammation. Cell 140:871–882
146. Hanahan D, Weinberg RA (2011) Hallmarks of cancer: the next generation. Cell 144:646–674
147. Ho EA, Piquette-Miller M (2006) Regulation of multidrug resistance by pro-inflammatory cytokines. Curr Cancer Drug Targets 6:295–311
148. Hanahan D, Weinberg RA (2000) The hallmarks of Cancer. Cell 100:57–70

149. Day C-P, Merlino G, Van Dyke T (2015) Preclinical mouse Cancer models: a maze of opportunities and challenges. Cell 163:39–53
150. Hwang C-I, Boj SF, Clevers H, Tuveson DA (2016) Preclinical models of pancreatic ductal adenocarcinoma. J Pathol 238:197–204
151. Sakai S, Iwata C, Tanaka HY, Cabral H, Morishita Y, Miyazono K, Kano MR (2016) Increased fibrosis and impaired intratumoral accumulation of macromolecules in a murine model of pancreatic cancer co-administered with FGF-2. J Control Release 230:109–115
152. Domcke S, Sinha R, Levine DA, Sander C, Schultz N (2013) Evaluating cell lines as tumour models by comparison of genomic profiles. Nat Commun 4:2126
153. Daniel VC, Marchionni L, Hierman JS et al (2009) A primary Xenograft model of small-cell lung Cancer reveals irreversible changes in gene expression imposed by culture in vitro. Cancer Res 69:3364–3373
154. Piaskowski S, Bienkowski M, Stoczynska-Fidelus E et al (2011) Glioma cells showing IDH1 mutation cannot be propagated in standard cell culture conditions. Br J Cancer 104:968–970
155. Voskoglou-Nomikos T, Pater JL, Seymour L (2003) Clinical predictive value of the in vitro cell line, human xenograft, and mouse allograft preclinical cancer models. Clin Cancer Res 9:4227–4239
156. Johnson JI, Decker S, Zaharevitz D et al (2001) Relationships between drug activity in NCI preclinical in vitro and in vivo models and early clinical trials. Br J Cancer 84:1424–1431
157. Sharpless NE, DePinho RA (2006) The mighty mouse: genetically engineered mouse models in cancer drug development. Nat Rev Drug Discov 5:741–754
158. Becher OJ, Holland EC (2006) Genetically engineered models have advantages over Xenografts for preclinical studies. Cancer Res 66:3355–3359
159. Izeradjene K, Combs C, Best M et al (2007) KrasG12D and Smad4/Dpc4 Haploinsufficiency cooperate to induce mucinous cystic neoplasms and invasive adenocarcinoma of the pancreas. Cancer Cell 11:229–243
160. Hingorani SR, Wang L, Multani AS, Combs C, Deramaudt TB, Hruban RH, Rustgi AK, Chang S, Tuveson DA (2005) Trp53R172H and KrasG12D cooperate to promote chromosomal instability and widely metastatic pancreatic ductal adenocarcinoma in mice. Cancer Cell 7:469–483
161. Hingorani SR, Petricoin EF, Maitra A et al (2003) Preinvasive and invasive ductal pancreatic cancer and its early detection in the mouse. Cancer Cell 4:437–450
162. Ijichi H, Chytil A, Gorska AE, Aakre ME, Fujitani Y, Fujitani S, Wright CVE, Moses HL (2006) Aggressive pancreatic ductal adenocarcinoma in mice caused by pancreas-specific blockade of transforming growth factor-β signaling in cooperation with active Kras expression. Genes Dev 20:3147–3160
163. Hsu PD, Lander ES, Zhang F (2014) Development and applications of CRISPR-Cas9 for genome engineering. Cell 157:1262–1278
164. Joung JK, Sander JD (2013) TALENs: a widely applicable technology for targeted genome editing. Nat Rev Mol Cell Biol 14:49–55
165. Urnov FD, Rebar EJ, Holmes MC, Zhang HS, Gregory PD (2010) Genome editing with engineered zinc finger nucleases. Nat Rev Genet 11:636–646
166. Walrath JC, Hawes JJ, Van Dyke T, Reilly KM (2010) Genetically engineered mouse models in cancer research. Adv Cancer Res 106:113–164
167. Houghton JA, Taylor DM (1978) Growth characteristics of human colorectal tumours during serial passage in immune-deprived mice. Br J Cancer 37:213–223
168. Tentler JJ, Tan AC, Weekes CD, Jimeno A, Leong S, Pitts TM, Arcaroli JJ, Messersmith WA, Eckhardt SG (2012) Patient-derived tumour xenografts as models for oncology drug development. Nat Rev Clin Oncol 9:338–350
169. Hidalgo M, Amant F, Biankin AV et al (2014) Patient-derived Xenograft models: an emerging platform for translational Cancer research. Cancer Discov 4:998–1013

170. Morelli MP, Calvo E, Ordoñez E, Wick MJ, Viqueira B-R, Lopez-Casas PP, Bruckheimer E, Calles-Blanco A, Sidransky D, Hidalgo M (2012) Prioritizing phase I treatment options through preclinical testing on personalized Tumorgraft. J Clin Oncol 30:e45–e48
171. Hidalgo M, Bruckheimer E, Rajeshkumar NV, Garrido-Laguna I, De Oliveira E, Rubio-Viqueira B, Strawn S, Wick MJ, Martell J, Sidransky D (2011) A pilot clinical study of treatment guided by personalized Tumorgrafts in patients with advanced Cancer. Mol Cancer Ther 10:1311–1316
172. Gao H, Korn JM, Ferretti S et al (2015) High-throughput screening using patient-derived tumor xenografts to predict clinical trial drug response. Nat Med 21:1318–1325
173. Nardella C, Lunardi A, Patnaik A, Cantley LC, Pandolfi PP (2011) The APL paradigm and the "co-clinical trial" project. Cancer Discov 1:108–116
174. Chen Z, Cheng K, Walton Z et al (2012) A murine lung cancer co-clinical trial identifies genetic modifiers of therapeutic response. Nature 483:613–617
175. Pergolini I, Morales-Oyarvide V, Mino-Kenudson M et al (2017) Tumor engraftment in patient-derived xenografts of pancreatic ductal adenocarcinoma is associated with adverse clinicopathological features and poor survival. PLoS One 12:e0182855
176. John T, Kohler D, Pintilie M et al (2011) The ability to form primary tumor Xenografts is predictive of increased risk of disease recurrence in early-stage non-small cell lung Cancer. Clin Cancer Res 17:134–141
177. Lai Y, Wei X, Lin S, Qin L, Cheng L, Li P (2017) Current status and perspectives of patient-derived xenograft models in cancer research. J Hematol Oncol 10:106
178. Yu J, Seldin MM, Fu K et al (2018) Topological arrangement of cardiac fibroblasts regulates cellular plasticity. Circ Res 123:73–85
179. Levinger I, Ventura Y, Vago R (2014) Life is three dimensional—as in vitro Cancer cultures should be. pp 383–414
180. Edmondson R, Broglie JJ, Adcock AF, Yang L (2014) Three-dimensional cell culture systems and their applications in drug discovery and cell-based biosensors. Assay Drug Dev Technol 12:207–218
181. Davies PF (2009) Hemodynamic shear stress and the endothelium in cardiovascular pathophysiology. Nat Clin Pract Cardiovasc Med 6:16–26
182. Shyy JY-J, Chien S (2002) Role of integrins in endothelial mechanosensing of shear stress. Circ Res 91:769–775
183. White CR, Frangos JA (2007) The shear stress of it all: the cell membrane and mechanochemical transduction. Philos Trans R Soc B Biol Sci 362:1459–1467
184. Yamada KM, Cukierman E (2007) Modeling tissue morphogenesis and Cancer in 3D. Cell 130:601–610
185. Bin KJ (2005) Three-dimensional tissue culture models in cancer biology. Semin Cancer Biol 15:365–377
186. Wu M, Swartz MA (2014) Modeling tumor microenvironments in vitro. J Biomech Eng 136:021011
187. Fong ELS, Harrington DA, Farach-Carson MC, Yu H (2016) Heralding a new paradigm in 3D tumor modeling. Biomaterials 108:197–213
188. Katt ME, Placone AL, Wong AD, Xu ZS, Searson PC (2016) In vitro tumor models: advantages, disadvantages, variables, and selecting the right platform. Front Bioeng Biotechnol. https://doi.org/10.3389/fbioe.2016.00012
189. Park KM, Lewis D, Gerecht S (2017) Bioinspired hydrogels to engineer Cancer microenvironments. Annu Rev Biomed Eng 19:109–133
190. Mehta G, Hsiao AY, Ingram M, Luker GD, Takayama S (2012) Opportunities and challenges for use of tumor spheroids as models to test drug delivery and efficacy. J Control Release 164:192–204
191. Weeber F, Ooft SN, Dijkstra KK, Voest EE (2017) Tumor Organoids as a pre-clinical Cancer model for drug discovery. Cell Chem Biol 24:1092–1100

192. Sachs N, Clevers H (2014) Organoid cultures for the analysis of cancer phenotypes. Curr Opin Genet Dev 24:68–73
193. Aboulkheyr Es H, Montazeri L, Aref AR, Vosough M, Baharvand H (2018) Personalized Cancer medicine: an Organoid approach. Trends Biotechnol 36:358–371
194. Matsusaki M, Case CP, Akashi M (2014) Three-dimensional cell culture technique and pathophysiology. Adv Drug Deliv Rev 74:95–103
195. Nishiguchi A, Yoshida H, Matsusaki M, Akashi M (2011) Rapid construction of three-dimensional multilayered tissues with endothelial tube networks by the cell-accumulation technique. Adv Mater 23:3506–3510
196. van Duinen V, Trietsch SJ, Joore J, Vulto P, Hankemeier T (2015) Microfluidic 3D cell culture: from tools to tissue models. Curr Opin Biotechnol 35:118–126
197. Huh D, Hamilton GA, Ingber DE (2011) From 3D cell culture to organs-on-chips. Trends Cell Biol 21:745–754
198. Ahn J, Sei Y, Jeon N, Kim Y (2017) Tumor microenvironment on a Chip: the Progress and future perspective. Bioengineering 4:64
199. Tsai H-F, Trubelja A, Shen AQ, Bao G (2017) Tumour-on-a-chip: microfluidic models of tumour morphology, growth and microenvironment. J R Soc Interface 14:20170137
200. Zhang YS, Zhang Y-N, Zhang W (2017) Cancer-on-a-chip systems at the frontier of nanomedicine. Drug Discov Today 22:1392–1399
201. McWhorter FY, Davis CT, Liu WF (2015) Physical and mechanical regulation of macrophage phenotype and function. Cell Mol Life Sci 72:1303–1316
202. Sica A, Mantovani A (2012) Macrophage plasticity and polarization: in vivo veritas. J Clin Invest 122:787–795
203. Noy R, Pollard JW (2014) Tumor-associated macrophages: from mechanisms to therapy. Immunity 41:49–61
204. Hosoya H, Kadowaki K, Matsusaki M et al (2012) Engineering fibrotic tissue in pancreatic cancer: a novel three-dimensional model to investigate nanoparticle delivery. Biochem Biophys Res Commun 419:32–37
205. Matsusaki M, Komeda M, Mura S, Tanaka HY, Kano MR, Couvreur P, Akashi M (2017) Desmoplastic reaction in 3D-pancreatic Cancer tissues suppresses molecular permeability. Adv Healthc Mater 6:1700057
206. Priwitaningrum DL, Blondé J-BG, Sridhar A, van Baarlen J, Hennink WE, Storm G, Le Gac S, Prakash J (2016) Tumor stroma-containing 3D spheroid arrays: a tool to study nanoparticle penetration. J Control Release 244:257–268
207. Gong X, Lin C, Cheng J, Su J, Zhao H, Liu T, Wen X, Zhao P (2015) Generation of multicellular tumor spheroids with microwell-based Agarose scaffolds for drug testing. PLoS One 10:e0130348
208. Khawar IA, Park JK, Jung ES, Lee MA, Chang S, Kuh H-J (2018) Three dimensional mixed-cell spheroids mimic Stroma-mediated Chemoresistance and invasive migration in hepatocellular carcinoma. Neoplasia 20:800–812
209. Nishiguchi A, Matsusaki M, Kano MR, Nishihara H, Okano D, Asano Y, Shimoda H, Kishimoto S, Iwai S, Akashi M (2018) In vitro 3D blood/lymph-vascularized human stromal tissues for preclinical assays of cancer metastasis. Biomaterials 179:144–155
210. Liu C-Y, Matsusaki M, Akashi M (2016) Three-dimensional tissue models constructed by cells with nanometer- or micrometer-sized films on the surfaces. Chem Rec 16:783–796
211. Hsu Y-C, Acuña M, Tahara SM, Peng C-A (2003) Reduced phagocytosis of colloidal carriers using soluble CD47. Pharm Res 20:1539–1542
212. Rodriguez PL, Harada T, Christian DA, Pantano DA, Tsai RK, Discher DE (2013) Minimal "self" peptides that inhibit phagocytic clearance and enhance delivery of nanoparticles. Science 339:971–975
213. Lynch I, Dawson KA (2008) Protein-nanoparticle interactions. Nano Today 3:40–47
214. del Pino P, Pelaz B, Zhang Q, Maffre P, Nienhaus GU, Parak WJ (2014) Protein corona formation around nanoparticles – from the past to the future. Mater Horiz 1:301–313

215. Alexiou C, Arnold W, Klein RJ, Parak FG, Hulin P, Bergemann C, Erhardt W, Wagenpfeil S, Lübbe AS (2000) Locoregional cancer treatment with magnetic drug targeting. Cancer Res 60:6641–6648
216. Takae S, Akiyama Y, Otsuka H, Nakamura T, Nagasaki Y, Kataoka K (2005) Ligand density effect on biorecognition by PEGylated gold nanoparticles: regulated interaction of RCA 120 Lectin with lactose installed to the distal end of tethered PEG strands on gold surface. Biomacromolecules 6:818–824
217. Elias DR, Poloukhtine A, Popik V, Tsourkas A (2013) Effect of ligand density, receptor density, and nanoparticle size on cell targeting. Nanomed Nanotechnol Biol Med 9:194–201
218. Kumagai M, Kano MR, Morishita Y, Ota M, Imai Y, Nishiyama N, Sekino M, Ueno S, Miyazono K, Kataoka K (2009) Enhanced magnetic resonance imaging of experimental pancreatic tumor in vivo by block copolymer-coated magnetite nanoparticles with TGF-β inhibitor. J Control Release 140:306–311
219. Stirland DL, Matsumoto Y, Toh K, Kataoka K, Bae YH (2016) Analyzing spatiotemporal distribution of uniquely fluorescent nanoparticles in xenograft tumors. J Control Release 227:38–44
220. Wang T-H, Hsia S-M, Shieh T-M (2016) Lysyl oxidase and the tumor microenvironment. Int J Mol Sci 18:62
221. Olive KP, Jacobetz MA, Davidson CJ et al (2009) Inhibition of hedgehog signaling enhances delivery of chemotherapy in a mouse model of pancreatic cancer. Science 324:1457–1461
222. Lee JJ, Perera RM, Wang H et al (2014) Stromal response to hedgehog signaling restrains pancreatic cancer progression. Proc Natl Acad Sci 111:E3091–E3100
223. Rhim AD, Oberstein PE, Thomas DH et al (2014) Stromal elements act to restrain, rather than support, pancreatic ductal adenocarcinoma. Cancer Cell 25:735–747
224. Özdemir BC, Pentcheva-Hoang T, Carstens JL et al (2014) Depletion of carcinoma-associated fibroblasts and fibrosis induces immunosuppression and accelerates pancreas cancer with reduced survival. Cancer Cell 25:719–734
225. Öhlund D, Handly-Santana A, Biffi G et al (2017) Distinct populations of inflammatory fibroblasts and myofibroblasts in pancreatic cancer. J Exp Med 214:579–596
226. Gascard P, Tlsty TD (2016) Carcinoma-associated fibroblasts: orchestrating the composition of malignancy. Genes Dev 30:1002–1019
227. Chronopoulos A, Robinson B, Sarper M et al (2016) ATRA mechanically reprograms pancreatic stellate cells to suppress matrix remodelling and inhibit cancer cell invasion. Nat Commun 7:12630
228. Sherman MH, Yu RT, Engle DD et al (2014) Vitamin D receptor-mediated stromal reprogramming suppresses pancreatitis and enhances pancreatic cancer therapy. Cell 159:80–93
229. Cox TR, Bird D, Baker A-M, Barker HE, Ho MW-Y, Lang G, Erler JT (2013) LOX-mediated collagen crosslinking is responsible for fibrosis-enhanced metastasis. Cancer Res 73:1721–1732
230. Thompson CB, Shepard HM, O'Connor PM et al (2010) Enzymatic depletion of tumor hyaluronan induces antitumor responses in preclinical animal models. Mol Cancer Ther 9:3052–3064
231. Bala V, Rao S, Boyd BJ, Prestidge CA (2013) Prodrug and nanomedicine approaches for the delivery of the camptothecin analogue SN38. J Control Release 172:48–61
232. Gébleux R, Stringhini M, Casanova R, Soltermann A, Neri D (2017) Non-internalizing antibody-drug conjugates display potent anti-cancer activity upon proteolytic release of monomethyl auristatin E in the subendothelial extracellular matrix. Int J Cancer 140:1670–1679
233. Rajendran L, Knölker H-J, Simons K (2010) Subcellular targeting strategies for drug design and delivery. Nat Rev Drug Discov 9:29–42
234. Varkouhi AK, Scholte M, Storm G, Haisma HJ (2011) Endosomal escape pathways for delivery of biologicals. J Control Release 151:220–228
235. Ma D (2014) Enhancing endosomal escape for nanoparticle mediated siRNA delivery. Nanoscale 6:6415

236. Dominska M, Dykxhoorn DM (2010) Breaking down the barriers: siRNA delivery and endosome escape. J Cell Sci 123:1183–1189
237. Mura S, Nicolas J, Couvreur P (2013) Stimuli-responsive nanocarriers for drug delivery. Nat Mater 12:991–1003
238. Ganta S, Devalapally H, Shahiwala A, Amiji M (2008) A review of stimuli-responsive nanocarriers for drug and gene delivery. J Control Release 126:187–204
239. Dams ET, Laverman P, Oyen WJ, Storm G, Scherphof GL, van Der Meer JW, Corstens FH, Boerman OC (2000) Accelerated blood clearance and altered biodistribution of repeated injections of sterically stabilized liposomes. J Pharmacol Exp Ther 292:1071–1079
240. Ishida T, Ichihara M, Wang X, Kiwada H (2006) Spleen plays an important role in the induction of accelerated blood clearance of PEGylated liposomes. J Control Release 115:243–250
241. Ishida T, Maeda R, Ichihara M, Irimura K, Kiwada H (2003) Accelerated clearance of PEGylated liposomes in rats after repeated injections. J Control Release 88:35–42
242. Zhang P, Sun F, Liu S, Jiang S (2016) Anti-PEG antibodies in the clinic: current issues and beyond PEGylation. J Control Release 244:184–193
243. Engler AC, Ke X, Gao S, Chan JMW, Coady DJ, Ono RJ, Lubbers R, Nelson A, Yang YY, Hedrick JL (2015) Hydrophilic polycarbonates: promising degradable alternatives to poly(ethylene glycol)-based stealth materials. Macromolecules 48:1673–1678
244. Barz M, Luxenhofer R, Zentel R, Vicent MJ (2011) Overcoming the PEG-addiction: well-defined alternatives to PEG, from structure–property relationships to better defined therapeutics. Polym Chem 2:1900
245. Ilinskaya AN, Dobrovolskaia MA (2016) Understanding the immunogenicity and antigenicity of nanomaterials: past, present and future. Toxicol Appl Pharmacol 299:70–77
246. Bhabra G, Sood A, Fisher B et al (2009) Nanoparticles can cause DNA damage across a cellular barrier. Nat Nanotechnol 4:876–883
247. Sood A, Salih S, Roh D et al (2011) Signalling of DNA damage and cytokines across cell barriers exposed to nanoparticles depends on barrier thickness. Nat Nanotechnol 6:824–833
248. Hawkins SJ, Crompton LA, Sood A et al (2018) Nanoparticle-induced neuronal toxicity across placental barriers is mediated by autophagy and dependent on astrocytes. Nat Nanotechnol 13:427–433

Part II
Antibody Drug Conjugates (ADC)

Recent Progress in Linker Technology for Antibody-Drug Conjugates: Methods for Connection and Release

Shino Manabe

Abstracts Linkers connect antibodies to payloads in antibody-drug conjugates (ADCs). The choice of linker controls the conjugation-site and antibody-drug ratios as well as the releasing mechanism and the rate of release. In order to optimize the efficacy of ADCs, improvements in linker technology, especially site-selective conjugation methods, are being explored. Site-specific conjugation methods involving non-natural amino acid incorporation, glycan utilization, and enzymatic addition methodologies are discussed in this chapter. In addition to conjugation methodology, payload-releasing strategy is a very important consideration for delivering maximum focused toxicity to the vicinity of the tumor cells. Although protease-cathepsin-cleavable dipeptide Val-Cit sequence is currently the most common choice for an enzymatic cleavable linker, novel approaches for releasing payloads, such as light stimulation and chemical reaction, have been reported. In this chapter, advances in development of linker technology will be reviewed.

Keywords *p*-aminobenzyl carbamate · Anti-IL-7R antibody · Anti-insoluble fibrin (IF) antibody · Bio-orthogonal reaction · Cancer stromal targeting (CAST) therapy · Cathepsin · Diketopiperazine-releasing spacer · Drug-to-antibody ratio (DAR) · 1,6-elimitation reaction · Endo-β-*N*-acetylglucosaminidase · Enhanced permeability and retention (EPR) effect · Formylglycine-generating enzyme · Galactosyl transferase · β-glucuronidase-cleavable linker · Glycan · Inverse-electron-demand Diels-Alder reaction · Linker · Matrix assisted laser desorption/ionization time of flight mass spectrometry (MALDI-TOF MS) · Polyethylene glycol (PEG) · Maleimide · Monomethyl auristatin E (MMAE) · MS imaging · Nucleophilic aromatic substitution (S_NAr) · Plasmin · Sialyl transferase · SN-38 · Sortase A · Strain-promoted azide-alkyne cycloaddition reaction · Therapeutic index · THIOMAB · Transglutaminase · Val-Cit

---

S. Manabe (✉)
Cluster for Pioneering Research, RIKEN, Wako, Japan

Research Center for Pharmaceutical Development, Tohoku University, Sendai, Japan
e-mail: smanabe@riken.jp

Y. Matsumura, D. Tarin (eds.), *Cancer Drug Delivery Systems Based on the Tumor Microenvironment*, https://doi.org/10.1007/978-4-431-56880-3_5

Introduction

Antibody-drug conjugates (ADCs) are a new class of highly potent pharmaceutical drugs designed for targeted therapy; they are expected to be the next generation of antibody medicine [1–3]. Cytotoxic drugs are covalently linked to monoclonal antibodies for tumor antigen-specific antigen via linkers (Fig. 1). When a cytotoxic agent is bound to an antibody, the toxicity of the cytotoxic agent can be much reduced and thus tolerated by healthy cells. Antibodies specific to antigens that are overexpressed on cancer cells recruit conjugates carrying cytotoxic agents to the aberrant region and bind them to antigen-decorated cells. Antibodies target antigens only on the cell surface and exhibit very high specificity. In contrast, traditional chemotherapeutic agents are permeable and go inside cells, but have relatively low specificity, opening the possibility of collateral damage. Many ADCs are believed to be internalized by cells after attachment to surface receptors, and cytotoxic agents are released in lysosomes in their active form. ADCs can expand the narrow therapeutic index (i.e., the ratio of the toxic dose to effective dose) of traditional chemotherapeutic agents, because agents are selectively delivered to cancer region. In recent years, significant progress has been made in this area, and four ADCs have been approved, Mylotarg® (gemtuzumab ozogamicin), Adcetris® (brentuximab vedotin), Kadcyla® (ado-trastuzumab emtansine), and Besponsa® (inotuzumab ozogamicin) [4–7].

ADCs consist of three components, antibody, payload, and linker. The linker connects the payload to the antibody. Linker technology is a key challenge in ADC development. The linker is further divided into two portions, the enzyme recognition moiety and a spacer. An effective linker should (1) have sufficient stability under circulation in the bloodstream, (2) release payloads at the desired region at a suitable rate, (3) not interfere with binding to target antigens, and (4) be applicable to other subclasses of ADCs.

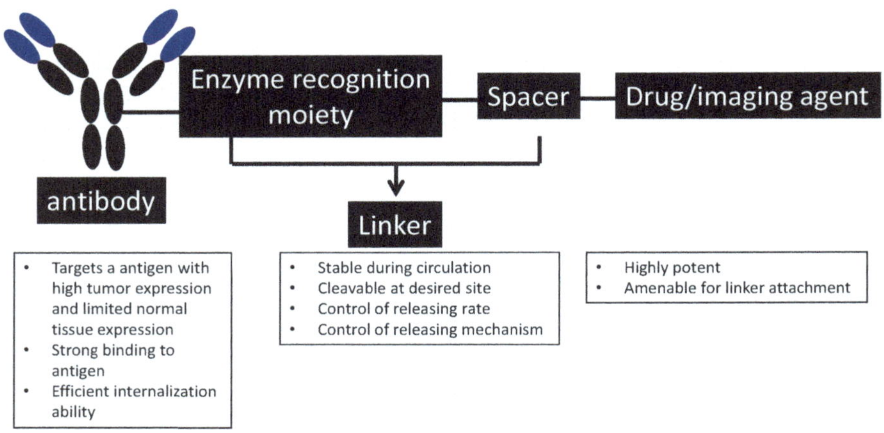

Fig. 1 Structure of antibody-drug conjugates

In many cases, classical methods for protein conjugation via Lys and Cys residues are applied for ADC preparation. However, classical conjugation methods usually generate heterogeneous ADCs due to their lack of chemical site-specificity. Heterogeneous ADCs may differ in the drug-antibody ratio (DAR) and conjugation-site, and this can result in a narrow therapeutic window with major pharmacokinetic implications. Although the FDA approves heterogeneous ADCs at present, non-selective approaches are now considered sub-optional for the development of next-generation ADCs [8, 9]. In addition to the conjugation-site and DAR, the linker controls the drug release rate, and release mechanism, which are known to exert a strong influence on pharmacokinetics/pharmacodynamics of the therapeutic agent.

History of Linker Development

In 1994, an ADC with a primitive linker was reported [10]. Doxorubicin was linked to a BR96 antibody by a hydrazone linkage between the ketone group of doxorubicin and the hydrazine group of an antibody (Fig. 2), known as attachment moieties. The hydrazone linker is expected to be cleaved in the acidic environment of lysosomes and endosomes. However, hydrazone linkages are less used for recent ADCs because they are hydrolyzed gradually during the circulation in blood and new linkers have replaced this approach.

A successful example of a linker is brentuximab vedotin (Fig. 3) [1], in which the potent tubulin inhibitor monomethyl auristatin E (MMAE) is linked via a cathepsin-cleavable linker. The linker consists of a Val-Cit (citrulline) dipeptide sequence as an enzyme recognition moiety and a *p*-aminobenzyl carbamate group as a spacer. The linker is stable during the circulation in blood, but is cleavable by protease cathepsin B once the ADC internalizes in cells. Cathepsin cleaves the carboxylic acid site of the Phe-Lys sequence with $t_{1/2}$ = 8 min. This half-time is too short for

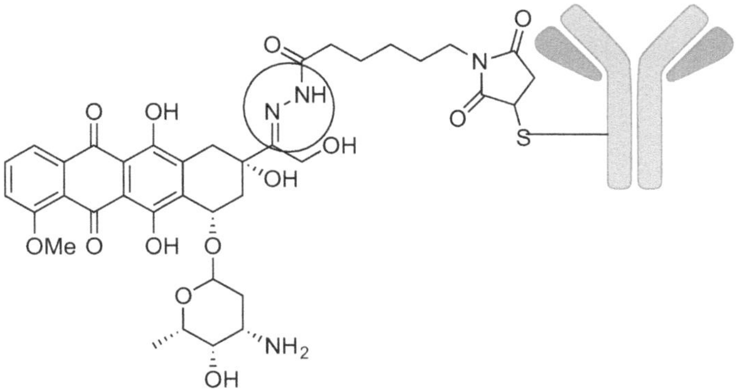

Fig. 2 Unstable linker at initial stage

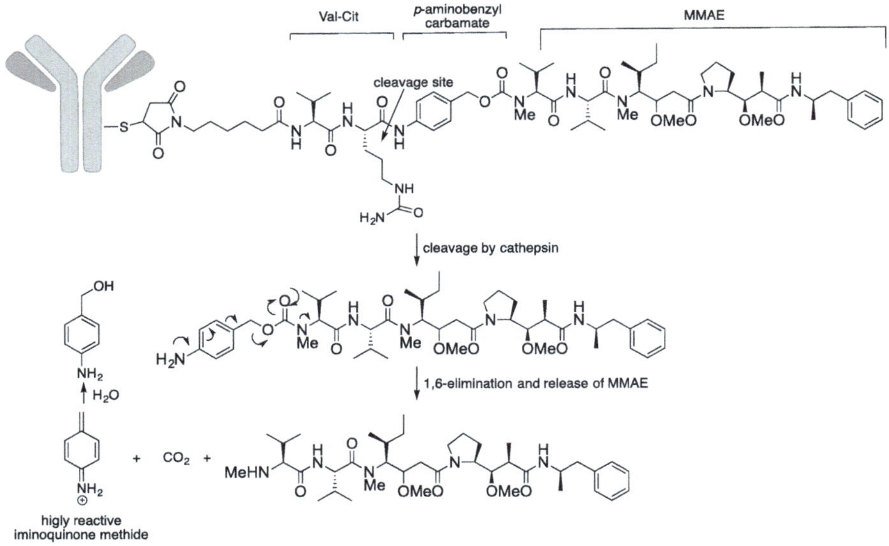

Fig. 3 Structure of brentuximab vedotin and MMAE releasing mechanism

Table 1 Doxorubicin releasing rate from dipeptide conjugates

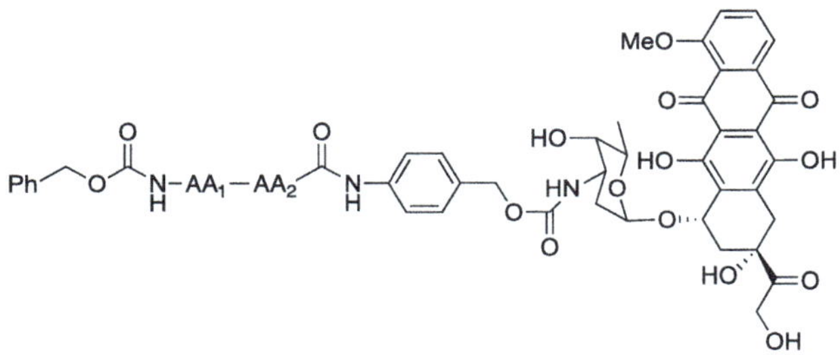

	AA1	AA2	PAB carbamate	$t_{1/2}$
1	Phe	Lys	-	-
2	Phe	Lys	+	8 min
3	Val	Cit	+	630 min

ADC therapy because a slow payload release is an important factor. Dubowchik et al. (2002) systematically investigated the cathepsin cleavage rate of doxorubicin-conjugated dipeptides (Table 1) [11]. They found that the Val-Cit sequence substantially enhances the half-time of cathepsin cleavage to 630 min. The replacement of a side-chain amino group of Lys to urea retards the cleavage rate; accordingly, a sequence change from Phe-Lys to Val-Cit is effective for sustained payload release. The *p*-aminobenzyl carbamate group is essential as a spacer. In the absence of the *p*-aminobenzyl carbamate group, doxorubicin is not released from the conjugate due to steric hindrance of doxorubicin itself. The *p*-aminobenzyl group plays a role in maintaining the distance between the payload and antibody to facilitate cleavage. Once the bond between the peptide carboxylic termini and *p*-aminobenzyl group is cleaved, the payload is released in a self-immolative 1,6-elimitation reaction.

During degradation, a highly reactive iminoquinone methide intermediate is formed. Generated iminoquinone methide may interact with cellular nucleophiles, such as glutathione. Some endopeptidases, such as prostate-specific antigen, can only cleave amide bonds, therefore, the carbamate site in the *p*-aminobenzyloxycarbonyl spacer is not cleaved by endopeptidases [12]. In order to avoid generating highly reactive species, several improvements have been achieved. One example is diketopiperazine-releasing spacer (Fig. 4) [13]. Although most peptide sequences have energetically favorable *trans*-conformation, Proline (Pro) prefers *cis*-conformation, and the *cis*-conformation of Pro easily forms diketopiperazine, so Gly-Pro, Pro-Pro, and Pro-Gly dipeptides are known to form cyclic diketopiperazine. When Gly-Pro spacer was introduced, payload was released as efficient as *p*-aminobenzyloxycarbonyl spacer.

Recently, cytotoxic agents were cleaved even in CRISPER-Cas9-reduced cathepsin B cell-lines [14]. The Val-Cit linker was cleaved by cathepsin S and carboxyesterase 1c [15]. In order to avoid hydrolysis of linker by carboxyesterase 1c, Tsuchikama et al. developed Glu-Val-Cit linker [16].

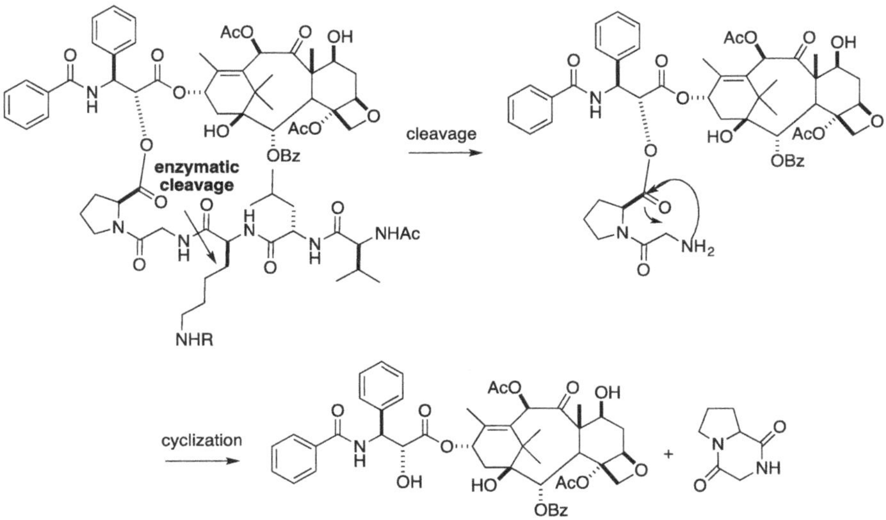

Fig. 4 Payload release by cyclization mechanism without highly reactive iminoquinone methide intermediate

Classical Conjugation Methods and Their Disadvantages

Most classical methods for the conjugation of potent cytotoxic agents to antibodies rely on the modification of the sulfhydryl group of cysteine residues and the amine group of lysine residues [17]. Typically, an antibody contains 90 accessible Lys residues, and a hydrophilic amino group is located on the antibody surface. The conjugation reaction at Lys residues proceeds without forcing conditions. Since the DARs of ADCs are typically 2–4, up to 10^6 distinct forms from one ADC could be generated. Such heterogeneous mixtures of ADCs may differ in both drug-loading ability and conjugation site, and this variation can result in a narrow therapeutic window, with major pharmacokinetic implications [18]. Furthermore, hydrophobic payloads added to hydrophilic amino groups alter the properties of antibodies. At present, the sulfhydryl group of cysteine is a preferred conjugation site. Four disulfide bonds are cleaved under reductive conditions, yielding eight SH groups. Cys residues are less abundant than Lys residues, but this methodology still generates ~15 distinct forms.

Classical conjugate methodology results in various forms of ADCs, and the binding ability to antigens may be reduced when the payload is conjugated at various regions. A simplified bioanalysis of a single, homogeneous drug substance with a very well-defined structure, loading, and impurity profile will facilitate regulatory approval. For these reasons, considerable interest in the field has shifted to the development of site-specific conjugation methods to obtain homogeneous ADCs (Fig. 5). Figure 5 shows outline of strategies for homogeneous ADCs.

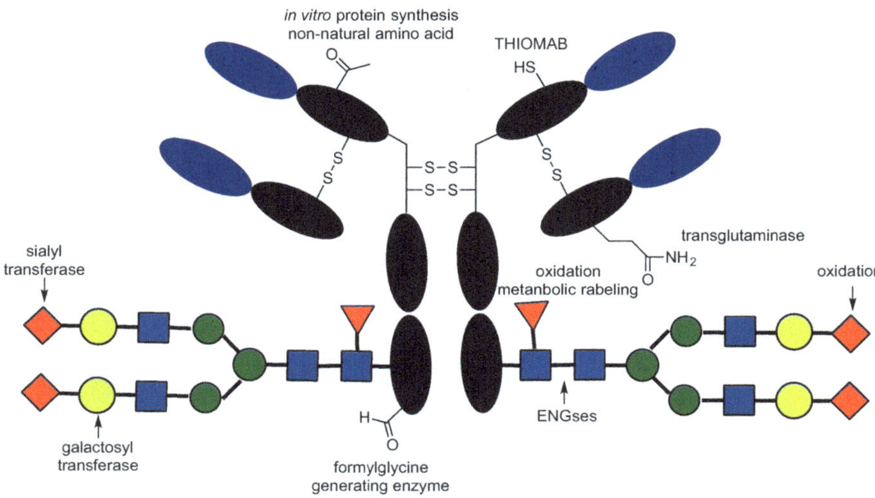

Fig. 5 Possible Conjugation site in antibody

Progress in Homogeneous ADC Preparation

Conjugation at Engineered Cysteines

The thiol group of cysteine residues has the highest nucleophilicity among the 20 amino acids under physiological conditions. This makes it a useful target for the selective and site-specific modification of antibodies. Junutula et al. (2008) introduced additional Cys residues to antibodies using site-directed mutagenesis; this methodology, referred to as THIOMAB, is now widely applied [19]. Although engineered Cys residues have been successfully introduced, they are mostly found as mixed disulfides with glutathione. To enable conjugation via introduced Cys, both the mixed and native interchain disulfides were reduced, followed by mild re-oxidation of the interchain disulfides with a gentle oxidant. From these processes, two reduced engineered Cys residues of the antibody were available for conjugation without changing all native disulfide bonds. Using this methodology, MMAE was conjugated with a DAR of 1.9, greater than 90% homogeneity, and activity *in vitro* and *in vivo*.

Amino Acid-Specific Conjugation

In addition to methods focusing on Lys and Cys, various additional amino acid-specific conjugation methods have been reported. Ala, Gly, Ile, Leu, Phe, and Pro are difficult to use for payload conjugation owing to their lack of relevant functional groups.

Trp is less abundant than other amino acids and therefore conjugation at Trp sites would be helpful for the generation of homogeneous ADCs. Koshland's reagent (2-hydroxy-5-nitrobenzyl-bromide) [20] or chlorosulfonium ions [21] have been used for Trp modification, but these reagents display cross-reactivity with His and Tyr.

Antos et al. (2009) reported that the combination of $Rh(OAc)_4$ and *N*-(*tert*-butyl) hydroxylamine promotes diazoester addition to Trp with 40–60% conversion at pH 6–7 (Fig. 6) [22]. Hansen et al. (2016) reported the chemo- and regioselective ethynylation of Trp by applying a hypervalent iodine reagent and $[AuCl(SMe_2)]$ to peptides and small proteins, like apomyoglobin [23]. Oxidative conjugation to Trp residues without transition metals has also been reported. 9-Azabicyclo[3,3,1] nonane-3-one *N*-oxyl (keto-ABNO) was added to Trp residues in the presence of $NaNO_2$ [24]. Small molecules, e.g., fluorescein, SN-38, and biotin, were conjugated via ketone groups. Although examples are limited to small proteins at present, this methodology has the potential to be applied to ADCs.

Ban et al. (2013) reported that 4-phenyl-3*H*-1,2,4-riazoline-3,5(4*H*)-diones (PTADs) react with aromatic rings of Tyr residues (Fig. 7) [25]. Phenols react with highly reactive electrophiles, such as diazodicarboxylates, in organic solvents in the

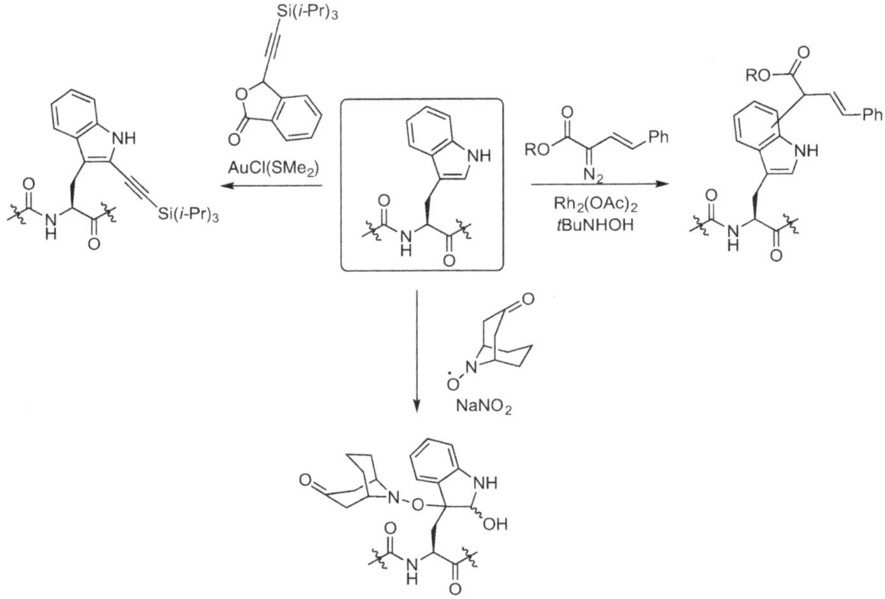

Fig. 6 Modification at Trp residue

presence of an acid. However, highly reactive diazaodicarboxylate regents decompose rapidly in physiological conditions. Cyclic diazodicarboxylate reagents, e.g., PTAD, are relatively stable. Unfortunately, once isocyanate is generated from PTADs, the reaction between the amino group and isocyanate gives urea as a side-product. Sato et al. (2015) reported an improved version of Tyr modification [26]. Cyclic hydrazide compounds are oxidized by hemin (an iron-protoporphyrin IX complex) and H_2O_2 *in situ*, generating reactive cyclic diazodicarboxyamide. This modification resulted in a better yield and selectivity when angiotensin II was employed as a substrate. Tyr was oxidized by *mushroom tyrosinase* to obtain highly reactive 1,2-quinone. Although 1,2-quinone receives a nucleophilic attack by amine or thiols, van Delft reported that quinone and strain-promoted alkyne react to give a cyclo-adduct for protein labeling [27]. In this technology, precise reaction conditions are required (reaction temperature, <16 °C) to avoid competitive side-reactions caused by side chains of Lys and Cys.

Compared with Cys conjugation, Met labeling under physiological conditions remains largely underdeveloped. Lin et al. (2017) reported that oxaziridine reacts with Met to give sulfur imidate (Fig. 8). A native Met on its light chain and two Met residues on its heavy chain did not react with oxaziridine because hydrophobic Met residues are buried in the antibody binding region [28]. They incorporated Met residues at *C*-termini of light chains of trastuzumab. Replaced Met residues reacted successfully with the oxaziridine reagent to give adducts.

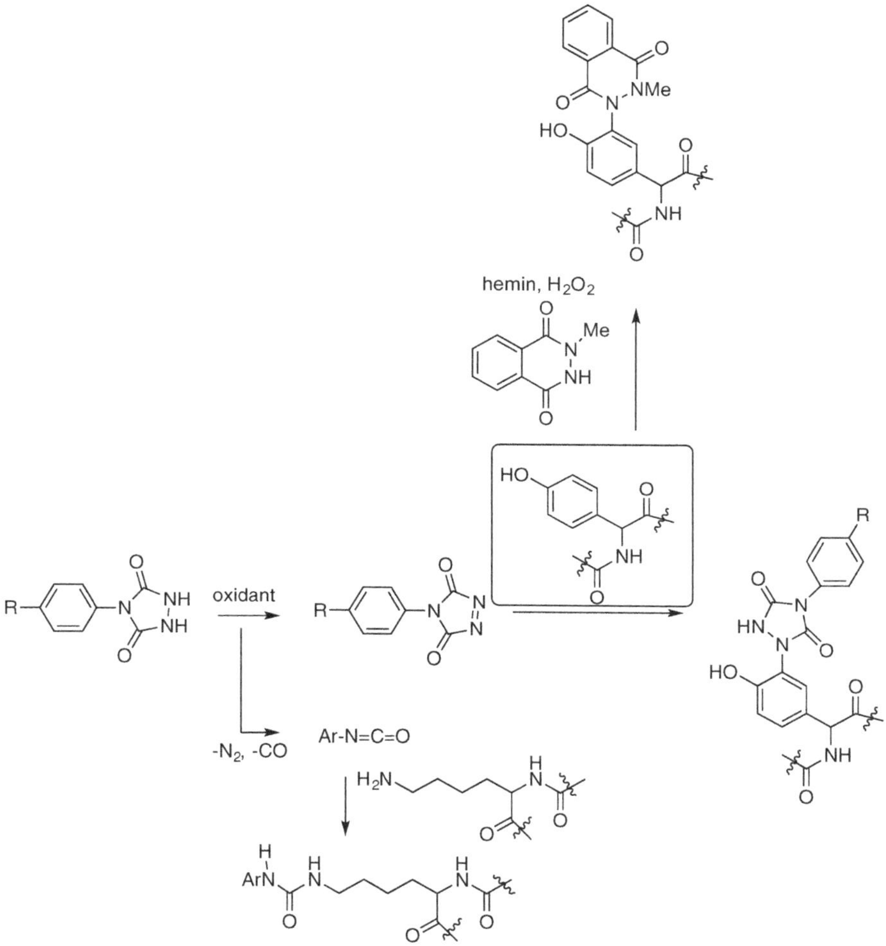

Fig. 7 Modification at Tyr residue

Fig. 8 Modification at Met residue

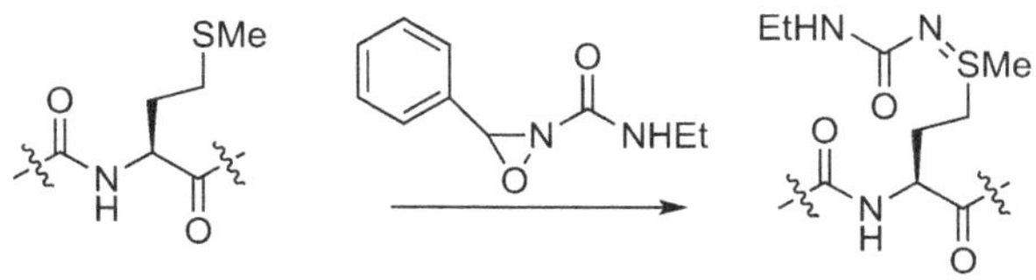

Conjugation at Non-natural Amino Acids Installed by in vitro *Protein Synthesis*

In vitro protein synthesis is a representative example of the broad potential for installing functional groups on a protein (Fig. 9) [29]. With this technique, non-natural amino acids carrying a handle for conjugation, such as an aldehyde, ketone, or azide functional group, can be installed. Carbonyl and azide groups are not found in the 20 amino acids specified by DNA, and their unique reactivities allow bio-orthogonal conjugation reactions. However, these technologies have the potential to lengthen the drug development time by requiring protein sequence re-engineering of the antibody, which can affect protein expression and stability, or by changing cell culture conditions in the bioreactor. In addition, potential immunogenicity of non-natural amino acids is not fully understood.

Axup et al. (2012) reported the incorporation of *p*-acetylphenylglycine in a constant region in trastuzumab [30]. MMAD (monomethyl auristatin D) was conjugated by oxime-ligation to generate nearly homogeneous ADCs with a DAR of 2.0. The prepared ADCs had *in vivo* efficacy and stability in serum. Zimmmerman et al. (2014) similarly reported *p*-azidomethyl phenylalanine in trastuzumab, and MMAF (monomethyl auristatin F) was conjugated by a strain-promoted azide-alkyne cycloaddition reaction [31].

Recently, many examples of bio-orthogonal reactions have been reported. Bio-orthogonal reactions were originally developed so that chemical reactions can occur in living systems without interfering with native biochemical processes [32]. Proteins contain various functional groups, including those that are sensitive to oxidative conditions, such as Met and Trp. The bio-orthogonal reaction can be applied to protein conjugation.

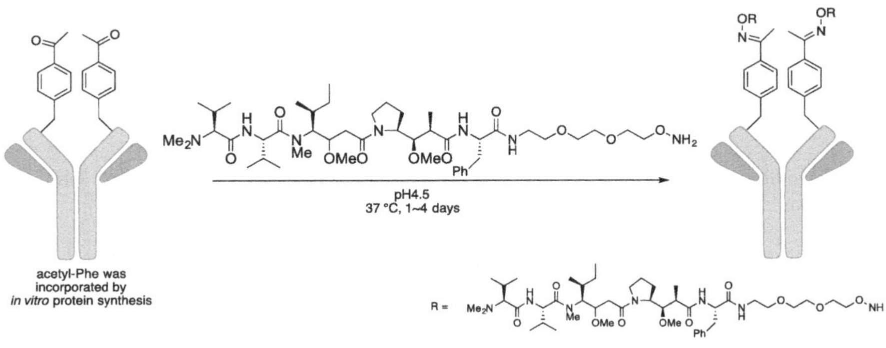

Fig. 9 Conjugation to non-natural amino acid by *in vitro* protein synthesis

Sequence-Selective Conjugation by Enzymes

Site-specific conjugation at specific amino acid selection is normally difficult. Because the same amino acid is present at multiple sites in an antibody, usual conjugation methods cannot recognize amino acids at a specific site. Sequence-selective conjugation has the potential to generate homogeneous ADCs. Protein modification by enzymes is one potential method. Formylglycine-generating enzyme converts the specific cysteine in CXPXR (X = Ser, Thr, Ala, or Gly in many cases) to formylglycine, whose aldehyde can participate in various orthogonal chemical couplings [33]. To create a carbonyl group in an antibody, formylglycine-generating enzyme is effective for ADC preparation. Though aldehyde group was generated by formylglycine generating enzyme, payloads were conjugated via oxime ligation. A disadvantage of formylglycine in water is the generation of the hydrated form, which reduces the reactivity. Drake et al. (2014) reported that the hydrazino-Pictet-Spengler reaction under neutral conditions between aldehyde function and hydrazine carrying indole yields homogeneous ADCs with stable linkage (Fig. 10) [34].

Transglutaminase isolated from *Streptoverticillium mobaraense* forms an amide bond between the glutamine side chain in the recognition sequence and primary amine. Strop et al. (2013) incorporated the recognition sequence "LLQG" at surface-accessible regions of an anti-EGFR antibody, anti-HER2 antibody, and anti-M1S1 antibody [35]. Then, MMAD was conjugated with a DAR of 1.9. They found that glutamine-tagged ADCs showed similar activity *in vitro* and *in vivo* and were better tolerated *in vivo* when compared to those with a higher drug loading (average DAR 3.6). These data suggest that either the selected position of the drug and/or the lower loading offers the possibility for improvement of the therapeutic index.

The transpeptidase sortase A is produced by gram-positive bacteria and cleaves an amide bond between Thr and Gly in the recognition peptide Leu-Pro-X-Thr-Gly

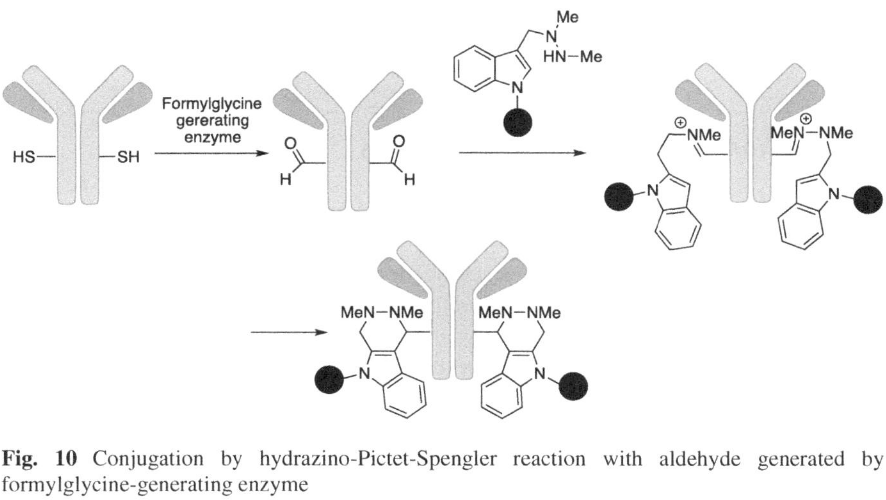

Fig. 10 Conjugation by hydrazino-Pictet-Spengler reaction with aldehyde generated by formylglycine-generating enzyme

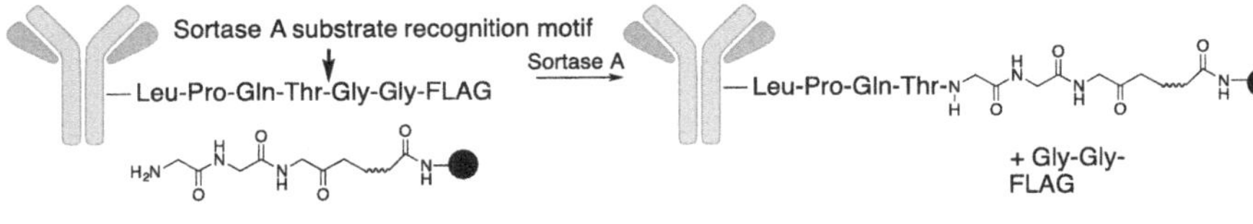

Fig. 11 Conjugation by sortase A

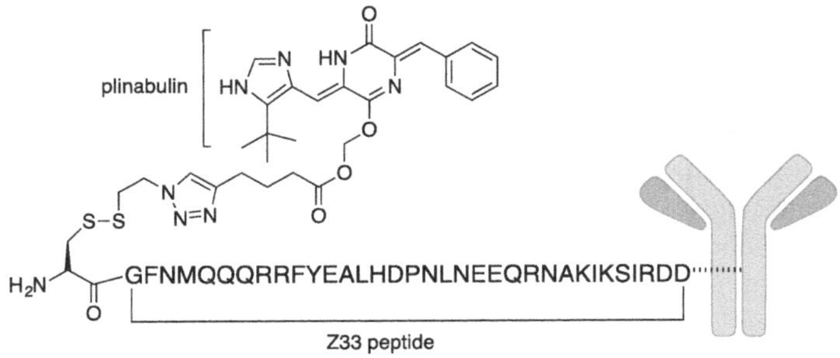

Fig. 12 ADC by non-covalent bond hybrid using Z33 peptide

(X = any amino acid) [36]. Cleavage occurs between Thr and Gly. It should be noted that the product also contains the Leu-Pro-X-Thr-Gly sequence, which is cleaved by sortase A, and the reaction mediated by sortase A is reversible. Transglutaminase directly catalyzes site-specific coupling in the LLQGA motif (Fig. 11).

Muguruma et al. (2016) utilized peptide Z33 from protein A for non-covalent type ADC preparation (Fig. 12) [37]. Z-peptide was derived from B-domain of protein A, and it has strong and site-selective binding affinity to Fc region in a noncovalent manner [38]. They conjugated plinabulin, originally found diketopiperazine-type microtuble depolymerization agent. Plinabulin-trastuzumab hybrid showed cytotoxicity against HER-2 overexpressed MCF-7 cells. Furthermore, it also showed cytotoxicity to Herceptin-resistant SKBR-3 cells *in vitro*.

Conjugation of Payloads at Glycan Sites

Conjugation to Aldehyde Generated by Fucose and Sialic Acid Oxidation

It is well-documented that protein glycosylation is one of the most prevalent post-translational modifications and can affect the stability, recognition process, and trafficking of proteins. In particular, the glycan structure affects antibody function [39, 40]. For example, the lack of a core fucose significantly enhances the affinity of an antibody to the FcγIIIa receptor, which is responsible for antibody-dependent cytotoxicity of up to 100-fold [41]. The Fc region of monoclonal immunoglobulin G (IgG) carries a conserved *N*-linked glycan chain at Asn297. This oligosaccharide is

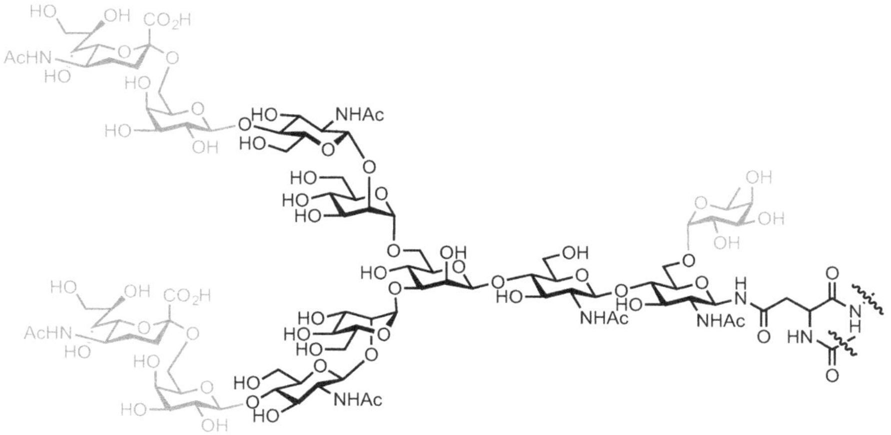

Fig. 13 Structure of *N*-glycan of antibody. Gray moiety is heterogeneous portion

characterized by considerable microheterogeneity owing to its complex biosynthetic process. Typically, *N*-linked oligosaccharide has a heterogeneous form with respect to its core fucose, sialic acid, and galactose (gray part in Fig. 13). The major isomers of biantennary glycans are composed of the G0, G1, and G2 isoforms (i.e., 0, 1, or 2 terminal galactose residues, respectively). In general, the portion of sialylated oligosaccharide is very small, i.e., only <10–14% of glycans are sialylated. These oligosaccharide structures could be used for site-specific chemical modification strategies. Since the *N*-glycan is located far from the antigen-binding Fab region, any modification of this glycan is not expected to affect the antigen-binding ability of the antibody and is apparently site-selective.

Antibody Modification by Fucose and Sialic Acid Oxidation

Because fucose and sialic acid have *cis*-diol, selective oxidative cleavage by $NaIO_4$ gives aldehyde functional group. Cytogen Corporation reported $NaIO_4$ treatment of an anti-phosphocholine IgM, HPCM2 [42]. The ^{125}I and ^{111}In -conjugated conjugated antibodies via glycan retained their binding affinity. In contrast, conjugation via lysine, aspartic acid, or glutamic acid lowered the affinity. *In vivo* biodistribution and tumor localization data suggest that antibodies radiolabeled via oligosaccharides might be beneficial for immunoscintigraphy. Zuberbühler et al. (2012) reported aldehyde generation by the oxidative cleavage of fucose and subsequent conjugation between aldehyde and hydrazide [43]. Matrix assisted laser desorption/ionization time of flight mass spectrometry (MALDI-TOF MS) showed a major fucosylated core heptaglycan and two minor glycans with one or two galactose residues (Fig. 14). The oxidative cleavage reaction can be monitored by liquid chromatography electrospray ionization (LC-ESI) MS based on a 14 Da increase in the MS peak.

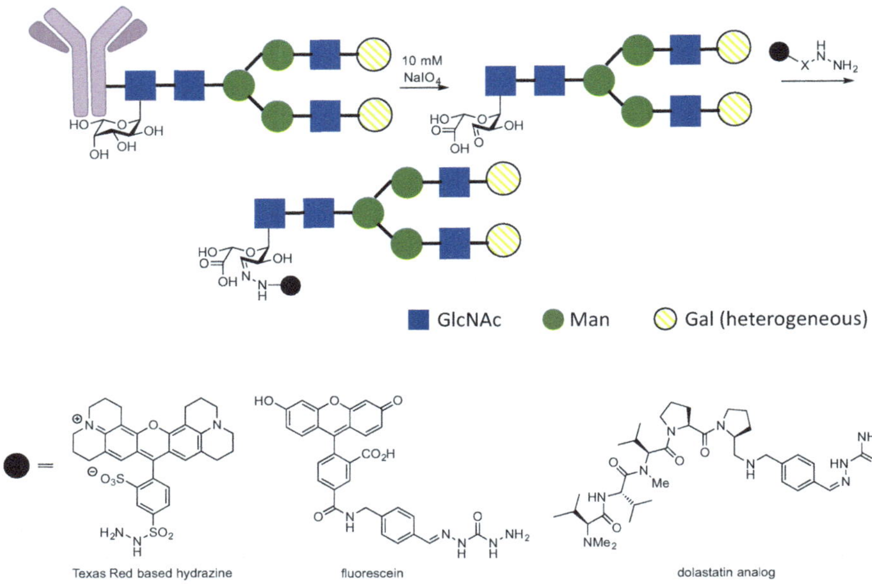

Fig. 14 Conjugation via oxidative cleavage of fucose

Overoxidation of one aldehyde to carboxylic acid was also observed. The fluorophores and dolastatin analog were conjugated via hydrazine linkage. The modified antibody showed comparable dissociation to that of the unmodified antibody based on a surface plasmon resonance experiment. The stability of hydrazone linkers depends on their structural characteristics, and the dolastatin analog was released slowly (half-time of 18 h).

Zhou et al. (2014) prepared ADCs by aldehyde-function generation via sialic acid oxidation [44]. The acyclic diol moiety of sialic acid is flexible and unhindered compared to those found in other monosaccharides; accordingly, sialic acid can be oxidized under relatively mild conditions. Normal antibodies contain low levels (<10–14%) of sialic acid. To increase sialic acid, the oligosaccharide was elongated by commercially available galactosyltransferase (Gal T) and α2,6-sialyltransferase (Sail T) using UDP (uridine diphosphate)-Gal and CMP (cytidine monophosphate)-sialic acid as glycosyl donors. MALDI-TOF MS analysis of PNGase (peptide-*N*-glucosidase) F-released *N*-glycans of control trastuzumab showed a mixture of glycan components, G0, G0F, G1, G1F, and G2F. After treatment with two transferases, HPAEC-PAD (high-performance anion exchange chromatography with pulsed amperometric detection) showed the presence of 2.2 sialic acids per antibody. The glycan-remodeled antibody was oxidized by $NaIO_4$ (0.25 to 2 mM) and as little as 0.5 mM $NaIO_4$ was enough for sialic oxidation. Antibody oxidation using 1 mM $NaIO_4$ caused partial methionine oxidation and a 15% reduction of binding affinity. After conjugation of the aminoxy drug-linker, HIC (hydrophobic interaction chromatography) showed 1.3–1.9 drug molecules per antibody, compared with 3.5–3.9

drugs per antibody when the drug was conjugated by thiol-maleimide-based conjugation. The *in vivo* efficacy of glycol ADCs was demonstrated in a HER2-positive tumor cell xenograft model.

As diagnosis and treatment are becoming increasingly linked, therapeutic monoclonal antibodies have been developed as molecular diagnostic tools. An antibody radiolabeled with short-lived isotopes can provide critical molecular information when used with single photon emission computed tomography or positron emission tomography imaging. Bejot et al. (2012) reported an aminooxy group tethered to chelator-conjugated Herceptin radiolabeled with the single photon emitter ^{111}In (172.247 keV, $t_{1/2}$ = 2.83 days) [45]. They prepared aminooxy-DOTA (tetraazacyclododecanetetraacetic acid)-conjugated trastuzumab via *N*-glycan oxidation by $NaIO_4$ and compared it to DOTA with an amide bond at lysine residues. The two radiolabeled antibodies showed nearly identical behaviors with respect to stability, immunoreactivity, and affinity both *in vitro* and *in vivo*. These methodologies require prolonged exposure of the antibody to low pH and severe oxidative conditions, and the modification of amino acid residues, such as cysteine, methionine, and tryptophan, is a potential issue.

Conjugation by Modified Monosaccharide Transfer by Transferase

Since galactosyl transferase tolerates a modified galactosyl donor structure, galactose with a handle for functionalization can be added by galactosyltransferase. Although the wild-type enzyme shows poor catalytic efficiency for substrates larger than galactose, Boeggeman et al. (2009) reported that C2-keto-galactose was incorporated by mutant galactosyltransferase β(1,4)Gal-T1-Y289L [46]. In addition to C2-keto-galactose, azide-GalNAc was also added by the enzyme. After the removal of native galactose by galactosidase, the modified sugar was transferred by the enzyme. Native galactose was replaced with modified galactose in this process. Then, the fluorescent reagents Alexa Fluor 488 and biotin with an aminooxy-group were conjugated via the chemical handles. This methodology has been applied to various antibodies, including rituximab, infliximab, and bevacizumab. Based on ELISA, modified infliximab did not show decreased binding to VEGF, and trastuzumab with Alexa Fluor 488 was useful to target HER-2-expressing cells.

Zeglis et al. (2014) used positron emission tomography and optical imaging with a site-specifically labeled immunoconjugate by a similar method [47]. *N*-acetylgalactosamine (GalNAc) with an azide group was transferred to a galactose-containing antibody by recombinant galactosyl transferase [48]. Then, a hydroxamic acid chelator was installed via strain-promoted click chemistry for diagnosis.

Sialyltransferases tolerate modifications at the C5- and C-9 positions of sialic acids. Li et al. (2014) used 9-azide-functionalized CMP-sialic acid to introduce a handle for conjugation (Fig. 15) [49]. After galactosyl transferase and recombinant α-2,6-sialyltransferase I treatment of heterogeneous *N*-glycan of anti-CD22 with UDP-galactose and CMP-sialic acid, sialic acid carrying an azide group was incorporated, and homogeneous *N*-glycan was obtained. High-pH anionic exchange

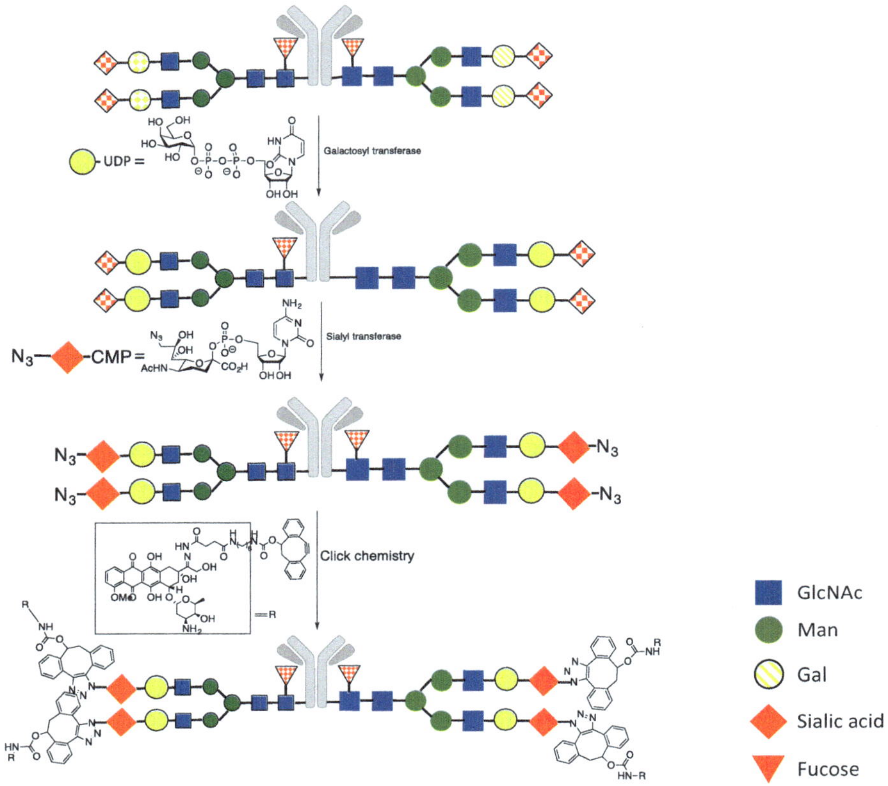

Fig. 15 Conjugation at glycan site by glycosyl transferases and bioorthogonal reaction

chromatography indicated that there were 4.3 azido-containing sialic acid residues per antibody. The use of α(2,3)-sialyltransferase of *Pasteurella multocida* or ST3GalIV resulted in the partial transfer of the donor. DIBO (dibenzylcyclooctynol)-FITC and DIBO-biotin conjugates reacted with the azide group of sialic acid by the Cu-free Huisgen reaction. Hydrazone-linked doxorubicin was also immobilized on the antibody. After the immobilization of doxorubicin via sialic acid, the binding ability to the Fc Receptor IIIA was not decreased, and *in vivo* cytotoxicity was observed in a dose-dependent manner.

Modified Glycan Transfer by an Oligosaccharide Transferase Mutant

In addition to monosaccharyl transferases, endo-β-*N*-acetylglucosaminidase (ENGase) mutants could be useful for site-specific conjugation. EndoS, one of endo-β-*N*-acetylglucosaminidases, is widely used for glycan trimming of antibodies to obtain single GlcNAc moieties [50]. Combinations of EndoS mutants with reduced

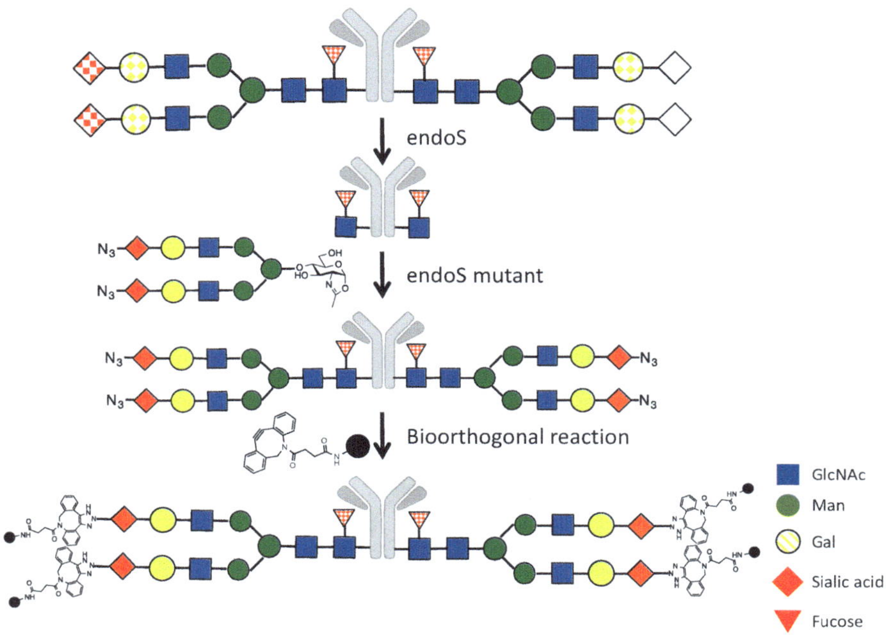

Fig. 16 Conjugation at glycan site by mutant endo-β-*N*-acetylglucosaminidases (ENGases)

hydrolytic activity and the transition-mimic sugar oxazoline are used for the transfer of oligosaccharides with functional groups (Fig. 16) [51]. Parsons et al. (2016) showed that the pH must be controlled for transglycosylation [52]. At pH 7.5, an MS analyses indicated that there were more than two glycans per antibody. At pH 6.5, the side-reaction was retarded to give two glycans per antibody. Suda et al. (2016) [53] and Wang (2016) [54] independently reported that amine in Lys reacts with oxazoline. An acidic pH reduces the side-reaction between the amino group and oxazoline by protonation of the amino group. Manabe et al. [55] reported stepwise addition of sugar oxazoline under pH 6.5 significantly suppressed side-reactions, and real-time monitoring or reactions was achieved by hydrophilic interaction chromatography (HILIC) column chromatography [55]. MMAE-conjugated trastuzumab at a glycan site showed enhanced killing of HER2(+) cells (SK-BR-3) compared with trastuzumab.

Metabolic Modification of Antibodies

Metabolic modification of *N*-glycan is possible. When fucose analogs are added outside of cells, fucosyltransferase adds fucose analogs to the core *N*-acetylglucosamine (GlcNAc). Usually, fucose analogs are added in the peracetylated form, as this modification is known to increase cellular uptake efficiency. Acetate is subsequently hydrolyzed in the cytosol and converted to GDP-sugar by

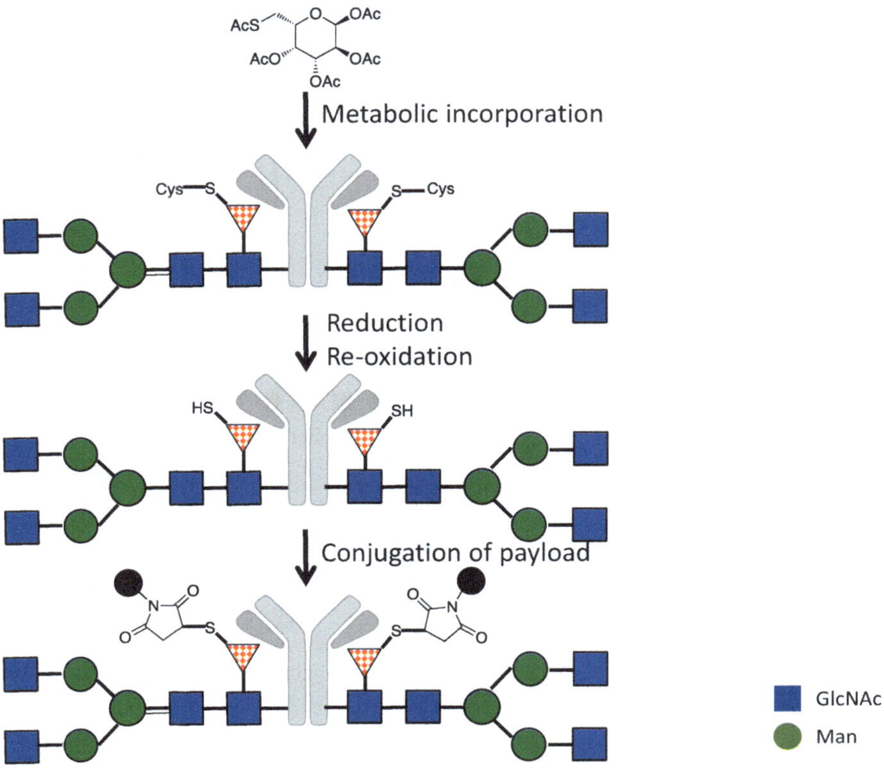

Fig. 17 Incorporation of thiol at glycan site by metabolic engineering

the fucose salvage pathway. Extensive studies have examined the application of the metabolic modification of *N*-glycan to site-specific anticancer conjugation techniques (Fig. 17) [56]. Thiol carrying a fucose derivative was incorporated in place of fucose with 70% efficacy. Anti-CD70 (h1F6) and anti-CD30 (cAc10) antibodies were used for generating ADCs via incorporated thio-fucose in Chinese Hamster Ovary (CHO) cells. In total, 1.2–1.4 molecules of thiol-modified fucose were incorporated per antibody, and thiol could be a useful handle for the conjugation of anticancer agents. However, the incorporated thiol of fucose was present in a disulfide form with a cysteine residue, so a reduction process was required for conjugation. After the reduction of disulfide bonds with dithiothreitol, 9.3 molecules of MMAE with a cathepsin-cleaved linker per antibody were conjugated. These ADCs showed potent cytotoxicity in CD30-positive cell lines *in vitro*.

The availability of the modified sugar moiety makes it possible to link cytotoxic agents and imaging agents at specific sites. Novel glycosyltransferases have been designed with broad specificities for the transfer of a sugar residue with a chemically reactive functional group to a specific terminal moiety of a glycoconjugate.

Improvement of Maleimide-Thiol Linker Stability

The addition of payloads via sulfhydryl group of Cys residues following the reduction of disulfide bond is common. However, because the reaction between the sulfhydryl group and maleimide is reversible, the stability of ADCs is a frequent issue [57]. Payloads are sometimes released in an unwanted region or in serum during circulation. This undesired reaction results in a narrow therapeutic window of ADCs. In phosphate-buffered saline, ADCs with maleimide-thiol linkage are rather stable, but they are not stable in blood or plasma. Half-lives of ADCs with maleimide-thiol linkage are usually 1–2 days. The released maleimoide-thiol linker conjugated with a payload is added to other proteins, such as albumin (Fig. 18). Because thiol has the strongest nucleophilicity among amino acid components. Owing to the development of THIOMAB technology, conjugation at Cys sites has become an attractive method. To overcome the instability of maleimide-thiol linkage, several linkages have been reported.

The bromoacetamide group and thiol react to form a stable amide bond (Fig. 19). Allery reported that the bromoacetamide-thiol linkage is stable, with a half-life of ~7 days. However, basic conditions are required when bromoacetamide-thiol linkages are prepared, while nearly neutral conditions are sufficient for maleimide-thiol addition.

To prevent reversible maleimide-thiol linkage, self-hydrolyzing maleimides have been developed (Fig. 20) [58]. Near maleimide moieties, an amino group was incorporate to enhance the hydrolysis of imide functional groups under neutral conditions. The linkage between linker and antibody was changed from imide to amide by hydrolysis. Once an amide bond is generated, the linkage becomes stable. Indeed, ADC using an anti-CD30 antibody conjugated with MMAE via the Val-Cit sequence showed excellent stability in mice, improved antitumor activity, and reduced neutropenia.

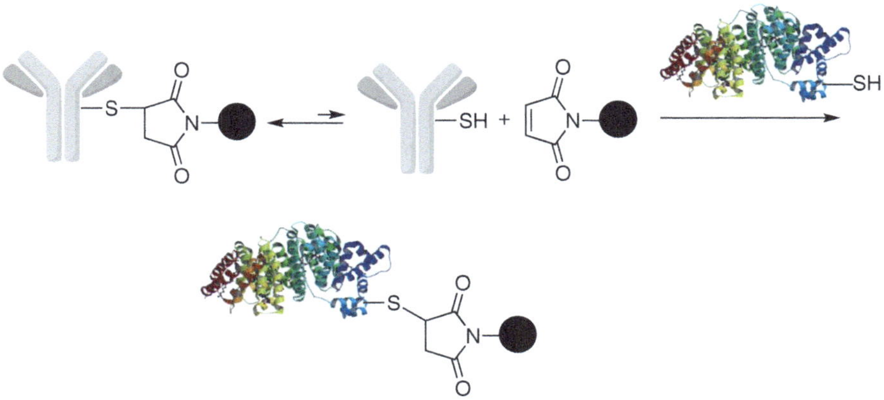

Fig. 18 scrambling of payloads with other proteins due to retro-Michael reaction

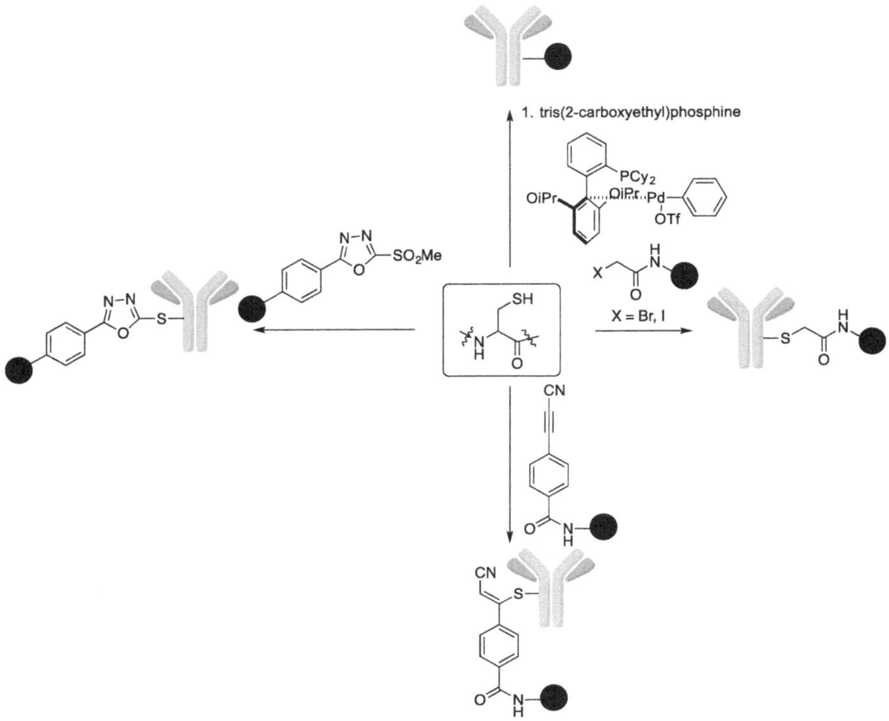

Fig. 19 Conjugation at Cys site with stable linkages

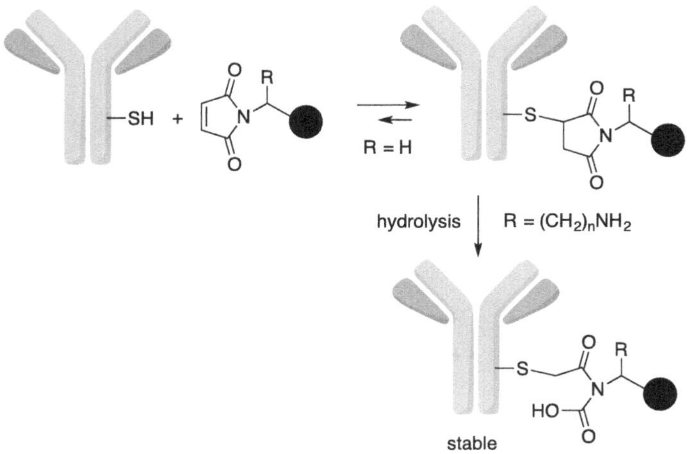

Fig. 20 Self-hydrolyzing maleimide linker

Pd-mediated allylation typically occurs at Tyr (*O*-allylation) sites over Lys (*N*-allylation) and cysteine (*S*-allylation). Buchward reported that the Pd-tolyl complex using 2-dicyclohexylphophino-2′,6′-diisopropxybiphenyl (RuPhos) as a ligand and trifluoromethanesulfonate as the counter-ion was an effective catalyst for Cys modification [59]. This unique reactivity is probably explained by the lower electrophilicity than that of typical Pd reagents. The reaction was performed under neutral to slightly basic conditions within a short period (<5 min), and the catalyst was air-stable. Using this strategy, the aryl payload vandetanib was directly conjugated to trastuzumab, and the DAR was 4.4. Unfortunately, selective releasing function of vandetanib may be difficult, because vandetanib was conjugated directly to the antibody.

Patterson et al. (2014) reported that a phenyloxadiazole sulfone linker improved antibody conjugate stability in human plasma [60]. Sulfone conjugation with selenocysteine also improved stability.

4-((4-(Cyanoethynyl)benzoyl)oxy)-2,3,5,6-tetrafluorobenzensulfonate was designed as a bifunctional group for conjugation [61]. Amino groups are easily replaced with electron-deficient eater groups, followed by the conjugate addition of a thiol group to the triple bond. These linkages are stable in human plasma for more than 72 h.

Traditional cysteine conjugation methodology is of limited value for site-specific conjugation, because it cannot recognize a specific Cys residue among multiple Cys residues. A sequence-specific conjugation method for Cys would, therefore, be useful, and Pentelute developed a nucleophilic aromatic substitution (S_NAr) substitution reaction with perfluo-aromatic biphenyl compound [62]. Perfluoro-aromatic biphenyl specifically recognizes the Cys residue in the Phe-Cys-Pro-Phe sequence for substitution. In order to prepare the ADC, the Phe-Cys-Pro-Phe sequence was incorporated at the C-terminus of trastuzumab. As a result, conjugation of biotin and MMAF were achieved specifically and only at Cys in the Phe-Cys-Pro-Phe sequence. No conjugation at other native Cys residues was observed in an LC-MS analysis. The ADC thus formed had the same binding affinity to HER as trastuzumab in an Octet binding assay and killed BT474 cells *in vitro* (ED_{50} 0.19 nM).

Bis-sulfone reagents undergo bis-alkylation to conjugate both thiols derived from two cysteine residues of reduced native disulfide bonds (Fig. 21) [63]. This approach results in covalent re-bridging of the disulfide bond via a three-carbon bridge, leaving the protein structurally intact. ADCs using trastuzumab conjugated with MMAE are stable after 120 h and the homogeneity is increased, with a DAR of 4. A similar bridging approach has been reported based on the thiomaleamic acid compound. This linker released the payload doxorubicin under acidic conditions (pH 4.5).

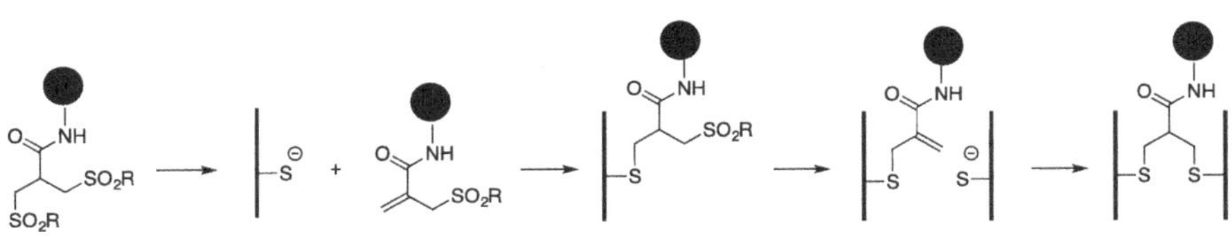

Fig. 21 Conjugation by three-carbon re-bridging of disulfides

Control of Payload Cleavage Manner

Cleavage of Payloads by Enzymatic Degradation

Although the preparation of homogeneous ADCs has been a focus of research, payload release is an important factor in linker design. Linker diversity is not sufficient at this time to permit a wide variety of choices for cleavage mechanisms. Cathepsin-cleavable Val-Cit and Val-Ala linkers are used in approximately half of pipeline ADCs [64]. The Val-Cit and Val-Ala linkers are also most common in research.

Lysosomal proteases cleave the Gly-Phe-Gly-Gly sequence (Fig. 22) [65]. Hydrophobic DNA topoisomerase I inhibitor exatecan derivatives were conjugated at Cys residues of trastuzumab. The average DAR was 7.7, which was remarkably higher than that of ordinary ADCs. After protease cleavage, hydroxy-linked exatecan derivative DX-8951f was released, instead of exatecan itself. The released high permeable DX-8951f shows cytotoxicity to HER2-negative cells as well as HER-2 positive cells *in vivo*. This ADC, called DS-8201a has potent antitumor activity in a broad selection of HER-2-positive models and great potential to respond to T-DM-1-insensitive HER-2 positive cancers.

Many examples of β-glucuronidase-cleavable linkers in ADCs have been reported because prodrugs using β-glucuronidase digestion have been developed. β-Glucuronic acid with *p*-aminobenzyl carbamate carrying payloads and antibodies attached at functional groups (Fig. 23) [66, 67]. Similar to cathepsin-cleavable link-

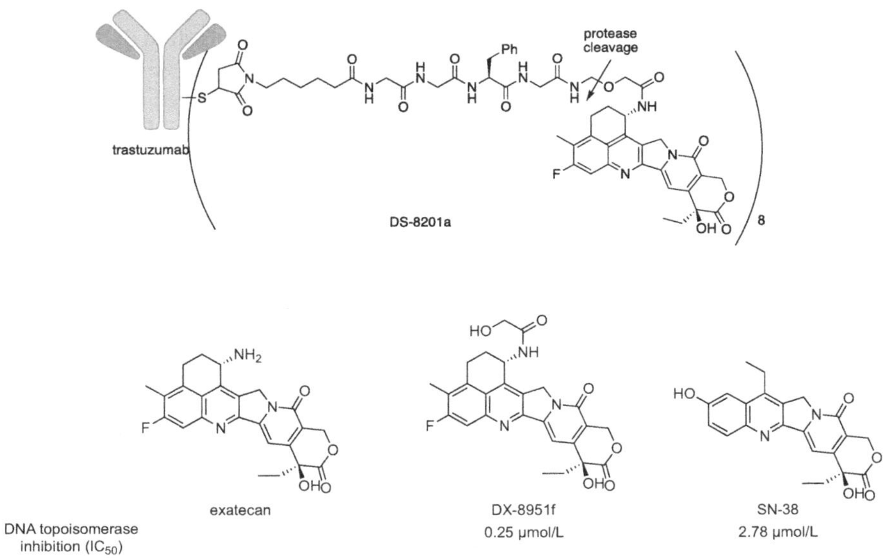

Fig. 22 Protease-release of topoisomerase inhibitors

ers, 1,6-elimination or cyclization occur after the cleavage of the glycosyl linkage to release payloads.

A pyrophosphate esterase cleavage linker has been reported [68]. Precision design is possible for pyrophosphate-cleavable linkers. Pyrophosphates have robust plasma stability, yet are cleaved in lysosomes. Kern et al. (2016) synthesized several alkyl-linked and aryl-linked pyrophosphate-linked glucocorticoids. The glucocorticoid linkers were linked by bio-orthogonal chemistry via azide groups in an anti-hCD-70 antibody incorporated by *in vitro* protein synthesis. Pyrophosphate and triphosphate diester linkers cleave much more rapidly than do monophosphate diesters, and this is correlated with the activity of their ADCs.

Matsumura successfully developed a monoclonal antibody, called anti-IF mAb, against insoluble fibrin (IF) following his intensive studies of cancer induced blood coagulation [69]. Tumor vascular porosity due to mural abnormalities in malformed capillaries in the neoplasm (see chapter "The Cancer Stroma and Its Relevance to Tumor Survival and Treatment" by Tarin in this book for pictures), is the major factor responsible for the enhanced permeability and retention (EPR) effect discovered by Matsumura and Maeda [70]. It is associated with followed by micro-hemorrhaging and tumor induced blood coagulation which produces IF in the interstitial space. Fuchigami et al. (2018) exploited this phenomenon to devise a novel onco-therapeutic regimen by creating an ADC using the anti-IF mAb conjugated with a chemotherapy payload [71]. The conjugate contains a linker severed specifically by

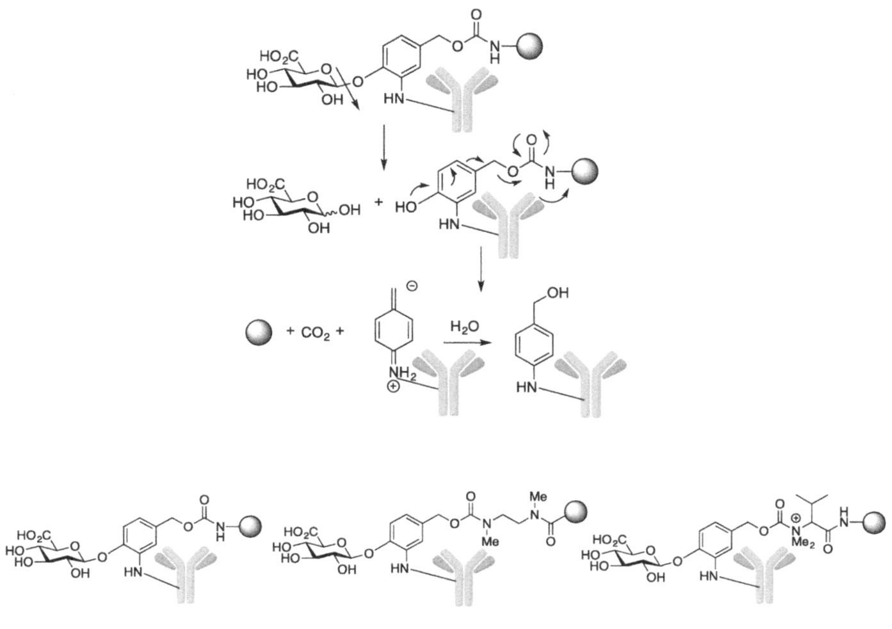

Fig. 23 Glucronidase cleavable linkers

plasmin, an enzyme which is activated only on binding to IF. Their strategic concept of cancer stromal targeting (CAST) therapy is that the IF-ADC selectively extravasates in the tumor rather than in other sites, due to the EPR effect. It can then access and bind to the specific pits it recognizes in the fibrin clot that are uncovered only when a fibrin clot forms. This utilizes the stroma to create a scaffold from which effective sustained release of free ACA may occur, as a result of cleavage of the Val-Leu-Lys linker by plasmin [71].

Payload Cleavage by Light

Light between 650–900 nm, often referred to as the near-IR window, is cytocompatible and has significant tissue penetration, with minimal toxicity. Light-based therapeutic modalities using phototoxic small molecules have an extensive history in the treatment of cancer and skin disorders. Heptamethine cyanide fluorophore linked payloads can be cleaved by near IR irradiation (690 nm) (Fig. 24) [72]. This strategy may be applied to the timed delivery of small molecules, if the target lesion is within range of penetration by the radiant energy.

Payload Cleavage by Chemical Reaction

The inverse-electron-demand Diels-Alder reaction between *trans*-cyclooctene and tetrazines triggers payload release [73]. After this reaction, resulting 4,5-dihydrapyridazines are tautomerized to 1,4-dihydrapyridazines, following which CO_2 and the payload are released (Fig. 25). Substituted *trans*-cyclooctenes have isolative diastereomers depending on the high energy barrier of the ring flip. It was shown that the reactivity of the axially oriented payload was 156 times lower than that of the equatorially located payload, probably due to steric hindrance and electronic effects [74]. To enhance the release rate of payloads, Rossin introduced the methyl group of cyclooctene to fix the conformation [75]. Using such a construct, Doxorubicin release was observed in mice by the addition of tetrazine. This "click to release" concept may circumvent endogenous intracellular and extracellular activation mechanisms in cancer biology. This strategy also expands the scope to non-internalizing cancer targets, and maximizes the by-stander effect in heterogeneous tumors, by attacking cancer cells that do not express the targeted receptor.

Non-internalizing ADCs are valuable for solid-cancer treatment [76]. Perrino et al. (2014) reported non-internalizing anti-fibronection antibody conjugated with maytansinoids via a disulfide linker show good therapeutic efficacy [77]. Yasunaga et al. (2011) reported ADCs toward to collagen 4 and fibrin were effective for localization of therapeutic agents in stroma-rich cancers. They linked SN-38 and

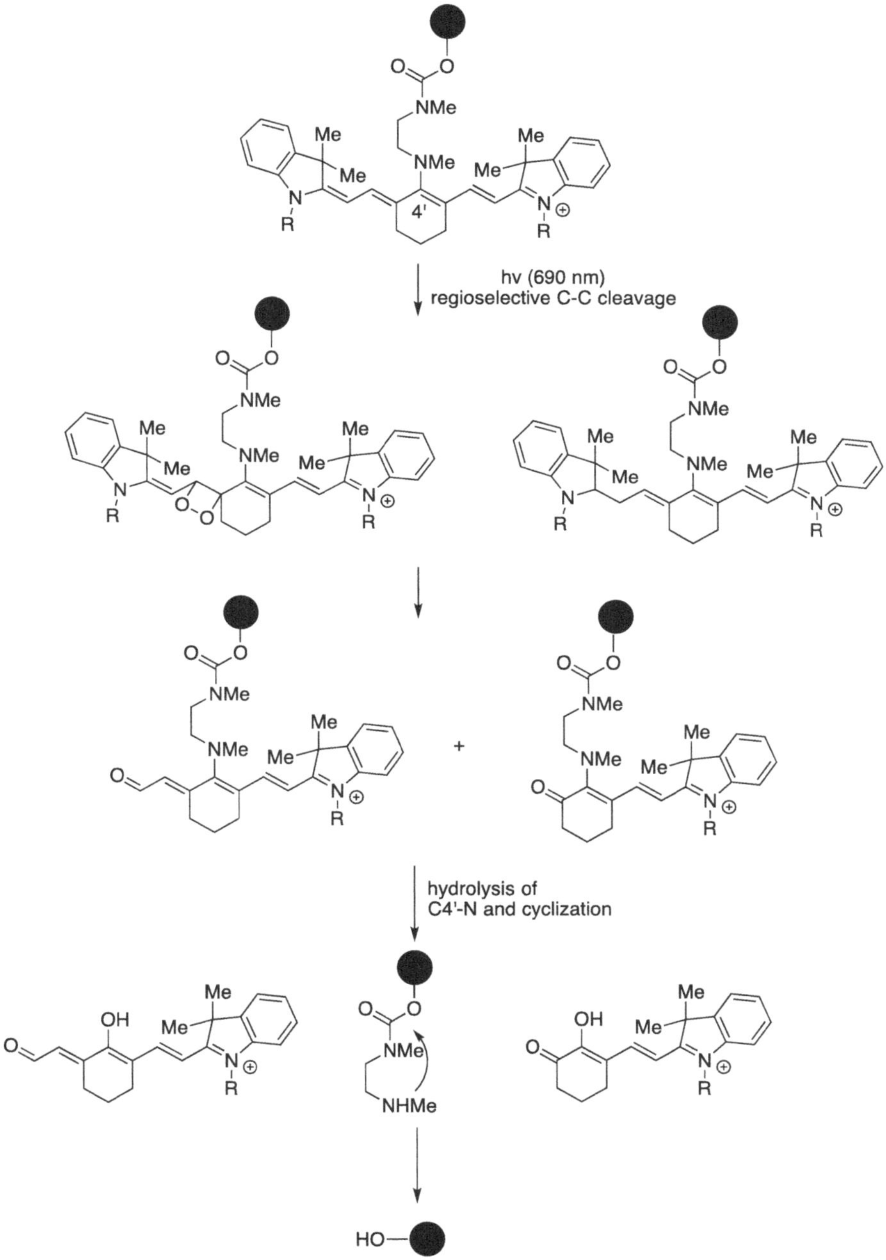

Fig. 24 Payload cleavable by near-IR light

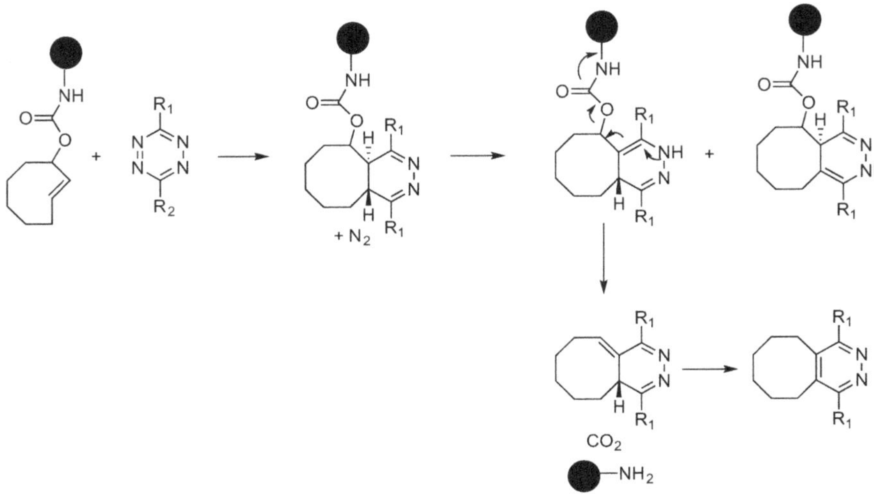

Fig. 25 Payload cleavage by chemical reaction

MMAE via an ester linkage to be hydrolyzed outside cells [78–80]. IMMU-132, sacituzumab govitecan, is an ADC derived from anti-Trop-2 and SN-38. SN-38 was linked via carbonate linkage in high DAR of 7.6, and it was gradually released by hydrolysis [81].

Detection of Released Cytotoxic Agents

Despite rapid progress in linker technology, additional logical linker designs are still needed. At present, methods for the evaluation of linker design, especially for the evaluation of cleavage function, are not available. Evaluations of linker design are usually focused only on detection of drug activity. When cytotoxicity is not observed, we cannot determine whether a linker is simply not cleaved or whether other factors contribute. Techniques have not yet been developed to measure the release rate and ACA distribution, but MS imaging is a potentially powerful method being explored for the detection of payload release. Fujiwara et al. (2016) [82] reported that MS imaging is a powerful tool for detecting the distribution of a released drug. MMAE itself is detectable in a semi-quantitative manner by MS imaging. Only released MMAE was detected by MS imaging, while antibody-bound MMAE was not detected owing to the high molecular weight of the ADC.

Genentic reported imaging of the linker cleavage process by using FRET (fluorescence resonance energy transfer) technology [83]. When payloads were released from antibody, fluorescence was recovered. They compared the internalization of ADC, the releasing rate, and the distribution of cytotoxic agents in the cell in

SK-BR-3 cells and TenB2-PC3 cells. Although a fluorescent reagent may change the distribution of cytotoxic agents, this methodology would be a powerful technique to understand complex cytotoxic releasing processes and aid the development more efficient linkers based on logical design.

The relationship between diseases and the up-regulation of metabolic enzymes is not clear. To release payloads in specific cells, cell type-specific and disease-related enzyme activity should be clarified. Onagi et al. (2017) used peptides with known sequences as substrates and analyzed metabolites by LC-MS. Enzymes were identified by zymography in a diced electrophoresis gel assay [84].

Other Roles of Linkers in ADCs

Pharmacokinetics, pharmacodynamics, and biodistribution are important factors in ADC design. Linkers can also control pharmacokinetics. High-loaded ADCs usually result in accelerated plasma clearance owing to the hydrophobic nature of payloads, although promising examples such as DS8201a and IMMU-132, have been identified. As the cleavable moiety of the linker changes from Val-Cit to glucuronic acid and with the introduction of PEG, hydrophobicity is reduced. The hydrophobicity is simply measured by HIC/HILIC column chromatography. As a result, the clearance speed is retarded, even for a DAR of 8 [85]. DAR can be increased by design of branched PEG linker [85].

The application of ADCs is not limited to cancer treatment. Lehar et al. (2015) reported that ADCs constructed from anti-*S. aureous* antibody and antibiotic rifalogue showed superior effects to vancomycin [86]. Yasunaga et al. (2017) identified a relationship between IL-7R signaling and steroid-resistance in lymphoid malignancy [87]. ADCs from anti-IR-7R and SN-38 showed significant efficacy. This result suggests that the use of ADCs could be a promising strategy to treat immune diseases as a novel alternative to steroid therapy.

Conclusion

In this chapter, recent advances in linker technology have been reviewed and evaluated The design of linkers is recognized to be crucial factor for efficient delivery of treatment. As of 2018, 570 clinical trials of ADCs are underway and more than 50,000 patients have been enrolled. However, there are many issues still to be solved in linker technology and it is essential to find a way by which a specific drug can be efficiently released at appropriate dosage, in the right place, at the most suitable time after administration. This chapter is contribution towards this clinical therapeutic mission.

References

1. Wu AM, Senter PD (2005) Arming antibodies: prospects and challenges for immunoconjugates. Nat Biotechol 23:1137–1146
2. Beck A, Goetsch L, Dumontet C et al (2017) Strategies and challenges for the next generation of antibody-drug conjugates. Nat Rev Drug Discov 16:315–337
3. Carter PJ, Lazar GA (2018) Next generation antibody drugs: pursuit of the high-hanging fruit. Nat Rev Drug Discov 17:197–223
4. Gualberto A (2012) Brentuximab Vedotin (SGN-35), an antibody–drug conjugate for the treatment of CD30-positive malignancies. Expert Opin Investig Drugs 21:205–216
5. Hamann PR, Hinman LM, Hollander I et al (2002) Gemtuzumab ozogamicin, a potent and selective anti-CD33 antibody-calicheamicin conjugate for treatment of acute myeloid leukemia. Bioconjug Chem 13:47–58
6. Verma S, Miles D, Gianni L et al (2012) Trastuzuab emtansine for HER2-positive advance breast cancer. N Engl J Med 367:1783–1791
7. Kantarjian H, Thomas D, Jorgensen J et al (2012) Inotuzumab ozogamicin, an antibody-CD22-calecheamicin conjugate, for refractory and relapsed acute lymphocytic leukaemia: a phase 2 study. Lancet Oncol 13:403–411
8. Hamblett KJ, Senter PD, Chace DF et al (2004) Effects of drug loading on the antitumor activity of a monoclonal antibody drug conjugate. Clin Cancer Res 10:7063–7070
9. McDonagh CF, Turcott E, Westendorf L et al (2006) Engineered antibody–drug conjugates with defined sites and stoichiometries of drug attachment. Protein Eng Des Sel 19:299–307
10. Trail PA, Willner D, Lasch SJ et al (1993) Cure of xenografted human carcinomas by BR96-doxorubicine immunoconjugates. Science 261:212–215
11. Dubowchik GM, Firestone RA, Padilla L et al (2002) Cathepsin B-labile dipeptide linkers for lysosomal release of doxorubicin from internalizing immunoconjugates: model studies of enzymatic drug release and antigen-specific in vitro anticancer activity. Bioconjug Chem 13:855–869
12. Denmaede SR, Nagy A, Gao J et al (1998) Enzymatic activation of a doxorubicin-peptide prodrug by prostate-specific antigen. Cancer Res 58:2537–2540
13. Manabe S, Machida H, Aihara Y et al (2013) Development of a diketopiperazine-forming dipeptidyl Gly-pro spacer for preparation of an antibody-drug conjugate. Med Chem Commun 4:792–796
14. Caculitan NG, Dela Cruz Chuh J, Ma Y et al (2017) Cathepsin B is dispensable for cellular processing of cathepsin B-cleavable antibody-drug conjugates. Cancer Res 77:7027–7037
15. Dorywalska M, Dushin R, Moine L et al (2016) Molecular bias of valine-citrulline-PABC linker instability in site-specific ADCs and its mitigation by linker design. Mol Cancer Ther 15:958–970
16. Anami Y, Yamazaki CM, Xiong W et al (2018) Glutamic acid-valine-citrulline linkers ensure stability and efficacy of antibody-drug conjugates in mice. Nat Commun 9:2512
17. Hermanson GT (2008) Bioconjugate Techniques, 2nd edn. Academic, San Diego
18. Junutula JR, Raab H, Clark S et al (2008) Site-specific conjugation of a cytotoxic drug to an antibody improves the therapeutic index. Nat Biotechnol 26:925–932
19. Junutula JR, Bhakta S, Raab H et al (2008) Rapid identification of reactive cysteine residues for site-specific labeling of antibody-Fabs. J Immunol Methods 332:41–52
20. Koshland DE, Karkhanis YD, Latham HG (1964) An environmentally-sensitive reagent with selectivity for the tryptophan residue in proteins. J Am Chem Soc 86:1448–1450
21. Savige WE, Fontana A (1980) Oxidation of tryptophan to phenylalanine by dimethylsulfoxide-hydrochloric acid. Selective modification of tryptophan containing peptides. Int J Pept Protein Res 15:287–297
22. Antos JM, McFarland JM, Iavarone AT et al (2009) Chemoselective tryptophan labelling with rhodium carbenoids at mild pH. J Am Chem Soc 131:6301–6308

23. Hansen MB, Hubálek F, Skrydstrup T et al (2016) Chemo- and regioselective ethynylation of tryptophan-containing peptides and proteins. Chem Eur J 22:1572–1576
24. Seki Y, Ishiyama T, Sasaki D et al (2016) Transition metal-fee tryptophan-selective bioconjugation of proteins. J Am Chem Soc 138:10798–10801
25. Ban H, Nagano M, Gavrilyuk J et al (2013) Facile and stable linkages through tyrosine: bioconjugation strategies with the tyrosine-click reaction. Bioconjug Chem 24:520–532
26. Sato S, Nakamura K, Nakamura H (2015) Tyrosine-specific chemical modification with in situ hemin-activated luminol derivatives. ACS Chem Biol 10:2633–2640
27. Bruins JJ, Westphal AH, Albada B et al (2017) Inducible, site-specific protein labeling by tyrosine oxidation-strain-promoted (4+2) cycloaddition. Bioconjug Chem 28:1189–1193
28. Lin S, Yang X, Jia S et al (2017) Redox-based reagents for chemoselective methionine bioconjugation. Science 355:597–602
29. Hallam TJ, Smider VV (2014) Unnatural amino acids in novel antibody conjugates. Future Med Chem 6:1309–1324
30. Axup JY, Bajjuri KM, Ritland M et al (2012) Synthesis of site-specific antibody-drug conjugates using unnatural amino acids. Proc Natl Acad Sci USA 109:16101–16106
31. Zimmerman ES, Heibeck TH, Gill A et al (2014) Production of site-specific antibody-drug conjugates using optimized non-natural amino acids in a cell-free expression system. Bioconjug Chem 25:351–336
32. Lang K, Chin JW (2014) Bioorthogonal reactions for labeling proteins. ACS Chem Biol 9:16–20
33. Appel MJ, Bertozzi CR (2015) Formylglycine, a post-translationally generated residue with unique catalytic capabilities and biotechnology applications. ACS Chem Biol 10:72–84
34. Drake PM, Albers AE, Baker J et al (2014) Aldehyde tag coupled with HIPS chemistry enables the production of ADCs conjugated site-specifically to different antibody regions with distinct in vivo efficacy and PK outcomes. Bioconjug Chem 25:1331–1341
35. Strop P, Liu SH, Dorywalska M et al (2013) Location matters: site of conjugation modulates stability and pharmacokinetics of antibody drug conjugates. Chem Biol 20:161–167
36. Paterson BM, Alt K, Jeffery CM et al (2014) Enzyme-mediated site-specific bioconjugation of metal complexes to proteins: Sortase-mediated coupling of Copper-64 to a single-chain antibody. Angew Chem Int Ed 53:6115–6119
37. Muguruma K, Yakushiji F, Kawamata R et al (2016) Novel hybrid compound of a Plinabulin prodrug with an IgG binding peptide for generating a tumor selective noncovalent-type antibody-drug conjugate. Bioconjug Chem 27:1606–1613
38. Braisted AC, Wells JA (1996) Minimizing a binding domain from protein A. Proc Natl Acad Sci USA 93:5688–5692
39. Kaneko Y, Nimmerjahn F, Ravetch JV (2006) Anti-inflammatory activity of immunoglobulin G resulting from Fc sialylation. Science 313:670–673
40. Ferrara C, Grau S, Jäger C et al (2011) Unique carbohydrate-carbohydrate interactions are required for high-affinity binding between FcγRIII and antibodies lacking core fucose. Proc Natl Acad Sci USA 108:12669–12674
41. Shields RL, Lai J, Keck R et al (2002) Lack of fucose on human IgG1 *N*-linked oligosaccharide improves binding to human FcγRIII and antibody-dependent cellar toxicity. J Biol Chem 277:26733–26740
42. Rodwell JD, Alvarez VL, Lee C et al (1986) Site-specific covalent modification of monoclonal antibodies: in vitro and in vivo evaluations. Proc Natl Acad Sci 83:2632–2636
43. Zuberbühler K, Casi G, Goncalo JL et al (2012) Fucose-specific conjugation of hydrazide derivatives to a vascular-targeting monoclonal antibody in IgG format. Chem Commun 48:7100–7102
44. Zhou Q, Stefano JE, Manning C et al (2014) Site-specific antibody-drug conjugation though glycoengineering. Bioconjug Chem 25:510–520

45. Bejot R, Goggi J, Moonshi SS et al (2012) Aminooxy-functionalized DOTA for radiolabeling of oxidized antibodies: evaluation of site-specific ^{111}In-labeled trastuzumab. Label Compd Radiopharm 55:346–353
46. Boeggeman E, Ramakrishnan B, Pasek M et al (2009) Site specific conjugation of fluoroprobes to the remodeled Fc *N*-glycans of monoclonal antibodies using mutant glycosyltransferases: application for cell surface antigen detection. Bioconjug Chem 20:1228–1236
47. Zeglis BM, Davis CB, Abdel-Atti D et al (2014) Chemoenzymatic strategy for the synthesis of site-specifically labeled immunoconjugates for multimodal PET and optical imaging. Bioconjug Chem 25:2123–2218
48. Zeglis BM, Davis CB, Aggeler R et al (2013) Enzyme-mediated methodology for the site-specific radiolabeling of antibodies based on catalyst-free click chemistry. Bioconjug Chem 24:1057–1067
49. Li X, Fang T, Boons GJ (2014) Preparation of well-defined antibody-drug conjugates through glycan remodeling and strain-promoted azide-alkyne cycloadditions. Angew Chem Int Ed 53:7179–7182
50. Trastoy B, Lomino JV, Pierce BG et al (2014) Crystal structure of *Streptococcus pyogenes* EndoS, an immunomodulatory endoglycosidase specific for human IgG antibodies. Proc Nat Acad Sci 111:6714–6719
51. Umekawa M, Li C, Higashiyama T et al (2010) Efficient glycosynthase mutant derived from *Mucor hiemalis* endo-β-*N*-acetylglucosaminidase capable of transferring oligosaccharide from both sugar oxazoline and natural *N*-glycan. J Biol Chem 285:511–521
52. Parsons TB, Struwe WB, Gault J et al (2016) Optimal synthetic glycosylation of therapeutic antibody. Angew Chem Int Ed 55:2361–2367
53. Suda M, Sumiyoshi W, Kinoshita T et al (2016) Reaction of sugar oxazolines with primary amines. Tetrahedron Lett 57:5446–5448
54. Wang N, Seko A, Daikoku S et al (2016) Non-enzymatic reaction of glycosyl oxazoline with peptides. Carbohydr Res 436:31–35
55. Manabe S, Yamaguchi Y, Matsumoto K et al (2019) Characterization of antibody products obtained through enzymatic and non-enzymatic glycosylation reactions with a glycan oxazoline and preparation of homogeneous antibody-drug conjugate via Fc N-glycan. Bioconjug Chem 30:1343–1355
56. Okeley NM, Toki BE, Zhang X et al (2013) Metabolic engineering of monoclonal antibody carbohydrates for antibody–drug conjugation. Bioconjug Chem 24:1650–1655
57. Alley SC, Benjamin DR, Jeffrey SC et al (2008) Contribution of linker stability to the activities of anticancer immunoconjugates. Bioconjug Chem 19:759–765
58. Lyon RP, Setter JR, Bovee TD et al (2014) Self-hydrolyzing maleimides improve the stability and pharmacological properties of antibody-drug conjugates. Nat Biotechnol 32:1059–1062
59. Vinogradova EV, Zhang C, Spokoyny AM et al (2015) Organomethallic palladium reagents for cysteine bioconjugation. Nature 526:687–691
60. Patterson JT, Asano S, Li X et al (2014) Improving the serum stability of site-specific antibody conjugates with sulfone linkers. Bioconjug Chem 25:1402–1407
61. Kolodych S, Koniev O, Baatarkhuu Z et al (2015) CBTF: new amine-to-thiol coupling reagent for preparation of antibody conjugates with increased plasma stability. Bioconjug Chem 26:197–200
62. Zhang C, Welborn M, Zhu T et al (2016) π-Clamp-mediated cysteine conjugation. Nat Chem 8:120–128
63. Castañeda L, Maruani A, Schumacher FF et al (2013) Acid-cleavable thiomaleamic acid linker for homogeneous antibody-drug conjugation. Chem Commun 49:8187–8189
64. Chari VJ (2016) Expanding the reach of antibody-drug conjugate. ACS Med Chem Lett 7:974–976
65. Ogitani Y, Aida T, Hagihara K et al (2016) DS-8201a, A novel HER-2 targeting ADC with a novel DNA topoisomerase I inhibitor, demonstrates a promising antitumor efficacy with differentiation from T-DM1. Clin Cancer Res 29:5097–5108

66. Jeffrey SC, De Brabander J, Miyamoto J et al (2010) Expanded utility of the β-glucuronide linker: ADCs that deliver phenolic cytotoxic agents. ACS Med Chem Lett 1:277–280
67. Kolalowski RV, Haelsig KT, Emmerton KK et al (2016) The methylene alkoxy carbamate self-immolative unit: utilization for the targeted delivery of alcohol-containing payloads with antibody-drug conjugates. Angew Chem Int Ed 55:7948–7951
68. Kern JC, Cancilla M, Dooney D et al (2016) Discovery of pyrophosphate diesters as tunable, soluble, and bioorthogonal linkers for site-specific antibody-drug conjugates. J Am Chem Soc 138:1430–1445
69. Hisada Y, Yasunaga M, Hanaoka S et al (2013) Discovery of an unexplored space in fibrin clots and its clinical significance. Sci Rep 3:2604
70. Matsumura Y, Maeda H (1986) A new concept for macromolecular therapeutics in cancer chemotherapy: mechanism of tumoritropic accumulation of proteins and the antitumor agent smancs. Cancer Res 46:6387–6382
71. Fuchigami H, Manabe S, Yasunaga M et al (2018) Chemotherapy payload of anti-insoluble fibrin Antibody-drug conjugate is released specifically upon binding to fibrin. Sci Rep 8:14211
72. Nani RR, Gorka AP, Nagaya T et al (2015) Near-IR light-mediated cleavage of antibody-drug conjugates using cyanine photocages. Angew Chem Int Ed 54:13635–13638
73. Li J, Chen PR (2016) Development and application of bond cleavage reactions in bioorthogonal chemistry. Nat Chem 12:129–137
74. Versteegen RM, Rossin R, ten Hoeve W et al (2013) Click to release: instantaneous doxorubicin elimination upon tetrazine ligation. Angew Chem Int Ed 52:14112–14116
75. Rossin R, van Duijnhoven SM, ten Hoeve W et al (2016) Triggered drug release from an antibody-drug conjugate using fast "click-to release" chemistry in mice. Bioconjug Chem 27:1697–1706
76. Matsumura Y (2012) Cancer stromal targeting (CAST) therapy. Adv Drug Deliv Rev 64:710–719
77. Perrino E, Steiner M, Krall N et al (2014) Curative properties of noninternalizing antibody-drug conjugates based on maytansinoids. Cancer Res 74:2569–2578
78. Yasunaga M, Manabe S, Matsumura Y (2011) New concept of cytotoxic immunoconjugate therapy targeting cancer-induced fibrin clots. Cancer Sci 102:1396–1402
79. Yasunaga M, Manabe S, Tarin D et al (2011) Cancer-stroma targeting therapy by cytotoxic immunoconjugate bound to the collagen 4 network in the tumor issue. Bioconjug Chem 22:1776–1783
80. Yasunaga M, Manabe S, Tarin D et al (2013) Tailored immunoconjugate therapy depending on a quantity of tumor stroma. Cancer Sci 104:231–237
81. Goldenberg DM, Cardillo TM, Govindan SV et al (2015) Trop-2 is a novel target for solid cancer therapy with sacituzumab govitecan (IMMU-132), an antibody-drug conjugate (ADC). Oncotarget 6:22496–22512
82. Fujiwara Y, Furuta M, Manabe S et al (2016) Imaging mass spectrometry for the precise design of antibody-drug conjugates. Sci Rep 6:24954
83. Lee BC, Chalouni C, Doll S et al (2018) FRET reagent reveals the intracellular processing of peptide-linked antibody–drug conjugates. Bioconjug Chem 29:2468–2477
84. Onagi J, Komatsu T, Ichihashi Y et al (2017) Discovery of cell-type-specific and disease-related enzymatic activity changes via global evaluation of peptide metabolism. J Am Chem Soc 139:3465–3472
85. Lyon RP, Bovee TD, Doronina SO et al (2015) Reducing hydrophobicity of homogeneous antibody-drug conjugates improves pharmacokinetics and therapeutic index. Nat Biotechnol 33:733–735
86. Lehar SM, Pillow T, Xu M et al (2015) Novel antibody-antibiotic conjugate eliminates intracellular *S. aureus*. Nature 527:323–328
87. Yasunaga M, Manabe S, Matsumura Y (2017) Immunoregulation by IL-7R-targeting antibody-drug conjugates: overcoming steroid-resistance in cancer and autoimmune disease. Sci Rep 6:10735

Preclinical Studies of ADC Therapy for Solid Tumors

Yoshikatsu Koga, Ryo Tsumura, and Yasuhiro Matsumura

Abstract Antibody engineering technologies that allow genetic modification of murine antibodies to produce chimeric mouse-human antibodies, or humanized antibodies, have led to the development of new therapeutic regimens using monoclonal antibodies (mAbs) for treatment of solid tumors. In some therapeutic designs, the antibody is itself the therapeutic agent, whereas in others it is the vehicle by which a cytotoxic agent is concentrated upon the tumor cells. The second approach diminishes the clearance of the therapeutic agent from the body and has yielded very promising results encouraging further study. Successful antibody drug conjugates (ADCs) have three important components; the antibodies, the linkers, and the payloads, which are chemotherapeutic drugs. The antibody is selected to bind to a specific component of the tumor tissue and fix the conjugate to it, thereby concentrating the effect of the drug. The linkers are categorized into cleavable and non-cleavable linkers and serve to bind the drug payload to the antibody. Using this strategy the payloads can achieve 100–1000 fold higher cytotoxic potency than conventional anti-cancer agents because comparable therapeutic effects can be achieved when the number of payloads conjugated to each antibody molecule is as few as 10 drug molecules. However, in some circumstances the efficacy of the ADC technique can be diminished by the composition of intercellular stroma of tumor, which can impede access of the ADC to the target tumor cells after extravasation from the leaky tumor vasculature.

There are six major issues to optimize in the treatment of solid tumors with ADCs, which are:

- extravasation of the conjugate from tumor vessels,
- traversing cancer stroma barrier,

Y. Koga
Department of Strategic Programs, Exploratory Oncology Research & Clinical Trial Center, National Cancer Center, Kashiwa, Japan

R. Tsumura · Y. Matsumura (✉)
Division of Developmental Therapeutics, Exploratory Oncology Research & Clinical Trial Center, National Cancer Center, Kashiwa, Japan
e-mail: yhmatsum@east.ncc.go.jp

Y. Matsumura, D. Tarin (eds.), *Cancer Drug Delivery Systems Based on the Tumor Microenvironment*, https://doi.org/10.1007/978-4-431-56880-3_6

- attaching to the antibody binding site,
- internalization into cancer cells,
- the drug releasing, and direct cytotoxic effect
- the bystander killing effect on cells adjacent to the site of release of the cytotoxic agent.

The ADCs which achieve these aims can significantly improve treatment outcomes for patients with solid tumors.

Keywords Antibody engineering · Linker technology · Payload · EPR effect · Cancer stroma barrier · Binding site barrier · Internalization · Releasing and visualization · Bystander killing effect

Introduction

Cancer is the main cause of death in developed countries, and many physicians and scientists are directing their efforts to conquer the disease. For cancer treatment, surgical or endoscopic therapy, radiation therapy, and chemotherapy are the main three pillars for curing and prolonging survival of cancer patients. Among these treatments, cancer chemotherapy plays a central role for cancer patients experiencing recurrence. Conventional cancer chemotherapy is based on the principle that rapidly proliferating cancer cells are killed by small molecule anti-cancer agents (ACAs). Unfortunately, the current ACAs, such as 5-FU, cisplatin, paclitaxel, doxorubicin and the others, lack the specificity to kill cancer cells without damaging proliferating normal cells, such as hematopoietic cells and mucosal cells of the gut. To resolve these drawbacks of ACAs, various formulations of drug delivery systems (DDS) have been developed to date. One of the main strategies in designing DDS is the "passive targeting" approach, which is based on differences between tumor blood vessels and normal vessels, and results in the so-called EPR (enhanced permeability and retention) effect [48]. High molecular weight prodrugs, such as ACAs incorporated in micelles or liposomes, can permeate through the gaps in the walls of the tumor blood vessels and accumulate in the tumor tissue, but do not extravasate from normal vessels. Another major strategy is "active targeting" which is based exploiting the specificity of antigen-antibody interactions. In the last two decades, several monoclonal antibodies (mAbs) have been approved for treatment of solid tumors. Examples include anti-human epidermal growth factor receptor 2 (HER2) antibody against HER2-positive breast cancer and gastric cancer, and anti-epidermal growth factor receptor (EGFR) antibody against head & neck cancer and colorectal cancer. The antitumor mechanisms of these mAbs are down regulation of signaling cascades, antibody-dependent cell mediated cytotoxicity (ADCC) or complement dependent cytotoxicity (CDC). In addition to these mechanisms, we can make good use of the mAbs by conjugating a cytotoxic payload to the mAb, making into an antibody drug conjugate (ADC). In this chapter, we will summarize the results of preclinical studies on ADCs for solid tumors.

Antibody Drug Conjugates (ADCs): Historical Aspects and General Principles

More than 120 years ago, Paul Ehrlich had a dream of the "magic bullet" that would seek out and specifically attack pathogens [68]. "Magic bullets" can identify their target molecules without harming other normal tissues. This concept was also applied to cancer treatment, and many anticancer agents were developed. Antibody based therapeutics has become a most valuable mode of therapy since the discovery of hybridoma technology for producing mAbs was first reported by Kohler G and Milstein C in 1975 [38]. After this method to obtain target specific mAbs easily became available, "missile therapy" using "immunoconjugates" was developed for clinical use in the late 1980's. Unfortunately, however, because murine antibodies were used for clinical trials in this era, these antibodies failed in clinical use, because they were found to be inactive and rapidly neutralized by the production of human anti-mouse antibodies (HAMA) [10]. At the same time, the idea of using these antibodies as tools for delivering ACAs, such as methotrexate [14] and doxorubicin [9], to cancer cells was investigated in order to expand the therapeutic window [54]. However, these mAbs were also murine antibodies, and thus, they had immunogenicity with rapid degradation of the conjugations. In the 1990's, antibody engineering technologies that allowed genetic modification of murine antibodies to produce chimeric mouse-human antibodies or humanized antibodies were developed. These mAbs are less likely to be recognized by the host immune system as a foreign antigen, and have half-lives similar to those of natural human Immunoglobulin G (IgG). These technologies led to advances in the treatment of solid tumors with mAbs, and several mAbs, such as trastuzumab (anti-HER2 humanized mAb) [67], bevacizumab (anti-VEGF humanized mAb) [29], and cetuximab (anti-EGFR chimeric mAb) [8], have been approved by the U.S. Food and Drug Administration (FDA). Following the improvements in clinical outcomes observed in using mAb therapy for solid tumors, the specific targeting capabilities of mAbs were used to further enhance therapeutic potency by coupling them with cytotoxic drugs in immunoconjugates. This strategy, now termed ADC, was designed to increase the concentration of toxic agents on the tumor cells. ADCs consist of three components: antibodies for drug carrier, linkers for conjugating antibody and drug, and payloads (cytotoxic drugs) for cancer cell toxicity (Fig. 1). This ADC strategy should be confined to highly toxic anticancer agents but not to ordinary ACA such as taxane, adriamycin and others because only less than three ACA molecules should be conjugated to the mAb, otherwise the affinity of the mAb is diminished if too many molecules of ACA are attached to the mAb. According to this principle, unrealistic amount of ADC should be administered if the mAbs are conjugated with an ordinary ACA. The linker should be stable in plasma to avoid releasing the payloads in the blood and be severed to

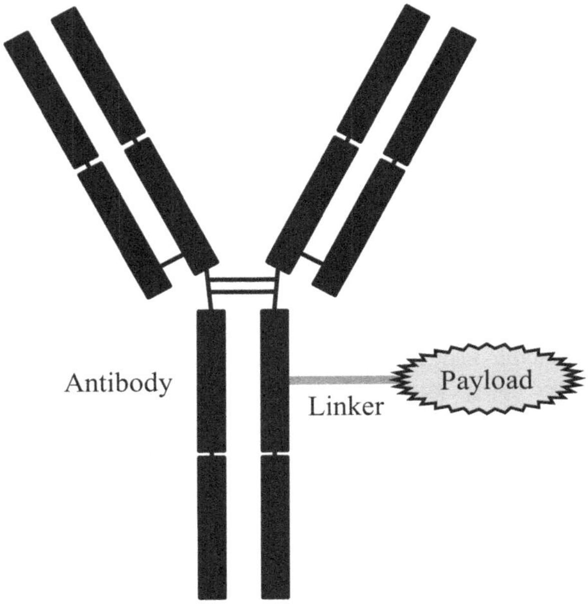

Fig. 1 Schema of the ADCs (antibody drug conjugates). ADCs are composed by antibodies for drug carriers, linkers for conjugation antibody and payload, and payloads for cancer cell toxicity

release the payloads in their active form only within cancer cells. To date, the ADC concepts for the cancer treatments are established as shown in (Fig. 2) [74]. Firstly, the ADCs bind to their target antigen on the cancer cell membrane. Then, the ADC-antigen complex is internalized into the early endosome. The endosome is merged to a lysosome with acidification and enzymatic reaction. The linker is cleaved only when the ADC internalize into cancer cells and the payloads are released in cancer cell cytosol in the acidic and enzymatic conditions of the lysosome.

Target Antigens for ADCs in Clinical Trials

To implement the ADC concept effectively, it is important to decide what mAbs, corresponding antigen and linker molecules should be selected. To date, four ADCs have been approved by the U.S. FDA (gemtuzumab ozogamicin was withdrawn in 2010 from the market because of off-target toxicity, but was approved again with a new dosing regimen in September 2017). Gemtuzumab ozogamicin [65], brentuximab vedotin [36], trastuzumab emtansine [80], and inotuzumab ozogamicin [35] target to CD33, CD30, HER2, and CD22 respectively. All these molecules are expressed higher on the surface of cancer cells than that of normal cells. Only trastuzumab emtansine is approved for treatment of non-hematological solid tumors, namely HER2-positive breast cancer. Now, 46 kinds of ADCs are being investigated in clinical trials for solid tumors (search as active, recruiting, or completed ADC trials in the ClinicalTrials.gov Web page on November 28, 2018). Since the payloads are released in the target cells and lead to cell death according to the ADC

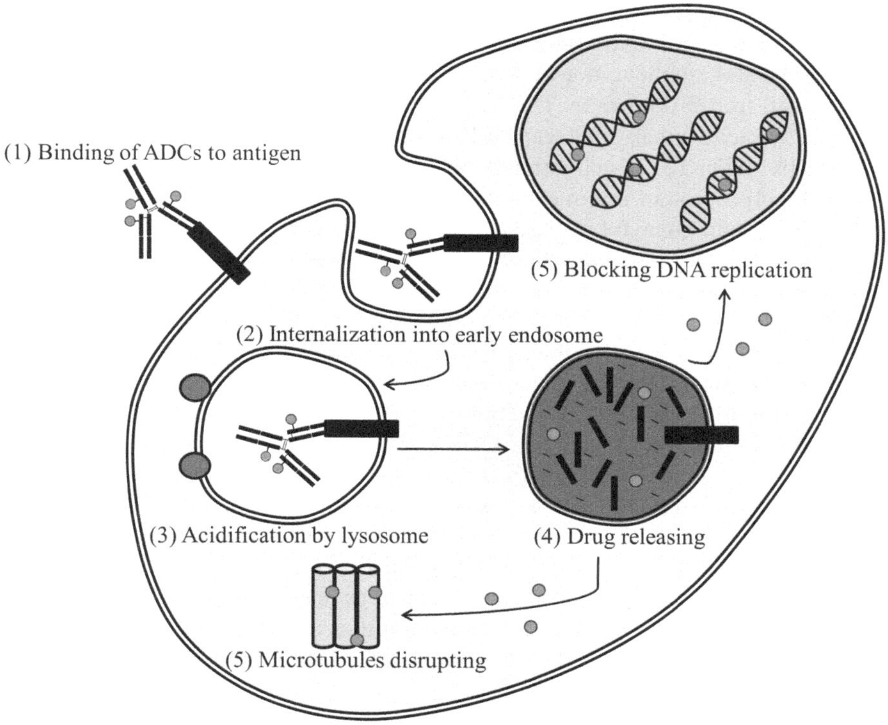

Fig. 2 Conventional concepts of ADCs for cancer treatment [74] (1) The ADCs bind to their target antigen on the cancer cell membrane. (2) The ADC-antigen complex internalize into the early endosome. (3) The endosome is merged to lysosome with acidification and enzymatic reaction. (4) The linker is cleaved and released payloads in their active form into cancer cell cytosol. (5) The payloads lead to apoptosis via microtubules disrupting the microtubule network or blocking replication of DNA minor groove

concept, most of ADCs are designed to target cancer cell surface antigens (98%; 45/46). One third of the targeted cancer cell molecules are classified into the receptor tyrosine kinases (RTKs). Because the RTKs are key molecules for the development and progression of many types of cancer via the intracellular signaling cascade, ADCs have high potent synergistic cytotoxic effects to the cancer cells not only by the cytotoxic payloads but also by mAbs composed in the ADCs which are capable of neutralizing and inhibiting the signals. RTKs are classified into 20 classes [63]. Same as trastuzumab emtansine, 9 ADCs target to RTK class I, EGF (epidermal growth factor) receptor family. Six ADCs, such as A166, ARX788 [31], DS-8201a [52], RC48-ADC, SYD985 [79], and XMT-1522, target to HER2, two ADCs, depatuxizumab mafodotin (ABT-414) [56] and AMG595 [26], target to EGFR and EGFR variant III (EGFRvIII), and U3-1402 targets to HER3. The other RTK classes are class IV, insulin receptor family, class V, fibroblast growth factor (FGF) receptor family, class VIII, HGF (hepatocyte growth factor) receptor family, and class X, AXL receptor family. W0101 targets to insulin growth factor 1 receptor (IGF-1R),

LY3076226 target to FGFR3, telisotuzumab vedotin (ABBV-399) [81] target to c-MET, and HuMax-AXL-ADC and CAB-AXL-ADC target to AXL. The other target molecules are the tumor-associated antigens; tissue factor (TF) by tisotumab vedotin [4], trop-2 (known as TACSTD2; tumor-associated calcium signal transducer 2) by sacituzumab govitecan (IMMU-132) [6], AG7 (novel tumor-associated antigen) by AbGn-107, T-cell immunoglobulin mucin receptor 1 (TIM-1) by CDX-014 [73], glycoprotein non-melanoma protein B (gpNMB) by glembatumumab vedotin [50], and guanylyl cyclase C (GCC) by MLN0264 [2] and TAK-164, the cell adhesion molecules; mesothelin by anetumab ravtansine (BAY94-9343) [23] and BMS-986148, nectin-4 by enfortumab vedotin (ASG-22ME) [7], and carcinoembryonic antigen related cell adhesion molecule 5 (CEACAM5) by labetuzumab govitecan (IMMU-130) [64], the transporter; folate receptor α (FRα) by mirvetuximab soravtansine (IMGN853) [1], solute carrier family 44 member 4 (SLC44A4) by ASG-5ME [49], and LIV-1 (zinc transporter) by SGN-LIV1A [70], and the cluster of differentiation (CD) number; CD27L, CD56, CD70, CD71, CD77, CD166, and CD205. Only one ADC (ABBV-085) in phase 1 clinical trial targets to cancer-associated fibroblast (CAF) that targets leucine rich repeat containing 15 (LRRC15). All of these mAbs target to cancer cells or cancer associated cells. In contrast to these ADCs, Matsumura Y proposed an alternative version called cancer stromal targeting (CAST) therapy by which ADCs can target cancer associated extracellular matrix within solid tumors [45, 82, 83] and use it as a scaffold for launching cytotoxic drugs. CAST therapy is described later in this chapter and in other chapters.

Three Components of ADCs

Antibodies

Immunoglobulin G (IgG) is a class of heterodimeric glycoproteins composed of two heavy chains (HC) and two light chains (LC) with approximately 150 kDa total molecular weight (Fig. 3a). Each HC is composed of four domains (one variable domain and three constant domains) and a hinge, aligned in the following order; V_H, C_H1, hinge, C_H2, and C_H3. Also each LC is composed of two domains (variable and constant domain), aligned as V_L and C_L. The variable domains (V_H and V_L) have a unique sequence and are important for binding to the target antigens with a high affinity and specificity. Six complementarity determining regions (CDRs), which are most important for antigen binding, are located within the variable domains (three CDRs in each V_H and V_L). The antigen-binding fragment (Fab) composed of V_H-C_H1 and V_L-C_L, and the crystallizable fragment (Fc) composed of C_H2 and C_H3 are connected by the hinge.

mAbs used for ADC construction, need to have the following properties: a high affinity and specificity to the target antigen (especially cancer cell surface antigen), efficient internalization capability, retention in the blood stream for a long time, and

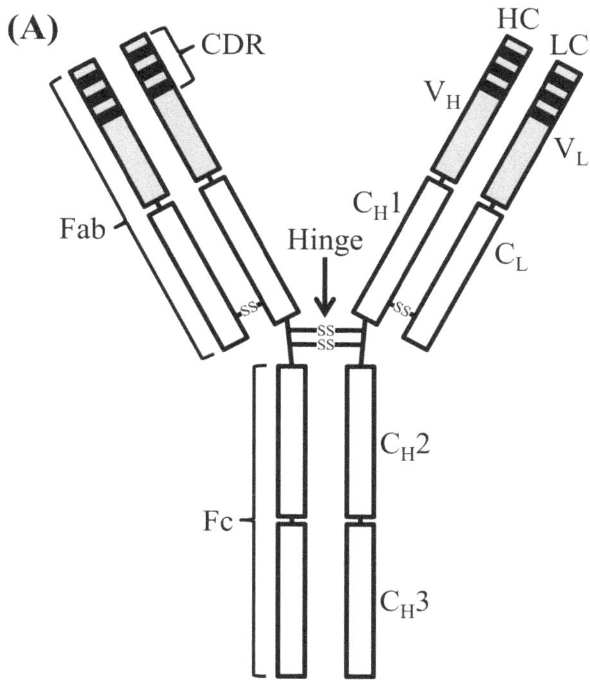

Fig. 3 Schema of the antibodies. (**a**) Schema of IgG. HC (Heavy chain) is composed of four domains (one variable domain, V_H, and three constant domains, C_H1, C_H2, and C_H3) and a hinge. LC (Light chain) is composed of a variable domain (V_L) and a constant domain (C_L). Six CDRs (complementarity determining regions) are located within the variable domains. The Fab (antigen-binding fragment) composed of V_H-C_H1 and V_L-C_L, and the Fc (crystallizable fragment) composed of C_H2 and C_H3 are connected by the hinge. (**b**) Several sizes of antibodies [60] The molecular weights of IgG consisted of two Fabs, the hinge, and the Fc, F(ab')$_2$ consisted of two Fabs and the hinge, and Fab are 150 kDa, 110 kDa, and 55 kDa, respectively. scFv-Fc (105 kDa, consisted of the scFv, the hinge, and the Fc), minibody (80 kDa), diabody (55 kDa), and scFv (25 kDa) can be obtained by the recombinant DNA technologies. (**c**) Therapeutic antibodies in clinical use [69] The chimeric antibodies are consisted of the variable regions from non-human species and the constant regions from human IgG. The humanized antibodies are developed by replacing the six CDRs from non-human variable domains into human variable domains. The fully human antibodies can be produced by the phage display technologies or using the transgenic mice with the human humoral immune system

a reduced immunogenicity. To obtain high efficacy in killing of cancer cells and low toxicity to normal cells, attention must be given not only to identifying specific target molecules but also to antibody engineering principles. Among several necessary conditions for constructing an effective ADC, the primary issue is the molecular weight or size of the antibody. To date, several sizes of antibodies can be produced by antibody engineering techniques (Fig. 3b). For example: IgG (150 kDa, consists of two Fabs, the hinge, and the Fc), F(ab')$_2$ (110 kDa, consists of two Fabs and the hinge) is produced from IgG by pepsin digestion, and Fab (55 kDa, can be produced

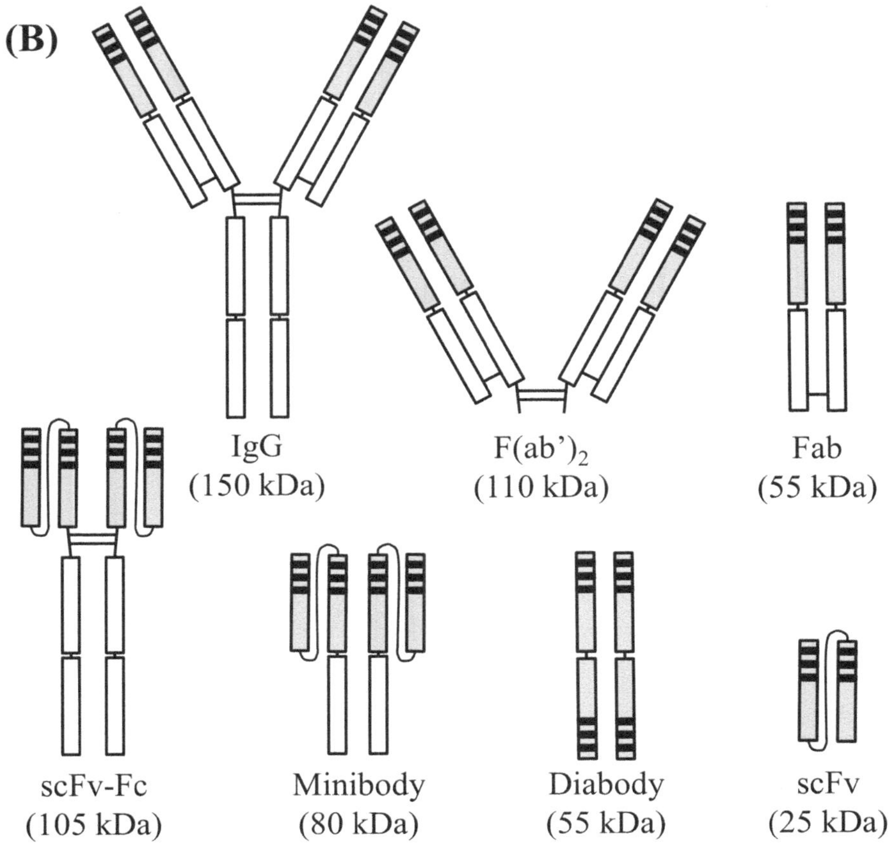

Fig. 3 (continued)

from IgG by papain digestion), scFv-Fc (105 kDa, consisting of the scFv, the hinge, and the Fc), minibody (80 kDa), diabody (55 kDa), and single-chain variable fragment (scFv, 25 kDa) can all be obtained by recombinant DNA technologies [60]. In order to enhance retention of the therapeutic agent in the blood stream, the full length IgG or an antibody with 100 kDa size, or more, is used for ADC construction, in order to prevent rapid urinary excretion.

To avoid destruction by the immune system, completely non-human antibodies were replaced by chimeric, humanized, and fully human antibodies for clinical use (Fig. 3c) [69]. The chimeric antibodies consist of the variable regions from non-human species and the constant regions from human IgG (contain 67% human sequences). The humanized antibodies are developed by replacing the six CDRs from non-human variable domains with human variable domains (contain 95% human sequences). Fully human antibodies are produced by the phage display technologies [28] or using transgenic mice with the human humoral immune system

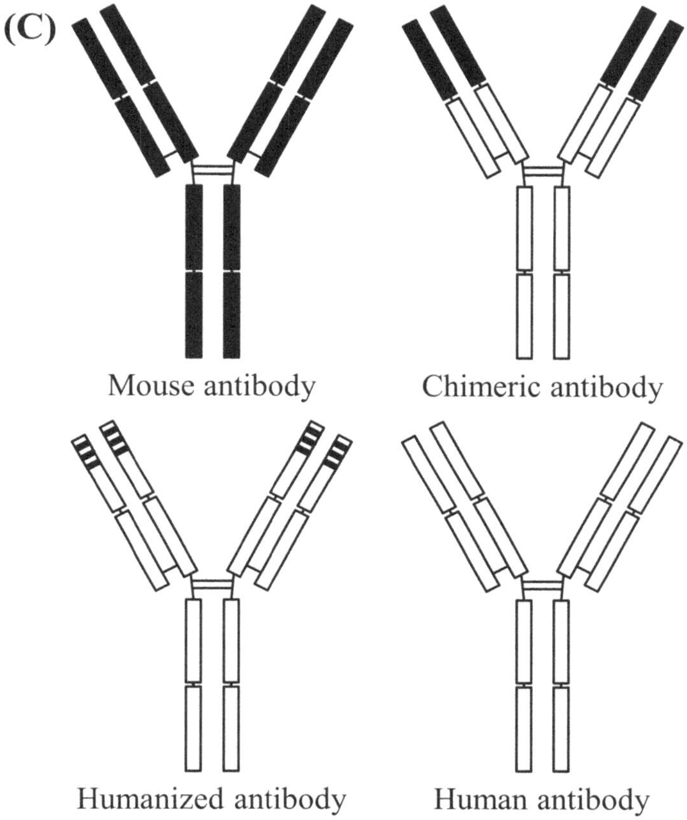

Fig. 3 (continued)

[32]. After the success of several therapeutic chimeric antibodies (cetuximab and rituximab [44]; anti-CD20 mAb), humanized (trastuzumab and bevacizumab), or fully human mAb (panitumumab [22]; anti-EGFR mAb) in the clinic, these engineered mAbs are widely used in preclinical and clinical studies.

Linkers

Linker technologies are closely related both to the capability of releasing payload after internalization of the ADCs into cancer cells and to the stability of ADCs in the blood stream. Successful linkers can have several mechanisms to connect the antibody and the payload; connection to the antibody, polarity, trigger, and self-immolative spacer [15]. The most commonly used amino acids that connect between the antibody and the linker-payloads are cysteine and lysine.

Cysteine residues are used for brentuximab vedotin [36] and lysine residues are used for trastuzumab emtansine [40]. These connection sites of the antibodies are related to the DAR (drug-to-antibody ratio) and stability. Because approximately 30 lysines on IgG can be used for conjugation, heterogeneous mixture of ADCs can be produced by random conjugation. This heterogeneity may affect to antigen-binding properties or blood stabitity of ADC. On the other hand, the cysteines on IgG are involved in intrachain and interchain disulfide bridges. Since there are only 4 interchain disulfide bridges (only 8 conjugation sites) in IgG (Fig. 3a), ADCs conjugated via cysteine residues are of lower heterogeneity those constructed via lysine residues. To reduce molecular aggregation, the polarity of the linkers has an important role in the pharmacokinetics of the ADC. Increasing the DAR of ADC makes aggregation because most of the payloads are hydrophobic. A polyethylene glycol (PEG) introduced into the polarity position of the linker makes ADC hydrophilic and can reduce this aggregation. The self-immolative spacer chemically undergoes a rapid and spontaneous intramolecular reaction that releases the free payload following activation by the trigger. For example, a para-aminobenzyl (PAB) group in the dipeptide linker is initially released by enzymatic cleavage and a 1, 6 elimination of the PAB group releases the free payload [76]. But the most important role of the linker for ADCs is that with the trigger (cleavable linker) or without the trigger (non-cleavable linker) (Table 1).

Table 1 Linkers used for ADCs in the clinic

Linkers	Characteristics
Cleavable linker: Cleaved by several specific mechanisms under the physiological environment	
Hydrazone	Designed for hydrolysis in the acidic environment: used in the gemtuzumab ozogamicin [65]
Carbonate	Designed for hydrolysis in the acidic environment: used in the sacituzumab govitecan (IMMU-132) [6], labetuzumab govitecan (IMMU-130) [64]
Disulfide	Designed for reducing by the glutathione: used in the anetumab ravtansine (BAY94-9343) [23], mirvetuximab soravtansine (IMGN853) [1]
Peptide	Designed for cleavage by specific intracellular proteases (cathepsin B): used in the brentuximab vedotin [36] (Val-Cit), rovalpituzumab tesirine [59] (Val-Ala), DS-8201a [52] (Gly-Gly-Phe-Gly)
Non-cleavable linker: Degraded in the lysosome without a specific releasing mechanism and increased plasma stability of ADCs and therapeutic window	
Thioether	Designed for intra-lysosomal proteolytic degradation: used in the trastuzumab emtansine [40], AMG595 [26]
Maleimide-caproyl	Designed for intra-lysosomal proteolytic degradation: used in the depatuxizumab mafodotin (ABT-414) [56], vorsetuzumab mafodotin (SGN-75) [72]

ADC antibody drug conjugate, *Val-Cit* valine-citrulline, *Val-Ala* valine-alanine, *Gly-Gly-Phe-Gly* glycine-glycine-phenilananine-glycine

Cleavable Linkers

Cleavable linkers can be severed by several specific mechanisms. The cleavable linkers are stable in the blood stream but are severed efficiently in the physiological environment, such as by the low pH environment in lysosomes, by high levels of glutathione concentrations, or by high levels of protease, such as cathepsin B [41]. Thus, the ADCs with cleavable linkers can act whenever the above conditions are encountered, and do not require strict internalization ability. There are three releasing mechanisms that are used for the ADCs entered in the clinical trials; a hydrazone or a carbonate linker lysed by hydrolysis under the acidic conditions, a disulfide linker reduced by glutathione, and a dipeptide or a tetrapeptide linker cleaved by lysosomal proteases. Regarding the approved ADCs, gemtuzumab ozogamicin [65] has a cleavable hydrazone linker and brentuximab vedotin [36] has a cleavable dipeptide (valine-citrulline [Val-Cit]) linker. Most of the ADCs entered in the clinical trials containe a cleavable linker (72%; 33/46). The peptide linkers, which are Val-Cit dipeptide [36], valine-alanine (Val-Ala) dipeptide [59, 75], or glycine-glycine-phenylalanine-glycine (Gly-Gly-Phe-Gly) tetrapeptide linker [52], are most commonly used in the clinic because the ADCs with peptide linkers have an excellent balance between plasma stability and intracellular protease cleavage [12].

Non-cleavable Linkers

Non-cleavable linkers do not mean "not cleaved in the cancer cells". But the greatest advantage of the non-cleavable linker compared to the cleavable linker is the stability in the plasma or normal tissues. The non-cleavable linkers do not have a specific releasing mechanism but depend on the catabolic degradation of the ADCs after internalization into the lysosomes. Thus, the ADCs with non-cleavable linkers require high potential internalization ability into the cancer cells. Regarding the approved ADCs, trastuzumab emtansine [40] has no releasing trigger and is constructed with MCC (maleimidomethyl cyclohexane-1-carbozylate), a non-cleavable thioether linker. The vorsetuzumab mafodotin (SGN-75) [72] and depatuxizumab mafodotin (ABT-414) [56] conjugates the antibodies and the payloads via non-cleavable maleimido-caproyl linker.

Payloads

The payloads have the most important roles in ADCs for killing the cancer cells directly. Usually, the payloads have 100–1000 fold high potent cytotoxicity than conventional ACAs and are cytotoxic at sub-nanomolar concentrations in cell lines, because the number of payloads conjugated to each antibody is less than 10

Table 2 Payloads used for ADCs in the clinic

Payloads	ADCs
Microtubule inhibitor: Binds to tubulin, blocks the microtubule polymerization and prevents the mitosis of the cells	
Auristatin	Used in the brentuximab vedotin [36] (MMAE), depatuxizumab mafodotin (ABT-414) [56] (MMAF)
Maytansine	Used in the trastuzumab emtansine [40] (DM1), anetumab ravtansine (BAY94-9343) [23] (DM4)
DNA damaging agent: Breaks the DNA strand or inhibits the DNA synthesis by the binding to the minor groove of DNA or inhibition of topoisomerase I or II	
Calicheamycin	DNA alkylating agent: used in the gemtuzumab ozogamicin [65]
Duocarmycin	DNA alkylating agent: used in the SYD985 [79]
PBD	DNA alkylating agent: used in the SGN-CD70A [33]
TOP1 inhibitor	Topoisomerase I inhibitor: used in the sacituzumab govitecan (IMMU-132) [6] (SN-38), DS-8201a [52] (Dxd)

ADC antibody drug conjugate, *MMAE* monomethyl auristatin E, *MMAF* monomethyl auristatin F, *DM1* maytansine derivative 1, *DM4* maytansine derivative 4, *PBD* pyrrolobenzodiazepines, *TOP1* topoisomerase I

molecules. The ADC strategy should be confined to highly toxic ACAs rather than ordinary ACAs because less than three ACA molecules should be conjugated to the mAb. Otherwise the affinity of the mAb is diminished, if too many molecules of ACA are attached to it. According to this principle, unrealistic amounts of ADC would need to be administered if the mAbs are conjugated with an ordinary ACA. The payloads used in ADCs are mainly categorized into two groups according to their mechanisms of action; namely microtubule inhibitors and DNA damaging agents (Table 2).

Microtubule Inhibitors

Regarding the ADCs in the clinical trials, most payloads (63%; 29/46) are microtubule inhibitors; auristatins and maytansines. The auristatins, such as MMAE (monomethyl auristatin E) and MMAF (monomethyl auristatin F), lead to G2/M phase cell cycle arrest and cell death through the induction of apoptosis [16]. The maytansinoids, such as derivative of DM1 (maytansine derivative 1) and DM4, bind directly to tubulin, inhibit microtubule assembly, and disrupt mitosis [40]. Both auristatins and maytansines interact with the Vinca alkaloid binding site on α-tubulin and block its polymerization and prevent the formation of the mitotic apparatus [3]. Regarding the approved ADCs, brentuximab vedotin [36] contains MMAE at DAR value of 4 and the trastuzumab emtansine [40] contains DM1 at DAR value of 3.5.

DNA Damaging Agents

The other types of payloads are the DNA damaging agents; such as calicheamicin, duocarmycin, pyrrolobenzodiazepines (PBD), SN-38, and Dxd (DX-8951 derivative). In the market, calicheamicin is contained in the gemtuzumab ozogamicin with a DAR value of 2 [65]. Calicheamicin is a DNA-alkylating agent which binds to the minor groove of DNA and breaks the DNA strands via generating a di-radical species and removing hydrogen atoms from the DNA backbone [66]. Duocarmycin and PBD bind to the minor groove of DNA at specific sites and cause irreversible alkylation of DNA [79]. SN-38 and Dxd inhibit topoisomerase I (TOP1) and induce DNA single-strand breaks during DNA replication, transcription, and chromatin remodeling [58]. Although the majority of the payloads of ADCs in current clinical trials consist of the microtubule inhibitor (auristatin or maytansine), some other DNA damaging agents, such as PBD dimers, duocarmycin, and Dxd have high potential for cytotoxicity for solid tumors. SGN-CD70A conjugated with PBD dimers has an impressive antitumor activity at 0.1 and 0.3 mg/kg in preclinical xenograft models [33]. Moreover, SYD985 (conjugated with duocarmycin) [79] and DS-8201a (conjugated with Dxd) [52] which targets HER2 have high antitumor effects in HER2 positive PDX mouse models, which are resistant to the trastuzumab emtansine treatment. These preclinical findings indicate that DNA damaging agents might be key candidates to use as payloads of ADCs for solid tumor treatment.

Hurdles to Overcome in the Preclinical Studies of ADCs

To develop effective cancer treatment, knowledge of the hallmarks of cancer and of the constitution of the tumor microenvironments is important for designing therapeutic targeting molecules [27]. The tumor microenvironment consists of two categories of cells, namely, the cancer cells and the surrounding non-neoplastic cancer stromal cells, and the extracellular matrix (ECM). In the case of leukemias or ascites tumors, blood plasma or ascites act as a fluid supporting medium, which also contains non-malignant cells. On the other hand, in solid cancers, the intercellular materials provide a firm scaffolding for the tumor cells and for the non-malignant stromal cells thereby restricting their mobility and creating a highly different microenvironment from that experienced by floating, circulating malignant cells. The non-malignant cells in the stroma of solid cancers, such as the tumor endothelial cells (TECs) [43], the CAFs [30], and the tumor-associated macrophages (TAMs) [39], are not "normal" cells, because they are educated and activated by the growth factors produced from the cancer cells and, retro-actively, promote the growth, invasion and metastasis of the cancer cells. Also, these cancer-associated stromal cells produce the collagen-rich ECM characteristic of solid tumors. Anti-cancer drugs act directly on cancer cells in the blood when administered via intravenous infusion, but ADCs used in the treatment of solid tumors must traverse several microenvironmental obstacles to act on the malignant cells (Fig. 4). In

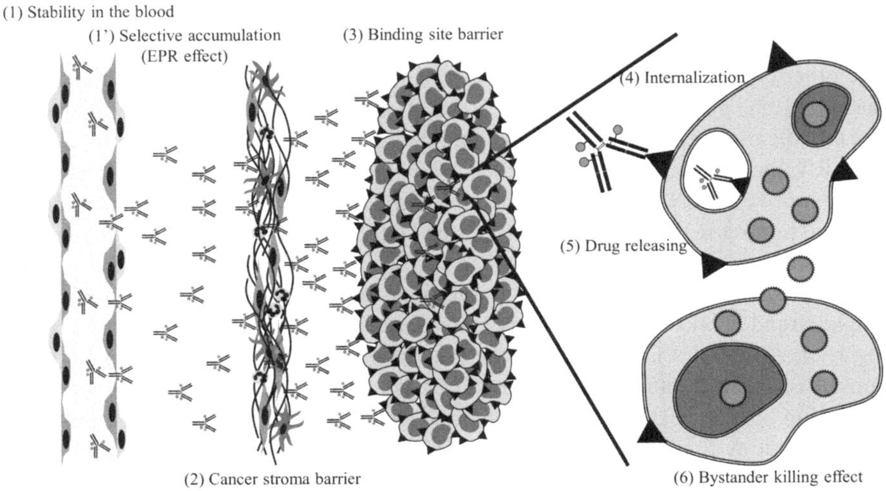

Fig. 4 Hurdles for ADCs in solid tumor. (1) First hurdle: blood stability and selective accumulation. Based on the EPR (enhanced permeability and retention) effect, macromolecules (particle size 10–1000 nm) are stable in the blood stream and are extravasated only from the tumor vessels. (2) Second hurdle: cancer stroma barrier. The cancer stroma cells, the ECMs (extracellular matrixes), and the collagen fibers are components of a physical blocking layer for reducing the drug penetration and diffusion. ADCs can penetrate and reach to the cancer cells because of the smaller size (15 nm). (3) Third hurdle: binding site barrier. The ADCs bind to the surface of the cancer nest via antigen-antibody interaction at the first contact to the antigen and are trapped to the surface of the cancer nest. The easily dissociation ADCs with high k_d (dissociation rate constant) value could penetrate into the cancer nest, reach to the center of the nest, and distribute homogenously in the nest. (4) Fourth hurdle: internalization into the cancer cells. The internalization ability via receptor mediated endocytosis depends on the types of the cells, the antigens, and the antibodies. (5) Fifth hurdle: drug releasing. The drug releasing mainly depends on the linker technology. (6) Last hurdle: bystander killing effect. The bystander killing effect can affect to not only the antigen positive cancer cells but also the adjacent antigen negative cancer cells via the transportation of the released drugs from the antigen positive cancer cells to the adjacent antigen negative cancer cells

particular, when an anti-cancer drug reaches a solid tumors through blood stream, the endogenous stroma of the cancer impedes the penetration of the drug.

Blood Stability and Selective Accumulation: EPR Effect

When the drugs are injected into a peripheral vessel, the first hurdle is the stability of the drugs in the blood stream and their specific extravasation from tumor vessels into the cancer tissues (Fig. 4). The conventional low molecular weight anti-cancer drugs are extravasated from both the normal and the tumor vessels and this causes several undesirable side effects in the patients. Since the drugs are also excreted from the urinary tract in a time dependent manner and a high concentration of the drugs is needed in order to achieve significant anti-tumor efficacy serious adverse events often occur. However, the EPR effect assists the extravasation and local accumulation of cancer therapeutic drugs (Fig. 5a) [48]. This concept relies on the pathophysiologic characteristics of solid tumor tissues; namely: hyper vascularity with angiogenesis, secretion of the vascular permeability factors (VPF) which appeared to be identical to vascular endothelial growth factor (VEGF) [13] and kinin [47] within cancer tissues, and the absence of effective lymphatic drainage systems. Many macromolecular anti-cancer drugs (nanoparticles) which benefit from the EPR effect for enhancing their action, such as ACA incorporated micelles, liposomes, and albumin-bound ACA, are approved or entered in clinical trials and the optimal size of the nanoparticles for these medications is defined as 10–100 nm [57]. For comparison it can be recalled that early studies on the EPR effect found that whole IgG, which had 150 kDa molecular weight and 15 nm particle size, was accumulated in the tumor tissue efficiently [48].

To compare the stability in the blood stream and the accumulation in tumor tissues of several sizes of antibodies, antibodies conjugated with fluorescence were injected into tumor bearing mice [62, 78]. Sato et al. genetically prepared anti- TF scFv conjugated with fluorescence, molecular size was 28 kDa (approximately 3 nm particle size) [62]. To study tumor specific accumulation, non-targeting scFv conjugated with same fluorescence was also produced. These scFv were injected into mice bearing spontaneous cutaneous tumors via the tail vein. Both types of scFv were excreted from the kidneys into the urine within 1 hr. after injection, and non-targeting scFv vanished from the mice body within 3 h after injection; however, the anti-TF scFv continued to accumulate selectively until 12 h after injection (Fig. 5b). The highest tumor to background ratio (TBR) in anti-TF scFv injection was observed at 3 h after injection. On the other hand, when anti-TF whole IgG, conjugated with same fluorescence, was injected in a same manner it was not excreted into the urine, and selectively accumulated in the tumor until 72 h after injection. The highest TBR in anti-TF IgG injection was observed at 24 h after injection. Tsumura et al. generated anti-TF Fab conjugated with fluorescence by digestion of whole IgG with papain, and its molecular size was 50 kDa (approximately 5 nm particle size) [78]. The non-targeting Fab, non-targeting whole IgG, anti-TF whole IgG conjugated with the same fluorescent marker were also produced. All antibodies were injected into mice bearing the tumors formed by the human pancreatic cancer cell line BxPC-3 tumor via the tail vein. Both Fab were excreted from the kidneys into the urine after injection, and non-targeting Fab vanished from the body within 12 h after injection; however, the anti-TF Fab was still retained at 24 h after injection (Fig. 5c). The highest TBR in anti-TF Fab injection was observed 12 h after injection. On the other hand, both IgG were not excreted in the urine, and non-targeting IgG accumulated in the tumor until 12 h after injection, due to the EPR effect (passive targeting); however, anti-TF IgG specifically accumulated in the tumor until 72 h after injection (passive plus active targeting). The highest TBR in anti-TF IgG injection was observed at 24 h after injection. These studies indicated that the smaller size of antibodies may be a suitable imaging tool for the cancer diagnosis because the smaller size antibodies can accumulate

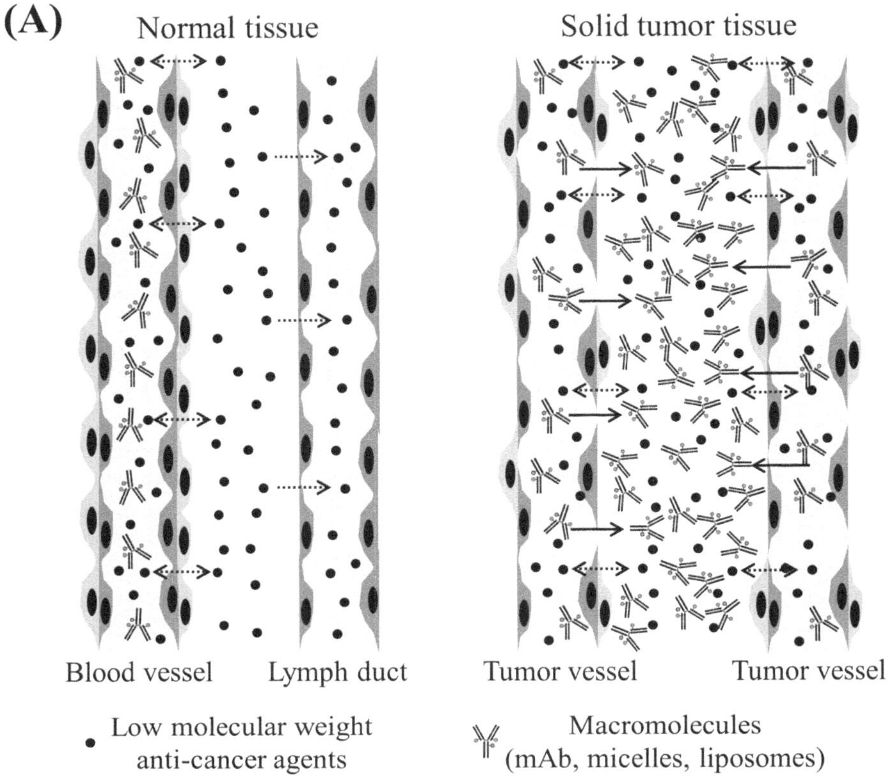

Fig. 5 EPR (enhanced permeability and retention) effect and stability of low molecule antibodies. (**a**) Mechanisms of the EPR effect [48]. The solid tumors have hyper vascularity with angiogenesis, irregular vascular architecture with opening gaps in the tumor vessels, potential for secretion of the vascular permeability factors from the cancer cells, and the absence of effective lymphatic drainage systems. The macromolecules including IgG can accumulate selectively in the tumor tissue. (**b**) scFv accumulation in the solid tumor [62]. scFv was excreted from the kidneys to the urine within 1 h after injection, and non-targeting scFv was vanished from the mice body within 3 h after injection; however, the anti-TF scFv was still accumulated selectively until 12 h after injection. The anti-TF whole IgG was not excreted to urine, and selectively accumulated to the tumor until 72 h after injection. The highest TBR (tumor to background ratio) in anti-TF scFv and anti-TF IgG injection was observed at 3 h and 24 h after injection, respectively. (**c**) Fab accumulation in the solid tumor [78]. Fab was excreted from the kidneys to the urine within 6 h after injection, and non-targeting Fab was vanished from the mice body within 12 h after injection; however, the anti-TF Fab was still accumulated selectively until 24 h after injection. IgG was not excreted to urine, and non-targeting IgG was accumulated to the tumor until 12 h after injection due to the EPR effect (passive targeting); however, anti-TF IgG was specifically accumulated to the tumor until 72 h after injection (passive plus active targeting). The highest TBR in anti-TF Fab and anti-TF IgG injection was observed at 12 h and 24 h after injection, respectively

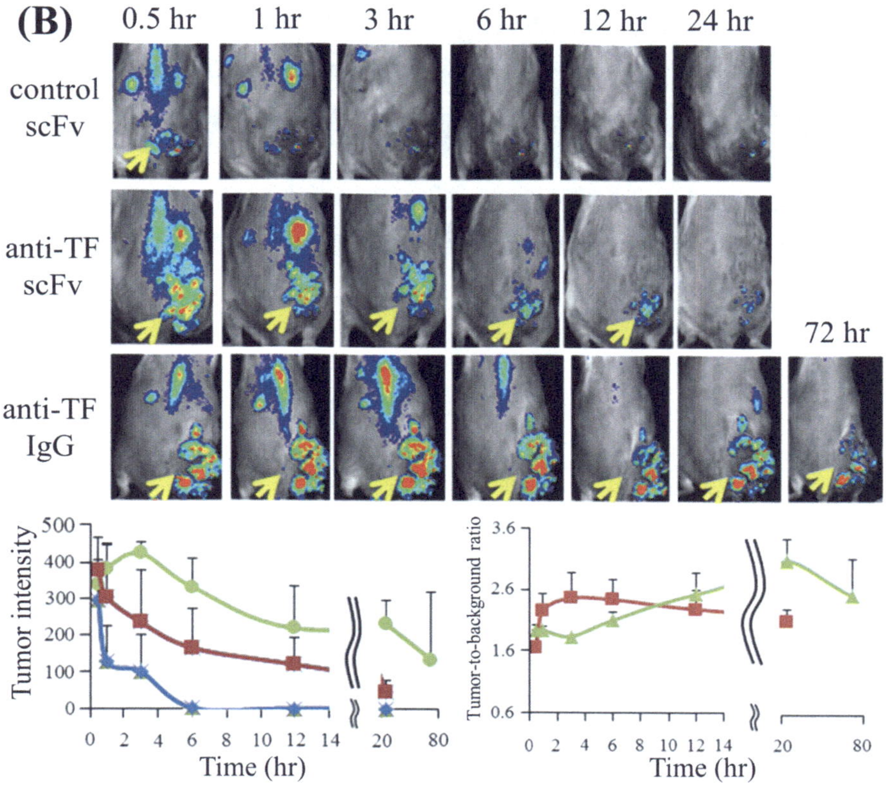

Fig. 5 (continued)

specifically in the tumor tissue and images with sufficiently high contrast can be rapidly obtained. In cancer therapeutics, the antibodies, as well as the ADCs, should accumulate efficiently in the cancer tissue for a long duration. Thus, IgG, the molecular size of which is 150 kDa (approximately 15 nm particle size), is suitable for the cancer treatment.

Cancer Stroma Barrier

After the extravasation of drugs from the tumor vessels, the second hurdle, the so-called the cancer stroma barrier is encountered. (Fig. 4) This is well illustrated by the following narrative. The agent MCC-465, categorized DDS consists of doxorubicin-incorporated liposomes conjugated with PEG and anti-GAH antibodies that specifically bind to stomach cancer cells was anticipated to be highly effective in clinic because the formulation utilizes both the EPR effect and an active

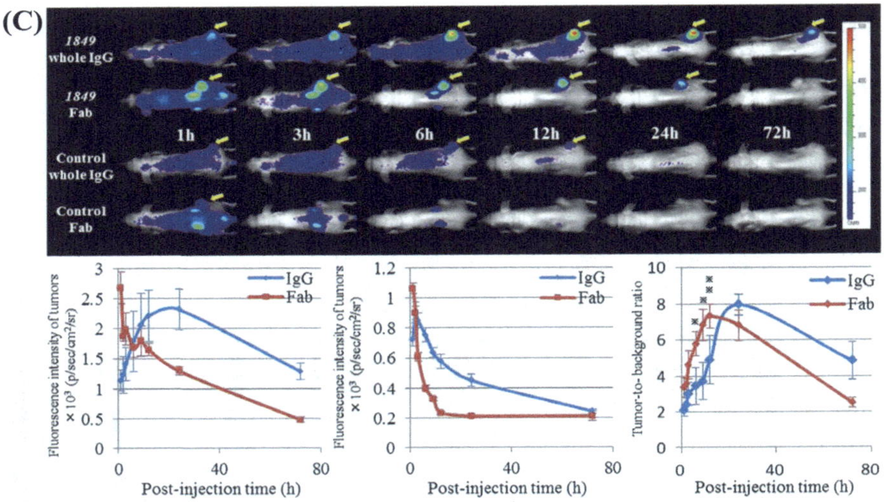

Fig. 5 (continued)

targeting by the anti-GAH antibodies. In fact, animal experiments revealed a remarkable antitumor effect in two different stomach cancer xenograft models. Unexpectedly, however, no antitumor response was observed in clinical trials [46].

After much consideration and various experiments, Matsumura Y concluded that DDS was effective in experimental tumor xenografts because of the paucity of stroma in such growths. On the other hand, DDS was clinically ineffective in human pancreatic cancer because of the tumor stromal barrier. To overcome this hurdle, he proposed a new concept, named CAST therapy [45] and produced several mAbs against cancer associated antigens in cancer stroma [83]. The cancer stroma barrier, which contains the cancer stroma cells, the ECMs, and the collagen fibers, is a physical blocking layer which impedes drug penetration and diffusion. If the physical barriers within the solid tumor are dense, the use of CAST therapy by which the ADCs can target and bind to the cancer stromal scaffolding, is highly appropriate. Yasunaga et al. and Fuchigami et al. reported several ADCs targeting the cancer stroma [18, 82–84]. In the formulation of ADCs used for the CAST therapy, the antibody and the linker are different from conventional ADCs. In this therapeutic agent design, the antibodies are selected to target collagen fibers, amorphous collagen (anti-collagen IV antibody) or fibrin clots (anti-fibrin antibody) in the cancer stroma, and the cleavable linkers do not need a strict internalization ability because the payloads are released outside the cells into the extracellular matrix. Thus, the anti-collagen IV ADC formed by conjugation with SN-38, TOP1 inhibitor, via non-enzymatically releasing linker (the ester linker) has an excellent tumor inhibition ability compared to the anti-EpCAM ADC conjugated with SN-38 via the pH sensitive cleavable carbamate linker [84]. The fibrin clot is another key molecule in the cancer stroma. In solid tumors, the malignant cancer cells invade the cancer stroma including the tumor vessels, resulting in microscopic hemorrhages in many places at different times. Following this, fibrin clots immediately form to stop the bleeding. Thus, fibrin clot formation in the cancer tissue lasts for as long as the cancer cells survive in the body. This whole process occurs silently and is called the "malignant cycle of blood coagulation" [45]. To make anti fibrin ADCs, the anti-fibrin antibodies are conjugated with SN-38 via the ester linker [82]. This ADC induces effective tumor suppression in mice with spontaneous cutaneous tumors. Since the ester linker cleaves in a serum gradually and the instability of the ADCs in a blood stream may induce a side effect. To overcome this shortcoming, the anti-fibrin antibody is conjugated with MMAE via tripeptide (Val-Leu-Lys) cleavable linker, which is cleaved by plasmin which is known to be activated only on the fibrin clots. The ADC was assessed in the transgenic mice with spontaneous pancreatic cancer, KPC (LSL-Kras^{G12D}/LSL-Trp53^{R172H}/Ptf1a-Cre) mice. This ADC produced a longer survival on the KPC mice compared to control ADC [18].

The other technique to penetrate the cancer stroma barrier is to select the optimal size of the macromolecules. The particle size of the most nanoparticles including micelles and liposomes is approximately 100 nm; regarding the approved nanomedicine, Doxil [51] and Abraxane [24] have 90–130 nm in diameters. Cabral et al. compared the accumulation and effectiveness of relatively smaller sizes of ACA incorporated micelles with diameters of 30, 50, 70, and 100 nm in tumors with a dense cancer stroma [5]. They found that the 30 nm size micelles could penetrate through the cancer stroma barrier, reach the cancer cells, and have the most efficient anti-cancer activity. But the sizes of the ADCs are 15 nm, less than the micelles, thus, the ADCs may more efficiently penetrate and reach to the cancer cells which are located far from the tumor vessels.

Binding Site Barrier (Fig. 4) [19, 34, 61]

The ADCs, which can bypass excretion in the urine, extravasate selectively from the tumor vessels and penetrate and diffuse through the cancer stroma, still encounter obstruction at the binding site barrier. Many cancer cells arranged in cancer nests in the stroma, and are not scattered as single cells, in the solid tissue of the tumor. The ADCs bind to the surfaces of the external cells in the cancer nest via antigen-antibody interaction at the first contact with the corresponding antigen. It is thought that specific and strong binding ability is a necessary requirement for good therapeutic antibodies. However, the ADCs formed using such antibodies become trapped at the surface of the cancer nest and their penetration and diffusion to the center of the cancer nest is retarded. This barrier can be seen in vivo, using fluorescence or radioisotope labeled antibodies, which bind to the cancer cell surface antigens when injected into tumor bearing mice. The anti-TF antibody clone 1849 has a high binding affinity to the human TF antigen with a K_D (binding affinity constant) value of 0.091 nM [37]. Although 1849 ADCs can reach the cancer nest through the cancer stroma barrier, they localize only at the surface of the cancer nest of the pancreatic cancer xenograft tumor. The same phenomenon is seen in similar studies on intracranial glioblastoma xenograft implants [71] Although the 1849 mAbs accumulate in the tumor, they are heterogeneously distributed in it despite the homogenous expression of TF in the tumor cells (Fig. 6a). The smaller size of antibodies, such as the 1849 Fab, can penetrate more efficiently to the center of the cancer nest compared to the 1849 whole IgG, although the majority of 1849 Fab are located at the surface of the cancer nest (Fig. 6b) [78]. However, as described before, the Fab, the size of which is 5 nm, cleared rapidly from the blood stream mainly due to excretion by the kidney. Different from diagnostic purpose, Fab form may not be suitable for therapeutic purpose because enough amount of the Fab may not reach to cancer tissues in the first place. In this context, Tsumura et al. conducted a biochemical and a pharmacological investigations using several anti-TF mAbs with different affinities [77]. The K_D value is generally recognized as the antigen binding ability for therapeutic antibodies and the higher avidity to the antigen, which means a smaller K_D value, leads to the success of the ADCs. The K_D value is calculated by the k_a (association rate constant) value and the k_d (dissociation rate constant) value according to the following formula; K_D (M) = k_d (s^{-1}) /k_a (M^{-1} s^{-1}). In their report, the K_D values of the 1849 ADC and the 1084 ADC were calculated as 0.54 nM and 34.96 nM, respectively, and therefore, the 1849 ADC had higher avidity to the antigen than the 1084 ADC. According to the conventional ADC theory, the 1849 ADC would have a higher cytotoxic effect in the TF expressing cancer cells compared to the 1084 ADC; however, same cytotoxic effects were observed in vitro between the 1849 ADC and the 1084 ADC. The k_a values of the 1849 ADC and 1084 ADC were 4.20×10^4 M^{-1} s^{-1} and 2.61×10^4 M^{-1} s^{-1}, respectively. The almost same k_a values of ADCs may affect to the similar cytotoxic effect in vitro. Also, the other characteristics between the two ADCs, such as the internalization ability into the cancer cells, the drug releasing ability under the presence of cathepsin B, and the DAR, were almost same. The only difference between the 1849 ADC and the 1084 ADC is the k_d value, which is the dissociation ability of the antibodies from the antigens. The k_d values of the 1849 ADC and 1084 ADC were 0.21×10^{-4} s^{-1} and 8.25×10^{-4} s^{-1}, respectively. This result means that the 1084 ADC can dissociate faster from the antigen than the 1849 ADC. Then, they conducted in vivo antitumor experiments of the two ADCs using the pancreatic cancer xenograft mice models. They started the treatment when the tumor size reached to 600 mm^3 the size of which were larger than that of ordinary in vivo experiments. Although the K_D value of the 1084 ADC was lower than that of the 1849 ADC, the tumor growth inhibition of the 1084 ADC was significantly higher than that of the 1849 ADC ($P < 0.01$). They also investigated the distribution of these ADCs in the tumor to clarify the reason why the 1084 ADC was superior to the 1849 ADC in antitumor studies *in vivo*. As in the former studies [37, 78], the 1849 ADC became located at the surface of the cancer nest (Fig. 6c). Surprisingly, the 1084 ADC could penetrate into the cancer nest, reach to the center of the nest, and in the end, distribute homogenously throughout it. They concluded that the ADCs with high potent dissociation ability from the antigens have high potent tumor penetration ability into the central region of the tumor, and then, also have highly potent antitumor activity in solid tumors.

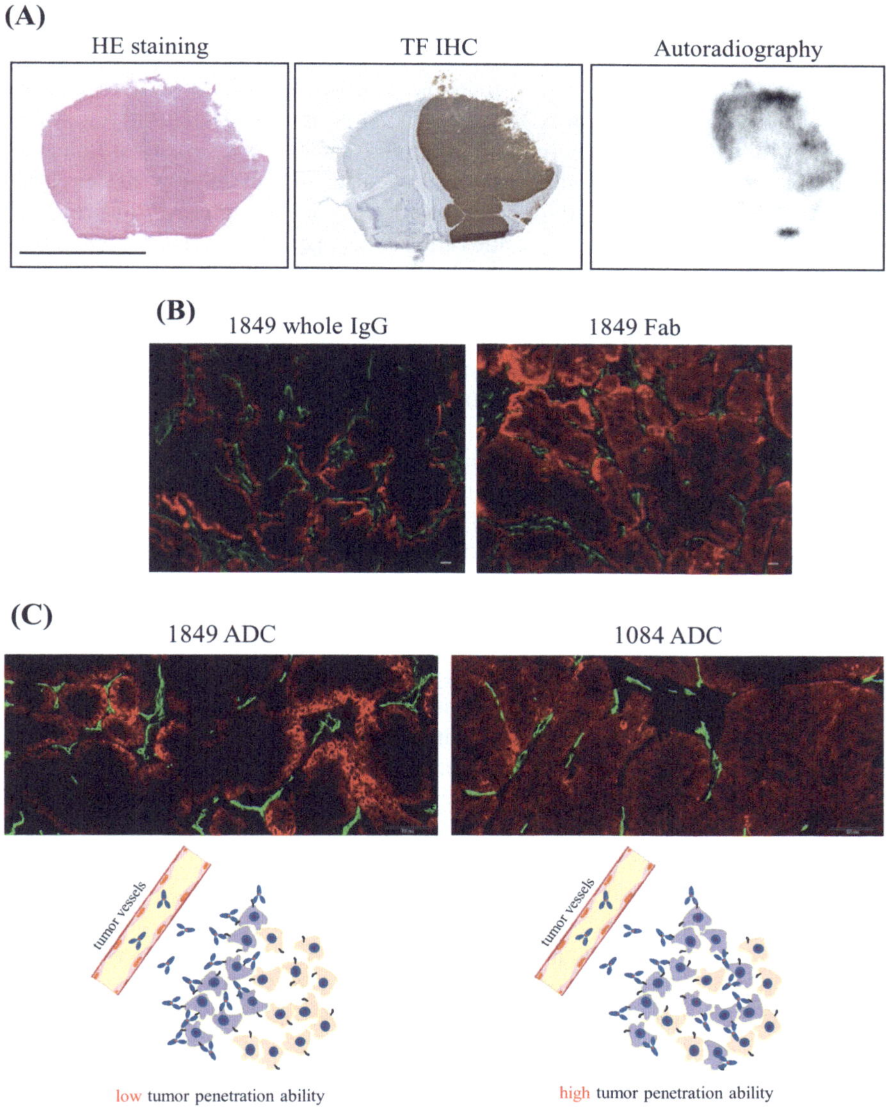

Fig. 6 Binding site barrier. (**a**) Heterogeneous accumulation of mAb [71]. The anti-TF IgG (clone 1849) are heterogeneously distributed in the intracranial tumor despite of the homogenously TF expression in the tumor. (**b**) Bio-distribution of several sizes of mAbs [78]. The smaller size of antibodies, the 1849 Fab, can be penetrated more efficiently to the center of the cancer nest compared to the 1849 whole IgG. (**c**) Bio-distribution of IgG with different dissociation ability [77]. Although the 1849 ADC is located in the surface of the cancer nest, the 1084 ADC with high potent dissociation ability from the antigen can penetrate into the cancer nest, reach to the center of the nest, and distribute homogenously in the nest

Internalization into the Cancer Cells

After the ADCs reach and bind to the cancer cells, internalization ability is the next hurdle for some kinds of ADCs (Fig. 4). Although the ADCs with cleavable linkers do not need strict internalization ability because the linkers are cleaved under prevailing conditions in the physiological environment (low pH environment, the high level of glutathione concentrations, or the high level of protease) even if the ADCs are located at outside of the cancer cells. Conversely, ADCs with non-cleavable linkers *are* required for strict internalization into the cancer cells, and especially into their lysosomes. In the internalization of nanoparticles into cancer cells, several well-known endocytosis pathways are utilized, such as receptor mediated endocytosis (also known as the clathrin mediated endocytosis), caveolae mediated endocytosis, micropinocytosis, and phagocytosis. However, a recent study demonstrated that, in the case of non-targeting nanoparticles, clathrin mediated endocytosis, caveolae mediated endocytosis and micropinocytosis are all involved in their internalization into non-phagocytic HeLa cells [25]. In that study, cylindrical particles with several sizes and several aspect ratios were applied to the HeLa cells. Although the low aspect ratio cylindrical particles (having a diameter = 200 nm and a height = 200 nm) and the high aspect ratio cylindrical particles (d = 150 nm, h = 450 nm) have similar volumes, the high aspect ratio particles were internalized into the HeLa cells faster than the low aspect ratio particles. Regarding same shape, the large particles with a diameter of 150 nm were internalized into the HeLa cells faster than the small particles with a diameter of 100 nm. Although the internalization ability of the nanoparticles into the non-phagocytic cells via several endocytosis pathways depends on the shape and the size of the nanoparticles, most sizes of the optimized therapeutic antibodies are homogenously 10–15 nm, and the internalization into the cells mainly depends on the receptor mediated endocytosis.

Internalization of agents into tumor cells depends upon the types of cells, antigens and the antibodies involved in the process. In studies on the role of the properties of the antigen, the cellular uptake and internalization of an anti-CD20 mAb against malignant lymphoma cells and of an anti-EpCAM mAb against pancreatic cancer cells were evaluated [84]. The anti-CD20 mAb was internalized efficiently into the cells and co-localized with the lysosomes at 12 h after the incubation. On the other hand, the anti-EpCAM mAb was still retained on the surface of the cells at the same time period. Generally, hematological tumor cells have more efficient internalization ability when compared to epithelial cancer cells. The internalization ability of another anti-EpCAM mAb was evaluated in three epithelial cancer cell lines manifesting high EpCAM expression on the cell surface [42]. Although the anti-EpCAM mAb could bind to all three cell lines, it was internalized and co-localized with lysosomes efficiently in BxPC-3, pancreatic cancer cells, moderately in WiDr, colon cancer cells, and rarely in MCF-7, breast cancer cells after 18 h incubation. These results indicate that internalization ability depends on the nature of the cells and the antigens as expected. In using these principles clinically with wild-type cancers, the cell types (non-phagocytic epithelial cells) and the cancer specific antigen cannot be changed. However, the above studies indicate that internalization of antibodies can be improved, by altering other parameters. Additionally, Friedman et al. found that some *combinations* of mAbs displayed highly potent internalization ability and synergistic down-regulation of RTKs [17]. Sym004, a mixture of two kinds of anti-EGFR antibodies, has higher anti-cancer efficacy compared to the already approved anti-EGFR mAbs in vitro and in vivo [21, 55]. The two kinds of antibodies in that study have different epitopes and the antibodies make receptor-antibody cross-linking. Thus, the Sym004 has excellent internalization ability and inhibition of EGFR signaling.

Releasing and Visualization of the Free Payloads

After internalization of ADCs into the cancer cells, drug releasing should be required. The drug releasing mainly depends on the linker technology described above. After the ADCs are internalized into the cells, the co-localization of the lysosome and the ADCs are important for both the cleavable and non-cleavable linkers. In preclinical studies, cancer cells incubated with the fluorescence labeled mAbs and another fluorescence labeled pH indicator are usually assessed using the fluorescence microscopy imaging (Fig. 7a) [37]. In such investigations, the ADCs have to be labeled with the fluorescence. Thus, the characteristics of the ADCs, such as molecular weight or polarity, might be changed and influence its stability and drug releasing ability in vivo. Also, the free payloads released from the ADCs cannot be detected. Fujiwara et al. reported that the free payloads released from non-labeled ADCs could be visualized by using the imaging mass spectrometry (IMS) [20]. The IMS could distinguish the free payloads from the ADCs (the payloads conjugated to mAbs). The ADCs of the anti-TF mAb conjugated with MMAE via Val-Cit cleavable linker, were injected into mice bearing TF high expressing pancreatic cancer xenografts via tail vein. The released MMAE was selectively located in the cancer nest compared to the cancer stroma at 24 h after injection (Fig. 7b). On the basis of observations like these, IMS is expected to be a powerful tool for the optimization of drug design, especially for antibody selection and the linker design, for ADCs in the preclinical stage.

Bystander Killing Effect

A cancer nest, situated in solid tumor tissue consists of many clonal cells. Moreover, the target antigen in the solid tumor is expressed heterogeneously. The "bystander killing effect", which can affect not only the antigen positive cancer cells, but also the adjacent antigen negative cancer cells, via the transportation of the released drugs from the antigen positive cancer cells, is known to be a promising approach

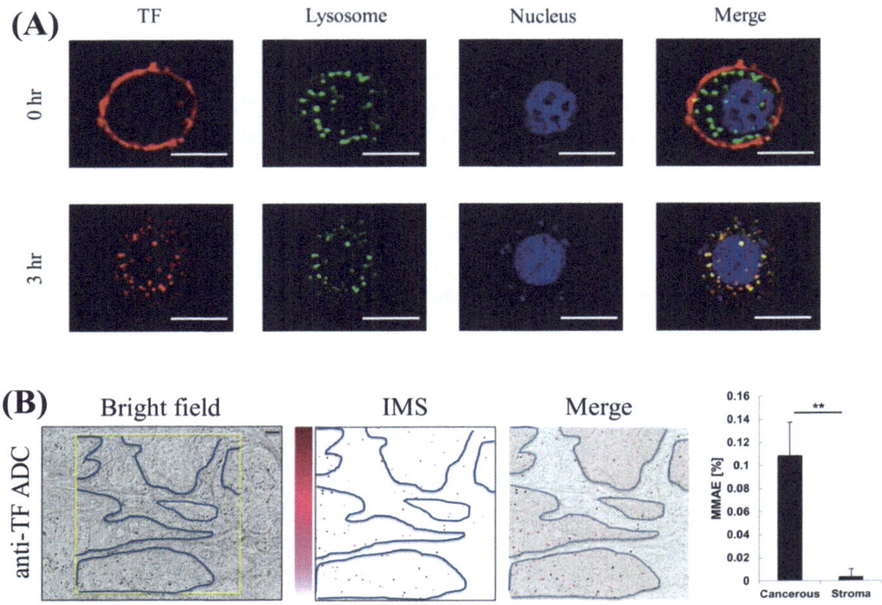

Fig. 7 Visual analyses of the internalization of the ADCs and the drug releasing from the ADCs. (**a**) Visualization of the internalized ADCs into the cancer cells [37]. The cancer cells incubated with the fluorescence labeled mAbs (red) and another fluorescence labeled pH indicator (green) are usually assessed using the fluorescent microscopy imaging. The mAbs are sufficiently internalized into the cytoplasm after a 3 h-incubation and localized in a lysosome. (**b**) Visualization of the free payloads released from the ADCs [20]. The IMS (imaging mass spectrometry) can distinguish the free payloads from the ADCs (the payloads conjugated to mAbs). The released free payloads are selectively located in the cancer nest compared to the cancer stroma at 24 h after injection

to obtaining effective cytotoxicity in solid tumors. Since the payloads which have bystander killing effects might not only exert therapeutic effects on the cancer cells but also be toxicity to nearby normal cells, the balance between efficacy and toxicity in vivo should be considered. To act on aggregates of cancer cells, payloads which diffuse through cell membranes easily should be selected. Generally, the payloads released from an ADC with a non-cleavable linker, such as the agent trastuzumab emtansine, have no bystander killing effect. These deductions are derived from the very thorough work of Ogitani et al. [53] who reported on a comparative study of the bystander killing effect of DS-8201a, the novel ADC now in clinical trials, anti-HER2-Dxd(2) and trastuzumab emtansine. The ADC DS-8201a consists of the anti-HER2 antibody, a tetrapeptide cleavable linker, and the Dxd TOP1 inhibitor. The ADC anti-HER2-Dxd(2) is similar to DS-8201a except for the payload (i.e. Dxd(2)) [53] It should be noted that the payloads of these three therapeutic agents differed as follows: the cell membrane permeability of Dxd(1) incorporated in DS-8201a, is higher than that of Dxd(2) and of Lys-SMCC-DM1 which is contained in trastuzumab emtansine. The cell growth inhibition abilities in vitro of DX-8201a, anti-HER2-Dxd(2), and of trastuzumab emtansine were almost same in any cells regardless of the level of HER2 expression. Also, the antitumor activity of these ADCs in vivo was almost same; tumor regression was observed in HER2-positive tumor xenografts but not in HER2-negative xenografts. On the other hand, in vitro *co-culture* killing assays for two cell lines treated with the ADCs indicated that only DS-8201a could kill both HER2-positive and HER2-negative cells growing together, but anti-HER2-Dxd(2) and trastuzumab emtansine killed only HER2-positive cells. Moreover, when these ADCs were injected into co-inoculated (mixture of the HER2-positive cells and the HER2-negative cells) xenograft mice, and the tumor growth inhibition was measured. The potent antitumor activity was observed in the DS-8201a group with the complete inhibition of both the HER2-positive and the HER2-negative cells. On the other hand, tumor reductions were not observed in the anti-HER2-Dxd(2) and the trastuzumab emtansine groups, because these ADCs could not inhibit HER2-negative cell growth. Interestingly, this bystander killing effect was observed only in the HER2-negative *cells* adjacent to the HER2-positive *cells*, not in HER2-negative *tumors* which were located far from the HER2-positive *tumors* (Fig. 8). This result indicates that the Dxd released from DS-8201a in the HER2-positive cells did not reach distant HER2-negative tumors. This means that only a limited amount of released Dxd entered and circulated in the blood. The ADCs with low permeable payloads might be useful for treating hematological tumors, such as leukemia because these tumor cells float on the blood stream as single units and the bystander killing effect is not required for eradicating circulating leukemia cells. More recently, Doi et al. reported that data of the phase 1 clinical

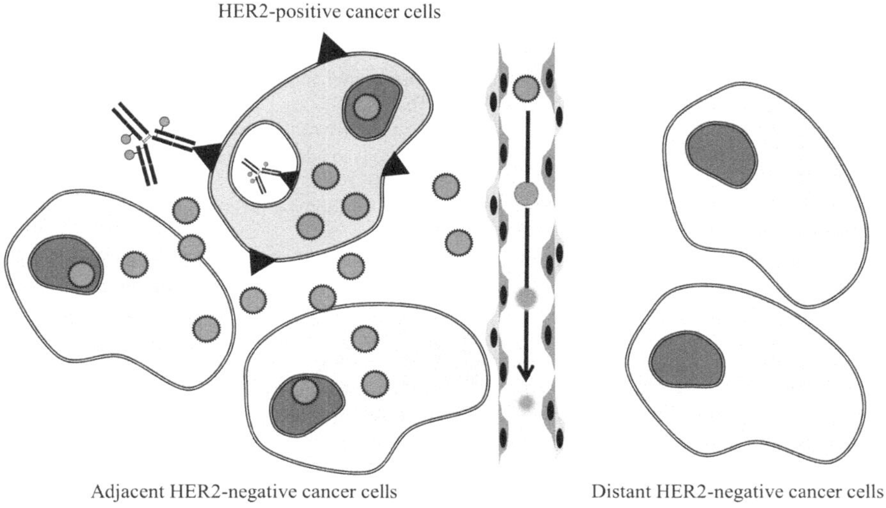

Fig. 8 Bystander killing effect to the antigen negative cancer cells [53]. The payloads which have high potent cell membrane permeability and are released in the antigen positive cancer cells affect to the antigen negative cells adjacent to the antigen positive cells, not to the antigen negative cells located far from the antigen positive cells

trial of the trastuzumab deruxtecan (same as DS-8201) in the patients with advanced breast and gastro-esophageal tumors was promising in terms of high antitumor effect and caused only limited, well-tolerated adverse effects. (See chapter "ADCs on the Market and in Clinical Development" by Abe) [11].

Conclusions

The successful ADCs have three important components; the antibodies, the linkers, and the payloads. For design of the ADCs, the characteristics of the tumor, especially the solid tumor, has to be well-informed. There are six hurdles to conquer in the solid tumors. First four hurdles, which are the stability and the extravasation from tumor vessels, the cancer stroma barrier, the binding site barrier, and the internalization into cancer cells, can be overcome by the antibody, the drug releasing can be controlled by the linker technology, and the bystander killing effect can be realized by the optimal payload. One hundred twenty years have passed since the "magic bullet" concept was proposed, and 40 years have passed after the first success of the monoclonal antibody technology. We are now seeing the advent of ADCs which have the potential to become the main therapeutic agents in medical oncology.

References

1. Ab O, Whiteman KR, Bartle LM et al (2015) IMGN853, a folate receptor-alpha (FRalpha)-targeting antibody-drug conjugate, exhibits potent targeted antitumor activity against FRalpha-expressing tumors. Mol Cancer Ther 14:1605–1613
2. Almhanna K, Kalebic T, Cruz C et al (2016) Phase I study of the investigational anti-guanylyl cyclase antibody-drug conjugate TAK-264 (MLN0264) in adult patients with advanced gastrointestinal malignancies. Clin Cancer Res 22:5049–5057
3. Bouchard H, Viskov C, Garcia-Echeverria C (2014) Antibody-drug conjugates-a new wave of cancer drugs. Bioorg Med Chem Lett 24:5357–5363
4. Breij EC, De Goeij BE, Verploegen S et al (2014) An antibody-drug conjugate that targets tissue factor exhibits potent therapeutic activity against a broad range of solid tumors. Cancer Res 74:1214–1226
5. Cabral H, Matsumoto Y, Mizuno K et al (2011) Accumulation of sub-100 nm polymeric micelles in poorly permeable tumours depends on size. Nat Nanotechnol 6:815–823
6. Cardillo TM, Govindan SV, Sharkey RM et al (2015) Sacituzumab govitecan (IMMU-132), an anti-Trop-2/SN-38 antibody-drug conjugate: characterization and efficacy in pancreatic, gastric, and other cancers. Bioconjug Chem 26:919–931
7. Challita-Eid PM, Satpayev D, Yang P et al (2016) Enfortumab vedotin antibody-drug conjugate targeting nectin-4 is a highly potent therapeutic agent in multiple preclinical cancer models. Cancer Res 76:3003–3013
8. Cunningham D, Humblet Y, Siena S et al (2004) Cetuximab monotherapy and cetuximab plus irinotecan in irinotecan-refractory metastatic colorectal cancer. N Engl J Med 351:337–345

9. Dillman RO, Shawler DL, Johnson DE et al (1986) Preclinical trials with combinations and conjugates of T101 monoclonal antibody and doxorubicin. Cancer Res 46:4886–4891
10. Dillman RO, Shawler DL, Mccallister TJ et al (1994) Human anti-mouse antibody response in cancer patients following single low-dose injections of radiolabeled murine monoclonal antibodies. Cancer Biother 9:17–28
11. Doi T, Shitara K, Naito Y et al (2017) Safety, pharmacokinetics, and antitumour activity of trastuzumab deruxtecan (DS-8201), a HER2-targeting antibody-drug conjugate, in patients with advanced breast and gastric or gastro-oesophageal tumours: a phase 1 dose-escalation study. Lancet Oncol 18:1512–1522
12. Dubowchik GM, Firestone RA, Padilla L et al (2002) Cathepsin B-labile dipeptide linkers for lysosomal release of doxorubicin from internalizing immunoconjugates: model studies of enzymatic drug release and antigen-specific in vitro anticancer activity. Bioconjug Chem 13:855–869
13. Dvorak HF (2002) Vascular permeability factor/vascular endothelial growth factor: a critical cytokine in tumor angiogenesis and a potential target for diagnosis and therapy. J Clin Oncol 20:4368–4380
14. Elias DJ, Hirschowitz L, Kline LE et al (1990) Phase I clinical comparative study of monoclonal antibody KS1/4 and KS1/4-methotrexate immunconjugate in patients with non-small cell lung carcinoma. Cancer Res 50:4154–4159
15. Flygare JA, Pillow TH, Aristoff P (2013) Antibody-drug conjugates for the treatment of cancer. Chem Biol Drug Des 81:113–121
16. Francisco JA, Cerveny CG, Meyer DL et al (2003) cAC10-vcMMAE, an anti-CD30-monomethyl auristatin E conjugate with potent and selective antitumor activity. Blood 102:1458–1465
17. Friedman LM, Rinon A, Schechter B et al (2005) Synergistic down-regulation of receptor tyrosine kinases by combinations of mAbs: implications for cancer immunotherapy. Proc Natl Acad Sci USA 102:1915–1920
18. Fuchigami H, Manabe S, Yasunaga M et al (2018) Chemotherapy payload of anti-insoluble fibrin antibody-drug conjugate is released specifically upon binding to fibrin. Sci Rep 8:14211
19. Fujimori K, Covell DG, Fletcher JE et al (1990) A modeling analysis of monoclonal antibody percolation through tumors: a binding-site barrier. J Nucl Med 31:1191–1198
20. Fujiwara Y, Furuta M, Manabe S et al (2016) Imaging mass spectrometry for the precise design of antibody-drug conjugates. Sci Rep 6:24954
21. Fukuoka S, Kojima T, Koga Y et al (2017) Preclinical efficacy of Sym004, novel anti-EGFR antibody mixture, in esophageal squamous cell carcinoma cell lines. Oncotarget 8:11020–11029
22. Gibson TB, Ranganathan A, Grothey A (2006) Randomized phase III trial results of panitumumab, a fully human anti-epidermal growth factor receptor monoclonal antibody, in metastatic colorectal cancer. Clin Colorectal Cancer 6:29–31
23. Golfier S, Kopitz C, Kahnert A et al (2014) Anetumab ravtansine: a novel mesothelin-targeting antibody-drug conjugate cures tumors with heterogeneous target expression favored by bystander effect. Mol Cancer Ther 13:1537–1548
24. Gradishar WJ, Tjulandin S, Davidson N et al (2005) Phase III trial of nanoparticle albumin-bound paclitaxel compared with polyethylated castor oil-based paclitaxel in women with breast cancer. J Clin Oncol 23:7794–7803
25. Gratton SE, Ropp PA, Pohlhaus PD et al (2008) The effect of particle design on cellular internalization pathways. Proc Natl Acad Sci USA 105:11613–11618
26. Hamblett KJ, Kozlosky CJ, Siu S et al (2015) AMG 595, an anti-EGFRvIII antibody-drug conjugate, induces potent antitumor activity against EGFRvIII-expressing glioblastoma. Mol Cancer Ther 14:1614–1624
27. Hanahan D, Weinberg RA (2011) Hallmarks of cancer: the next generation. Cell 144:646–674

28. Huls GA, Heijnen IA, Cuomo ME et al (1999) A recombinant, fully human monoclonal antibody with antitumor activity constructed from phage-displayed antibody fragments. Nat Biotechnol 17:276–281
29. Hurwitz H, Fehrenbacher L, Novotny W et al (2004) Bevacizumab plus irinotecan, fluorouracil, and leucovorin for metastatic colorectal cancer. N Engl J Med 350:2335–2342
30. Ishii G, Ochiai A, Neri S (2016) Phenotypic and functional heterogeneity of cancer-associated fibroblast within the tumor microenvironment. Adv Drug Deliv Rev 99:186–196
31. Jackson D, Atkinson J, Guevara CI et al (2014) In vitro and in vivo evaluation of cysteine and site specific conjugated herceptin antibody-drug conjugates. PLoS One 9:e83865
32. Jakobovits A, Amado RG, Yang X et al (2007) From XenoMouse technology to panitumumab, the first fully human antibody product from transgenic mice. Nat Biotechnol 25:1134–1143
33. Jeffrey SC, Burke PJ, Lyon RP et al (2013) A potent anti-CD70 antibody-drug conjugate combining a dimeric pyrrolobenzodiazepine drug with site-specific conjugation technology. Bioconjug Chem 24:1256–1263
34. Juweid M, Neumann R, Paik C et al (1992) Micropharmacology of monoclonal antibodies in solid tumors: direct experimental evidence for a binding site barrier. Cancer Res 52:5144–5153
35. Kantarjian HM, Deangelo DJ, Stelljes M et al (2016) Inotuzumab ozogamicin versus standard therapy for acute lymphoblastic leukemia. N Engl J Med 375:740–753
36. Katz J, Janik JE, Younes A (2011) Brentuximab vedotin (SGN-35). Clin Cancer Res 17:6428–6436
37. Koga Y, Manabe S, Aihara Y et al (2015) Antitumor effect of antitissue factor antibody-MMAE conjugate in human pancreatic tumor xenografts. Int J Cancer 137:1457–1466
38. Kohler G, Milstein C (1975) Continuous cultures of fused cells secreting antibody of predefined specificity. Nature 256:495–497
39. Komohara Y, Takeya M (2017) CAFs and TAMs: maestros of the tumour microenvironment. J Pathol 241:313–315
40. Lewis Phillips GD, Li G, Dugger DL et al (2008) Targeting HER2-positive breast cancer with trastuzumab-DM1, an antibody-cytotoxic drug conjugate. Cancer Res 68:9280–9290
41. Lu J, Jiang F, Lu A et al (2016) Linkers having a crucial role in antibody-drug conjugates. Int J Mol Sci 17:561
42. Lund K, Bostad M, Skarpen E et al (2014) The novel EpCAM-targeting monoclonal antibody 3-17I linked to saporin is highly cytotoxic after photochemical internalization in breast, pancreas and colon cancer cell lines. MAbs 6:1038–1050
43. Maishi N, Hida K (2017) Tumor endothelial cells accelerate tumor metastasis. Cancer Sci 108:1921–1926
44. Maloney DG, Grillo-Lopez AJ, White CA et al (1997) IDEC-C2B8 (Rituximab) anti-CD20 monoclonal antibody therapy in patients with relapsed low-grade non-Hodgkin's lymphoma. Blood 90:2188–2195
45. Matsumura Y (2012) Cancer stromal targeting (CAST) therapy. Adv Drug Deliv Rev 64:710–719
46. Matsumura Y, Gotoh M, Muro K et al (2004) Phase I and pharmacokinetic study of MCC-465, a doxorubicin (DXR) encapsulated in PEG immunoliposome, in patients with metastatic stomach cancer. Ann Oncol 15:517–525
47. Matsumura Y, Kimura M, Yamamoto T et al (1988) Involvement of the kinin-generating cascade in enhanced vascular permeability in tumor tissue. Jpn J Cancer Res 79:1327–1334
48. Matsumura Y, Maeda H (1986) A new concept for macromolecular therapeutics in cancer chemotherapy: mechanism of tumoritropic accumulation of proteins and the antitumor agent smancs. Cancer Res 46:6387–6392
49. Mattie M, Raitano A, Morrison K et al (2016) The discovery and preclinical development of ASG-5ME, an antibody-drug conjugate targeting SLC44A4-positive epithelial tumors including pancreatic and prostate cancer. Mol Cancer Ther 15:2679–2687

50. Naumovski L, Junutula JR (2010) Glembatumumab vedotin, a conjugate of an anti-glycoprotein non-metastatic melanoma protein B mAb and monomethyl auristatin E for the treatment of melanoma and breast cancer. Curr Opin Mol Ther 12:248–257
51. Northfelt DW, Dezube BJ, Thommes JA et al (1998) Pegylated-liposomal doxorubicin versus doxorubicin, bleomycin, and vincristine in the treatment of AIDS-related Kaposi's sarcoma: results of a randomized phase III clinical trial. J Clin Oncol 16:2445–2451
52. Ogitani Y, Aida T, Hagihara K et al (2016) DS-8201a, a novel HER2-targeting ADC with a novel DNA topoisomerase I inhibitor, demonstrates a promising antitumor efficacy with differentiation from T-DM1. Clin Cancer Res 22:5097–5108
53. Ogitani Y, Hagihara K, Oitate M et al (2016) Bystander killing effect of DS-8201a, a novel anti-human epidermal growth factor receptor 2 antibody-drug conjugate, in tumors with human epidermal growth factor receptor 2 heterogeneity. Cancer Sci 107:1039–1046
54. Panowski S, Bhakta S, Raab H et al (2014) Site-specific antibody drug conjugates for cancer therapy. MAbs 6:34–45
55. Pedersen MW, Jacobsen HJ, Koefoed K et al (2010) Sym004: a novel synergistic anti-epidermal growth factor receptor antibody mixture with superior anticancer efficacy. Cancer Res 70:588–597
56. Phillips AC, Boghaert ER, Vaidya KS et al (2016) ABT-414, an antibody-drug conjugate targeting a tumor-selective EGFR epitope. Mol Cancer Ther 15:661–669
57. Pillai G (2014) Nanomedicines for cancer therapy: an update of FDA approved and those under various stages of development. SOJ Pharm Pharm Sci 1:13
58. Pommier Y (2006) Topoisomerase I inhibitors: camptothecins and beyond. Nat Rev Cancer 6:789–802
59. Rudin CM, Pietanza MC, Bauer TM et al (2017) Rovalpituzumab tesirine, a DLL3-targeted antibody-drug conjugate, in recurrent small-cell lung cancer: a first-in-human, first-in-class, open-label, phase 1 study. Lancet Oncol 18:42–51
60. Saeed AF, Wang R, Ling S et al (2017) Antibody engineering for pursuing a healthier future. Front Microbiol 8:495
61. Saga T, Neumann RD, Heya T et al (1995) Targeting cancer micrometastases with monoclonal antibodies: a binding-site barrier. Proc Natl Acad Sci USA 92:8999–9003
62. Sato R, Obonai T, Tsumura R et al (2014) Preparation and characterization of anti-tissue factor single-chain variable fragment antibody for cancer diagnosis. Cancer Sci 105:1631–1637
63. Segaliny AI, Tellez-Gabriel M, Heymann MF et al (2015) Receptor tyrosine kinases: characterisation, mechanism of action and therapeutic interests for bone cancers. J Bone Oncol 4:1–12
64. Sharkey RM, Govindan SV, Cardillo TM et al (2018) Selective and concentrated accretion of SN-38 with a CEACAM5-targeting antibody-drug conjugate (ADC), labetuzumab govitecan (IMMU-130). Mol Cancer Ther 17:196–203
65. Sievers EL, Linenberger M (2001) Mylotarg: antibody-targeted chemotherapy comes of age. Curr Opin Oncol 13:522–527
66. Smith AL, Nicolaou KC (1996) The enediyne antibiotics. J Med Chem 39:2103–2117
67. Smith I, Procter M, Gelber RD et al (2007) 2-year follow-up of trastuzumab after adjuvant chemotherapy in HER2-positive breast cancer: a randomised controlled trial. Lancet 369:29–36
68. Strebhardt K, Ullrich A (2008) Paul Ehrlich's magic bullet concept: 100 years of progress. Nat Rev Cancer 8:473–480
69. Strohl WR (2014) Antibody discovery: sourcing of monoclonal antibody variable domains. Curr Drug Discov Technol 11:3–19
70. Sussman D, Smith LM, Anderson ME et al (2014) SGN-LIV1A: a novel antibody-drug conjugate targeting LIV-1 for the treatment of metastatic breast cancer. Mol Cancer Ther 13:2991–3000
71. Takashima H, Tsuji AB, Saga T et al (2017) Molecular imaging using an anti-human tissue factor monoclonal antibody in an orthotopic glioma xenograft model. Sci Rep 7:12341

72. Tannir NM, Forero-Torres A, Ramchandren R et al (2014) Phase I dose-escalation study of SGN-75 in patients with CD70-positive relapsed/refractory non-Hodgkin lymphoma or metastatic renal cell carcinoma. Investig New Drugs 32:1246–1257
73. Thomas LJ, Vitale L, O'neill T et al (2016) Development of a novel antibody-drug conjugate for the potential treatment of ovarian, lung, and renal cell carcinoma expressing TIM-1. Mol Cancer Ther 15:2946–2954
74. Thudium K, Bilic S, Leipold D et al (2013) American Association of Pharmaceutical Scientists National Biotechnology Conference Short Course: translational challenges in developing antibody-drug conjugates: May 24, 2012, San Diego, CA. MAbs 5:5–12
75. Tiberghien AC, Levy JN, Masterson LA et al (2016) Design and synthesis of tesirine, a clinical antibody-drug conjugate pyrrolobenzodiazepine dimer payload. ACS Med Chem Lett 7:983–987
76. Toki BE, Cerveny CG, Wahl AF et al (2002) Protease-mediated fragmentation of p-amidobenzyl ethers: a new strategy for the activation of anticancer prodrugs. J Org Chem 67:1866–1872
77. Tsumura R, Manabe S, Takashima H et al (2018) Influence of the dissociation rate constant on the intra-tumor distribution of antibody-drug conjugate against tissue factor. J Control Release 284:49–56
78. Tsumura R, Sato R, Furuya F et al (2015) Feasibility study of the Fab fragment of a monoclonal antibody against tissue factor as a diagnostic tool. Int J Oncol 47:2107–2114
79. Van Der Lee MM, Groothuis PG, Ubink R et al (2015) The preclinical profile of the duocarmycin-based HER2-targeting ADC SYD985 predicts for clinical benefit in low HER2-expressing breast cancers. Mol Cancer Ther 14:692–703
80. Verma S, Miles D, Gianni L et al (2012) Trastuzumab emtansine for HER2-positive advanced breast cancer. N Engl J Med 367:1783–1791
81. Wang J, Anderson MG, Oleksijew A et al (2017) ABBV-399, a c-met antibody-drug conjugate that targets both MET-amplified and c-met-overexpressing tumors, irrespective of MET pathway dependence. Clin Cancer Res 23:992–1000
82. Yasunaga M, Manabe S, Matsumura Y (2011) New concept of cytotoxic immunoconjugate therapy targeting cancer-induced fibrin clots. Cancer Sci 102:1396–1402
83. Yasunaga M, Manabe S, Tarin D et al (2011) Cancer-stroma targeting therapy by cytotoxic immunoconjugate bound to the collagen 4 network in the tumor tissue. Bioconjug Chem 22:1776–1783
84. Yasunaga M, Manabe S, Tarin D et al (2013) Tailored immunoconjugate therapy depending on a quantity of tumor stroma. Cancer Sci 104:231–237

ADCs on the Market and in Clinical Development

Yuki Abe, Kiyoshi Sugihara, Takashi Nakada, Javad Shahidi, Gilles J. A. Gallant, Takahiro Jikoh, and Toshinori Agatsuma

Abstract Antibody–drug conjugates (ADCs) represent a promising class of drugs with a wider therapeutic index (TI) than conventional chemotherapeutic agents due to their efficient and specific drug delivery. ADC payload is attached to antibody by chemical linkage. Two classes of cytotoxins – DNA damaging agents and microtubule inhibitors – have been validated as ADC payloads in clinical trials and four ADCs, MYLOTARG®, ADCETRIS®, KADCYLA®, and BESPONSA®, have been approved so far. There are many different ADCs in clinical development stage. Novel linker-payload technologies are evaluated and a new class of payload DNA topoisomerase I inhibitor is emerging. This review will focus on introduction of the four approved ADC drugs and promising novel ADCs with DNA topoisomerase I inhibitors, Sacituzumab Govitecan and [fam-] Trastuzumab Deruxtecan, in late stages of clinical development.

Keywords Antibody–drug conjugate, ADC · MYLOTARG® · BESPONSA® · ADCETRIS® · HER2 · KADCYLA® · [fam-] Trastuzumab deruxtecan · DS-8201a · Sacituzumab govitecan · IMMU-132 · Drug to antibody ratio, DAR · Bystander effect · DNA topoisomerase I inhibitor · Derivative of exatecan, DXd · Payload

Introduction

Antibody–drug conjugates (ADCs) represent a promising class of drugs with a wider therapeutic index (TI) than conventional chemotherapeutic agents due to their efficient and specific drug delivery [1–3]. Four ADCs have been approved [1, 4].

Y. Abe (✉)
Oncology Research Laboratories II, Daiichi Sankyo Co., Ltd., Tokyo, Japan
e-mail: abe.yuki.c3@daiichisankyo.co.jp

K. Sugihara · T. Nakada · T. Agatsuma
Oncology Research Laboratories I, Daiichi Sankyo Co., Ltd., Tokyo, Japan

J. Shahidi · G. J. A. Gallant · T. Jikoh
Oncology Research & Development, Daiichi Sankyo Inc., Basking Ridge, NJ, USA

Y. Matsumura, D. Tarin (eds.), *Cancer Drug Delivery Systems Based on the Tumor Microenvironment*, https://doi.org/10.1007/978-4-431-56880-3_7

Three of these agents are approved for hematological malignancies and one in HER2-positive breast cancer. Two classes of cytotoxins – DNA damaging agent and microtubule inhibitors – have been validated as ADC payloads in clinical trials. Gemtuzumab ozogamicin (MYLOTARG®) and inotuzumab ozogamicin (BESPONSA®) contain the same DNA binding cytotoxic antibiotic, a derivative of calicheamicin [4, 5]. Brentuximab vedotin (ADCETRIS®) has a tubulin polymerization inhibitor monomethyl auristatin E (MMAE) [6, 7]. Ado-trastuzumab emtansine (KADCYLA®) contains a different tubulin polymerization inhibitor, a derivative of maytansinoid DM1 [8]. Novel linker-payload technologies have been evaluated in clinical settings and a new class of payload DNA topoisomerase I inhibitor is emerging. This review will focus on introduction of the approved ADC drugs and promising novel ADCs with DNA topoisomerase I inhibitors in late stages of clinical development [9, 10].

ADC Constracts with DNA Damaging Agent

Gemtuzumab Ozogamicin (MYLOTARG®)

Gemtuzumab ozogamicin (GO) is an anti-CD33 ADC comprising of humanized IgG4 antibody, acid-labile hybrid 4-(4′-acetylphenoxy) butanoic acid linker and calicheamicin derivative (Fig. 1). GO is the firstly approved ADC by the United States Food and Drug Administration (US FDA) [11]. The average drug loading number of calicheamicin per antibody, drug to antibody ratio (DAR) is around 2–3 [11]. Calicheamicin is a potent antibiotic from the actinomycete *Micromonspora*

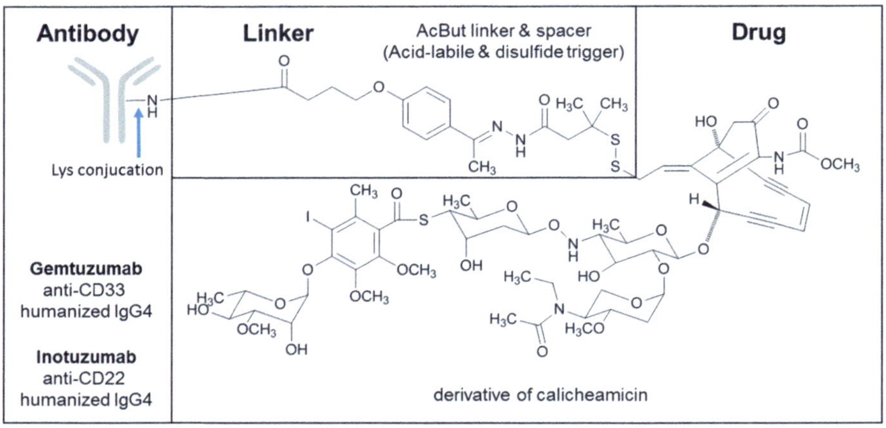

Average of drug-to-antibody ratio (DAR): gemtuzumab ozogamicin is 2 to 3, inotuzumab ozogamicin is 5 to 7
AcBut = 4-(4-acetylphenoxy) butanoic acid

Fig. 1 Schematic structure of gemtuzumab ozogamicin and inotuzumab ozogamicin

echinospora calichensis [12, 13]. Calicheamicin binds to the minor groove of DNA in a sequence specific manner and causes double-strand DNA breaks, resulting in cell death [14]. CD33 is expressed on leukemia blasts in most patients with acute myeloid leukemia (AML) [15, 16]. Once GO binds to the target, the ADC is internalized into AML cells and the linker gets degradation in acidic lysosomal environment. During the process, calicheamicin derivative is released and attacks DNA in AML cells [15, 16].

Marketing approval of GO was granted with phase II results by the US-FDA under the accelerated approval regulations in the year 2000. GO was indicated for the treatment of patients with CD33 positive AML in first relapse who are 60 years of age or older and who are not considered candidates for cytotoxic chemotherapy. The approved dose was 9 mg/m^2 IV over 4 h and repeated in 14 days with a 16% complete remission and a 30% overall response rate [17]. In contrast, in 14 phase II studies conducted after the approval, GO failed to show clinical benefit in first-line treatment in combination with standard chemotherapy. GO also showed fatal anaphylaxis, adult respiratory distress syndrome (ARDS), and hepatotoxicity, especially veno-occlusive disease (VOD) in AML patients [18]. The drug manufacturer then voluntarily withdrew the US New Drug Application in 2010, but GO remained available in Japan [19]. Recently, GO was re-approved for newly diagnosed and relapsed/refractory AML in US (2017) and Europe (2018), based on several investigator-led clinical trials including ALFA-0701 [20]. In the phase III ALFA-0701 trial, to reduce toxicity, investigators chose a new regimen based on the administration of GO at a lower dose of 3 mg/m^2 (maximum dose 5 mg) on days 1, 4, and 7 (3-3-3 split regimen). The rationale for lower doses was based on the responses to doses of 1–4 mg/m^2 in the initial phase I study and a saturation of more than 80% of the CD33 sites after dosing with 4 mg/m^2 or 6 mg/m^2 [20, 21]. The rationale for administration of fractionated doses was based on the rapid re-expression of CD33 molecules on the cell surface after a first exposure [15, 20, 22]. This new dosing schedule was associated with decreased incidence of hepatotoxicity, VOD and early mortality, suggesting selection of right dosing schedule considering pathophysiology of the target molecule is very important for ADC clinical development.

Inotuzumab Ozogamicin (BESPONSA®)

Inotuzumab ozogamicin is an anti-CD22 ADC composed of a humanized IgG4 antibody and the same drug-linker as Gemtuzumab ozogamicin (GO) (Fig. 1). DAR is 5–7 [23]. CD22 is a B-cell-specific transmembrane sialoglycoprotein and expressed on mature B cells and B cell malignancies including acute lymphoblastic leukemia (ALL), non-Hodgkin lymphoma (NHL), chronic lymphocytic leukemia, and hairy cell leukemia [4]. Inotuzumab ozogamicin showed dose-dependent efficacy in preclinical models [24, 25]. Compared to AML and NHL, ALL cell lines were more sensitive to inotuzumab ozogamicin, which indicated its potential use for treating relapsed ALL [4, 26].

The FDA approved inotuzumab ozogamicin based on data from an open-label, 2-arm, randomized global phase III INO-VATE ALL study [27]. In the trial, patients aged over 18 years with relapsed or refractory CD22 positive, Philladelphia chromosome (Ph)-positive or negative ALL (n = 326) were randomized (1:1) to inotuzumab ozogamicin or investigator's choice of standard therapies. Patients in the inotuzumab ozogamicin arm (n = 164) received a starting dose of 1.8 mg/m^2 per cycle (0.8 mg/m^2 on day 1; 0.5 mg/m^2 on days 8 and 15 of each cycle). Once complete remission (CR) or CR with incomplete hematologic recovery (CRi) was achieved, the day 1 dose was reduced to 0.5 mg/m^2. Patients with standard therapies (n = 162) received either FLAG (cytarabine, frudarabine, granulocyte-colony stimulating factor), cytarabine plus mitoxantrone, or high-dose cytarabine. Of the first 218 (inotuzumab ozogamicin, n = 109; standard, n = 109) of 326 patients randomized, complete remission rate was significant higher with inotuzumab ozogamicin than standard therapies (80.7% vs 29.4%; P < 0.001). Remission duration was longer (4.6 vs 3.1 months; P = 0.034) and progression-free survival was significantly longer (5.0 vs 1.8 months; P < 0.001) with inotuzumab ozogamicin versus standard. The median overall survival (OS) was 7.7 vs 6.7 months (P = 0.04) [4, 27].

The most frequent grade 3 or higher nonhematologic adverse events with inotuzumab ozogamicin were liver-related. VOD of any grade occurred in 15 patients (11%) who received inotuzumab ozogamicin and in 1 patient (1%) who received standard therapy [27]. VOD seems to be common risk by GO and inotuzumab ozogamicin. Further investigation is required to understand the mechanism of VOD and its relation to the calicheamicin payload.

ADC Constructs with Microtubule Inhibitor

Brentuximab Vedotin (ADCETRIS®)

Brentuximab vedotin is an anti-CD30 ADC conjugated with a tubulin polymerization inhibitor monomethyl auristatin E (MMAE) via a protease-cleavable dipeptide linker (Fig. 2) [6, 28]. DARis 4. Cathepsin B is one of the enzymes responsible for linker cleavage [29]. MMAE is a synthetic analog of the natural product dolastatin 10 from the marine shell-less mollusk *Dolabella auricularia* [30]. In cells, released MMAE induces G2/M phase growth arrest and apoptosis [6]. CD30 is a transmembrane glycoprotein highly expressed on several lymphoid neoplasms, such as Hodgkin's lymphoma (HL) and anaplastic large cell lymphoma (ALCL) [28].

Brentuximab vedotin was approved by the US FDA for the treatment of relapsed or refractory HL and systemic ALCL in 2011. In the phase III AETHERA trial [31], 329 patients were randomly assigned to the Brentuximab vedotin group (n = 165) or the placebo group (n = 164). PFS was significantly longer in patients in the brentuximab vedotin group. The median PFS was 42.9 months (95% CI: 30.4–42.9) and 24.1 months (95% CI: 11.5-NE) for the placebo (P = 0.0013). The most frequent

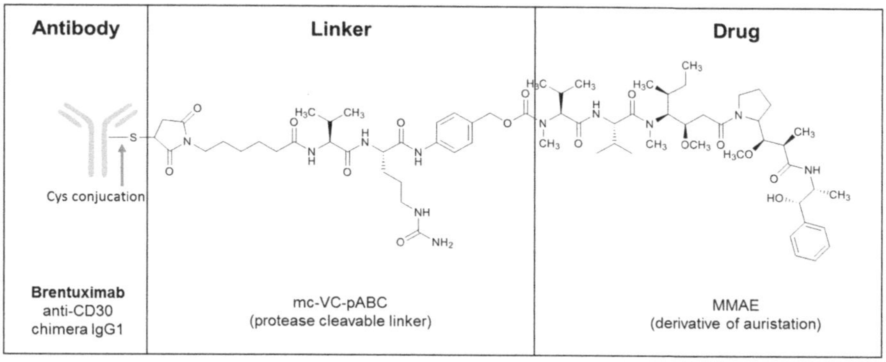

Average of drug-to-antibody ratio (DAR) is 4
MMAE = monomethyl auristatin E
mc-VC-pABC = maleimidocaproyl valine-citrulline *p*-aminobenzyloxycarbamoyl

Fig. 2 Schematic structure of brentuximab vedotin

adverse events in the brentxumab vedotin group were peripheral sensory neuropathy and neutropenia.

Recently, FDA further approved brentuximab vedotin to treat adult patients with previously untreated stage III or IV classical Hodgkin lymphoma (cHL) in combination with chemotherapy in 2018. Approval was based on a large, international, open-label, randomised, multicenter trial, ECHELON-1 [32], in which 664 were assigned to receive either Brentuximab vedotin plus doxorubicin, vinblastine, and dacarbazine (Adcetris + AVD) (n = 664) or doxorubicin, vinblastine, dacarbazine and bleomycin (ABVD) (n = 670). Patients were randomized to receive Brentuximab vedotin + AVD or ABVD on Days 1 and 15 of each 28-day cycle for up to 6 cycles. At a median follow-up of 24.9 months, 2-year modified PFS in the A + AVD and ABVD groups were 82.1% and 77.2%, respectively. The adverse events of A + AVD are more associated with neutropenia and neurotoxicity including peripheral sensory neuropathy than ABVD. The MMAE ADC is known to cause neurotoxicity in preclinical and clinical settings [33].

Ado-Trastuzumab Emtansine (KADCYLA®)

Ado-trastuzumab emtansine also known as T-DM1 (KADCYLA®) is an anti-HER2 ADC that is comprised of trastuzumab linked to a tubulin polymerization inhibitor maytansinoid DM1 *via* non cleavable SMCC (succinimidyl 4-(*N*-maleimidomethyl) cyclohexane-1-carboxylate) linker to lysine residue of the antibody (Fig. 3). DAR is 3.5 [8, 34].

In dose escalation of phase I study [35], maximum tolerated dose (MTD) of T-DM1 had been determined 3.6 mg/kg every three weeks (q3w) by dose-limiting toxicity (DLT) of thrombocytopenia. Two single-arm phase II study of T-DM1,

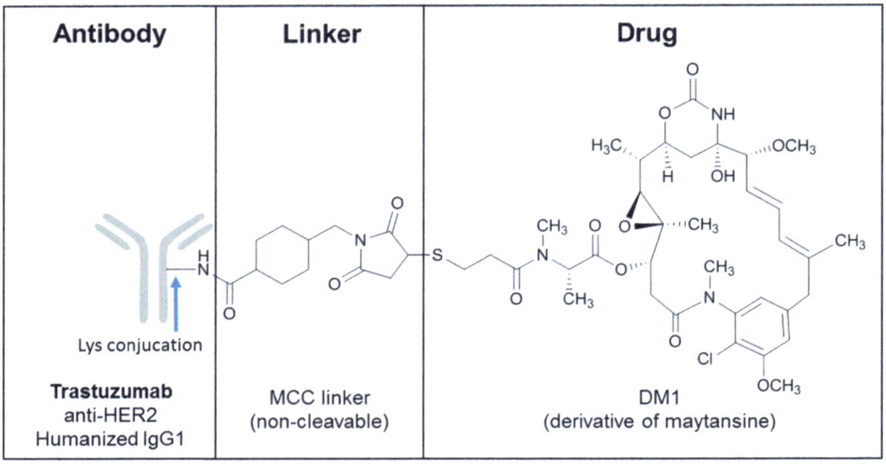

Average of drug-to-antibody ratio (DAR) is 3.5
DM1 = *N*2'-deacetyl-*N*2'-(3-mercapto-1-oxopropyl)-maytansine
MCC = 4-(*N*-maleimidylmethyl)cyclohexane-1-carbonyl

Fig. 3 Schematic structure of trastuzumab emtansine

TDM4258g [36] and TDM4374g [37], were conducted in patients with HER2-positive metastatic breast cancer (mBC). T-DM1 demonstrated antitumor efficacy in both studies. Among treated patients (n = 112 and 110), 29 (objective response rate (ORR) 25.9%) and 36 (ORR 34.5%) patients had an objective tumor response. And the median progression-free survival (PFS) as assessed by independent review was 4.6 and 6.9 months. The first ~~of~~ phase III trial was demonstrated superiority of T-DM1 compared with standard chemotherapy drug. The EMILIA [38], as pivotal study, was comparing T-DM1 to capecitabine plus lapatinib in nearly 1000 patients with HER2-positive advanced breast cancer, who had previously been treated with trastuzumab plus a taxane. The median PFS as assessed by independent review was 9.6 months with T-DM1 versus 6.4 months with lapatinib plus capecitabine (hazard ratio, 0.65; 95% confidence interval (CI), 0.55 to 0.77), and median overall survival (OS) at the second interim analysis crossed the stopping boundary for efficacy (30.9 months versus 25.1 months; hazard ratio, 0.68; 95% CI, 0.55 to 0.85). The ORR was higher with T-DM1 (43.6%, versus 30.8% with lapatinib plus capecitabine); results for all additional secondary end points favored T-DM1. Rates of adverse events of grade 3 or above were higher with lapatinib plus capecitabine than with T-DM1 (57% versus 41%). Grade 3 or 4 thrombocytopenia was more common toxicity in patients treated with T-DM1 (12.8%) but rarely led to hemorrhage. Transaminases were increased more frequently with T-DM1 as well, with grade 3 or 4 elevations occurring in less than 5% of study patients. Both thrombocytopenia and transaminase elevation were managed by dose interruptions and dose reductions, when necessary. Patients treated with T-DM1 had lower incidence of grade 3 or 4 diarrhea (1.6% versus 20.7%) and hand foot syndrome (0% versus 16.4%) than patients treated with lapatinib plus capecitabine. An application was submitted to the FDA for T-DM1 based on EMILIA data; T-DM1 was approved in 2013 as a treatment for patients with HER2-positive breast cancer.

ADCs in Late Stages of Clinical Development

There are more than 70 different ADCs in clinical development stage and different drug-linker systems from the approved ADCs have been applied to some ADCs. There are at least 6 ADCs in pivotal clinical trials (Table 1). Mirvetuximab soravtansine is targeting FOLR1 of ovarian cancer patients with ADC containing another microtubule inhibitor DM4 payload [39]. Rovalpituzumab tesirine is an anti-DLL3 ADC with DNA crosslinking agent pyrorrobenzodiazepin derivative for lung cancer patients [40]. Sacituzumab govitecan is an anti-TROP2 ADC for metastatic breast cancer patients with DNA topoisomerase I inhibitor SN-38, an active metabolite of irinotecan approved for some cancer indications [41]. Trastuzumab deruxtecan (DS-8201a) is a new HER2-targeting ADC with a novel DNA topoisomerase I

Table 1 ADCs which are commercialized or in pivotal studies

Name	Developer	Target	Linker-Drug	Status
Gemtuzumab Ozogamicin (MYLOTARG®/CMA-676)	Pfizer	CD33	Hydrazone–Calicheamicin	Launched
Brentuximab Vedotin (Adcetris®/SGN-35)	Seattle genetics/ Takeda	CD30	Vc–MMAE	Launched
Trastuzumab Emtansine (Kadcyla®/T-DM1)	Genentech/ Roche	HER2	SMCC-DM1	Launched
Inotuzumab Ozogamicin (Besponsa®/CMC-544)	Pfizer	CD22	Hydrazone–Calicheamicin	Launched
Mirvetuximab soravtansine (IMGN853)	ImmunoGen	FOLR1	Sulfo-SPDB–DM4	Phase3
Sacituzumab Govitecan (IMMU-132)	Immunomedics	TROP2	CL2A-SN38	Phase3
Polatuzumab Vedotin (RG7596)	Genentech/ Roche	CD79b	Vc–MMAE	Phase3
Rovalpituzumab tesirine (Rova-T, SC16LD6.5)	AbbVie	DLL3	PEG8-va–SG3199 (PBD)	Phase3
Trastuzumab duocarmazine (SYD985)	Synthon	HER2	Vc–seco-DUBA	Phase3
Trastuzumab deruxtecan (DS-8201a)	Daiichi Sankyo	HER2	Peptide-based linker-DXd	Phase3

HER2 human epidermal growth factor receptor 2, *FOLR1* folate receptor 1, *TROP2* Trophoblast cell surface antigen 2, *DLL3* Delta-like protein 3, *vc* valine-citrulline, *MMAE* monomethyl auristatin E, *SMCC* succinimidyl-4-(N-maleimidomethyl)cyclohexane-1-carboxylate, *DM1 and DM4* maytansine derivatives, *sulfo-SPDB* N-hydroxysuccinimidyl 4-(2-pyridyldithio)-2-sulfobutanoate, *SN-38* the active metabolite of irinotecan, *PEG8* polyethylene glycol 8, *PBD* pyrrolobenzodiazepine, *seco-DUBA* seco-duocarmycin-hydroxybenzamide-azaindole, *DXd* an exatecan derivative

inhibitor exatecan derivative (DXd) and proteinase sensitive cleavable linker [42, 43]. A different HER2 ADC trastuzumab duocarmycin (SYD985) has another DNA binding agent duocarmycin derivative as ADC payload [44]. Some of the ADCs are expected to be approved within a few years, widening the therapeutic options for cancer patients.

ADC Constructs with DNA Topoisomerase I Inhibitor

Sacituzumab Govitecan (IMMU-132)

Sactituzumab govitecan (IMMU-132) is an ADC targeting human trophoblast cell surface protein 2 (TROP2). IMMU-132 contains SN-38, the active metabolite of irinotecan (CPT-11) of DNA topoisomerase I inhibitor linked by pH sensitive cleavable linker technology (CL2A) (Fig. 4) [41]. Irinotecan was firstly approved in 1994 and has been used as cancer medicine for several target indications such as gastric cancer, colorectal cancer and lymphoma [45, 46]. TROP2 is a cell-surface receptor while over-expressed by a wide range of solid tumors such as breast, colon and lung, and has limited expression in normal human tissues [47]. A humanized anti-TROP2 monoclonal antibody (sactituzumab or hRS7) internalizes into cancer cells following binding to TROP2 for the delivery of cytotoxic drugs. The average DAR of IMMU-132 was 7 to 8 mixed with DARs of 6, 7, and 8 (average DAR is 7.6), that evaluated by hydrophobic interaction chromatography (HIC) (Fig. 5) [48, 49]. IMMU-132 releases SN-38 with a shorter half-life than one day in human serum by the cleavage at the pH-sensitive benzyl carbonate in CL2A linker [48, 49]. It is hypothesized that using a less stable linker allows for SN-38 to be released at the tumor site after the ADC targets the cells, making the drug accessible to surrounding tumor cells and not just cells directly targeted by the ADC. Therefore, it is hypothesized that IMMU-132 have antitumor activity against a wide range of tumor types.

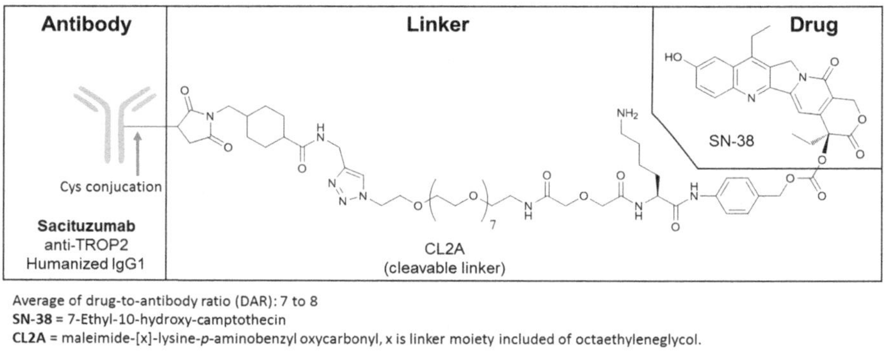

Fig. 4 Schematic structure of sacituzumab govitecan

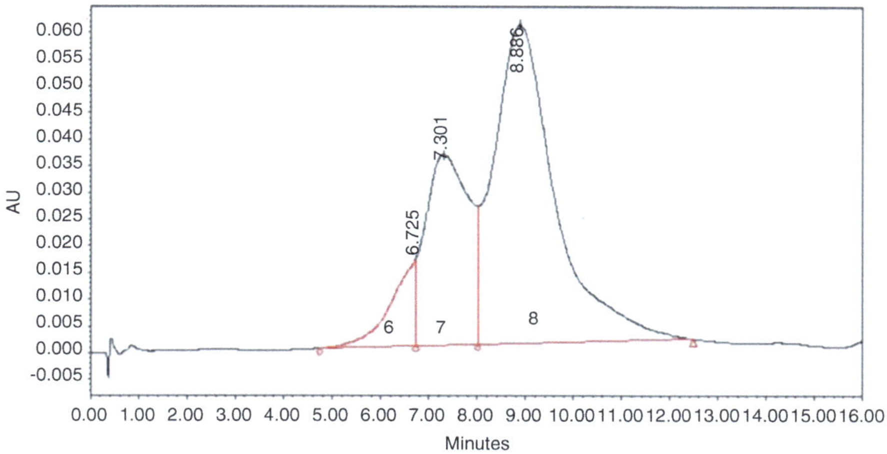

Fig. 5 The conjugated drug distribution by hydrophobic interaction chromatography. (Reproduced from Goldenberg et al. [49])

In dose-escalation phase I trial [50], IMMU-132 was administered on days 1 and 8 of 21-day cycles, with cycles repeated until dose-limiting toxicity or progression. Dose escalation followed a standard 3 + 3 scheme with 4 planned dose levels and dose delay or reduction allowed. Twenty-five patients were treated at dose levels of 8 (n = 7), 10 (n = 6), 12 (n = 9), and 18 (n = 3) mg/kg. Neutropenia was dose limiting, with 12 mg/kg the maximum tolerated dose (MTD) for cycle one, but too toxic with repeated cycles. Two patients achieved partial responses (triple-negative breast cancer, colon cancer) and 16 others had stable disease as best response. Twelve patients maintained disease control with continued treatment for 16–36 weeks; 6 survived 15 to over 20 months. No preselection of patients based on tumor TROP2 expression was done. IMMU-132 had acceptable toxicity and encouraging therapeutic activity in patients with difficult-to-treat cancers. The 8 mg/kg and 10 mg/kg doses were selected for further investigation.

In a single-arm, multicenter trial in patients with relapsed/refractory metastatic triple negative breast cancer (TNBC) who received a 10 mg/kg starting dose (69 patients) [51], the confirmed objective response rate (ORR) was 30% (partial response (PR), n = 19; complete response (CR), n = 2) (Fig. 6a), the median progression-free survival (PFS) was 6.0 (95% CI, 5.0 to 7.3) months (Fig. 6b), and median overall survival (OS) was 16.6 (95% CI, 11.1 to 20.6) months (Fig. 6c). Common grade 3 or above adverse events included neutropenia (39%), leukopenia (16%), anemia (14%), and diarrhea (13%); the incidence of febrile neutropenia was 7%. No neutralizing antibodies to the ADC or antibody were detected. IMMU-132 was well tolerated and induced early and durable responses in heavily pretreated patients with metastatic TNBC.

IMMU-132 has received Breakthrough Therapy designation from the FDA for the treatment of patients with triple-negative breast cancer (TNBC) who have failed prior therapies for metastatic disease. In July 2018, the US FDA grants Priority

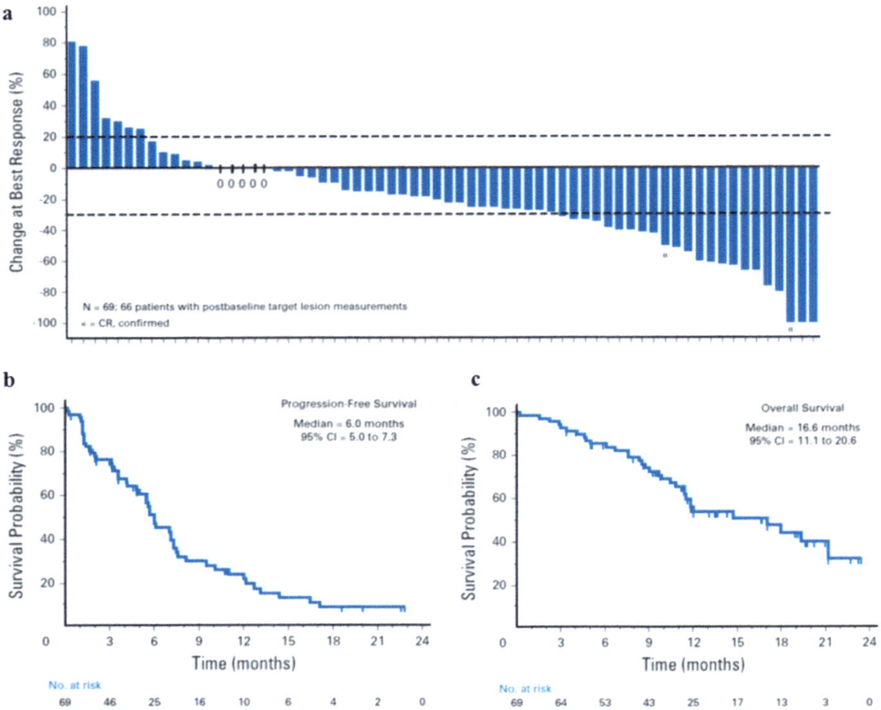

Fig. 6 (**a**) A waterfall plot that shows the percentage of change in sum of target lesions at the time of best response (**b**) Progression-free Survival b Overall survival. (Reproduced from Bardia et al. [7])

Review for sacituzumab govitecan for this population. The regulatory agency has also granted this ADC Fast Track designation for patients with TNBC, and for patients with small-cell lung cancer (SCLC) [52] or non-small-cell lung cancer (NSCLC) [53].

[Fam-] Trastuzumab Deruxtecan (DS-8201a)

[fam-] Trastuzumab deruxtecan (also known as DS-8201a, Fig. 7) is a HER2-targeting antibody drug conjugate (ADC) in development for HER2 expressing solid tumors [10, 42, 43]. DS-8201a is engineered through conjugation of a humanized anti-HER2 antibody with a topoisomerase I inhibitor payload using an enzymatically cleavable peptide (Gly-Gly-Phe-Gly)-based linker [42, 43]. The HER2 antibody is a humanized IgG1 monoclonal antibody produced with reference to the amino acid sequence of trastuzumab; thus DS-8201a binding to HER2 is similar to that of trastuzumab [42, 43]. The payload is the derivative of exatecan (DXd) which

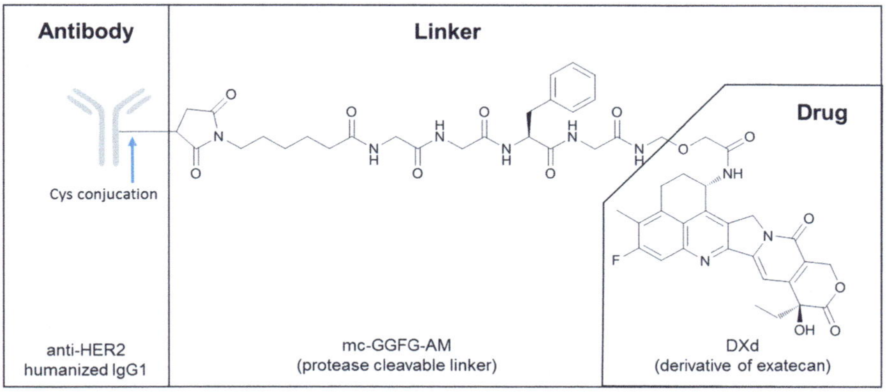

Average of drug-to-antibody ratio (DAR): 7 to 8
DXd = 2-hydroxyacetyl exatecan
mc-GGFG-AM= maleimidocaproyl glycine-glycine-phenylalanine-glycine-aminomethylene

Fig. 7 Schematic structure of trastuzumab deruxtecan

is almost 10-times more potent than the topoisomerase I inhibitor SN-38, the active metabolite of irinotecan [42]. Due to incorporation of the novel drug-linker, DS-8201a achieves a higher drug-to-antibody ratio (DAR) of 7 –8 with homogeneous conjugation compared to all currently approved ADCs [42]. Despite the high DAR, the drug-linker in DS-8201a is highly stable in plasma to confer favorable pharmacokinetic and safety profiles [42]. Furthermore, preclinical studies suggest a bystander effect for DS-8201a likely due to the fact that payload is cell membrane permeable after it is cleaved in HER2 expressing tumors; this effect is not observed with T-DM1 [43]. These features may contribute to potent antitumor activity in a broad range of tumors including those with low levels of HER2 expression and in a heterogeneous tumor environment. Compared to T-DM1 which is the only approved ADC against a solid tumor, HER2-positive breast cancer, DS-8201a has a different mechanism of cytotoxic action (topoisomerase I inhibitor vs. tubulin polymerization inhibitor), higher DAR (7 –8 vs. 3.5) with high plasma stability, and a bystander cytotoxic activity. Based on these differentiating features, the ADC DS-8201a is anticipated to be effective even in tumors that do not respond, or are resistant, to T-DM1.

Preclinical Studies

The DAR of DS-8201a was determined as approximately 8, which is theoretically the maximum drug loading number for conventional interchain cysteine conjugation. Therefore, drug distribution was homogeneous (Fig. 8) [42]. Despite the high DAR, DS-8201a showed a favorable pharmacokinetic profile in cynomolgus monkeys [42]. When DS-8201a was administered at 3 mg/kg intravenously, there was no clear difference between DS-8201a and the total antibody and DXd was detected

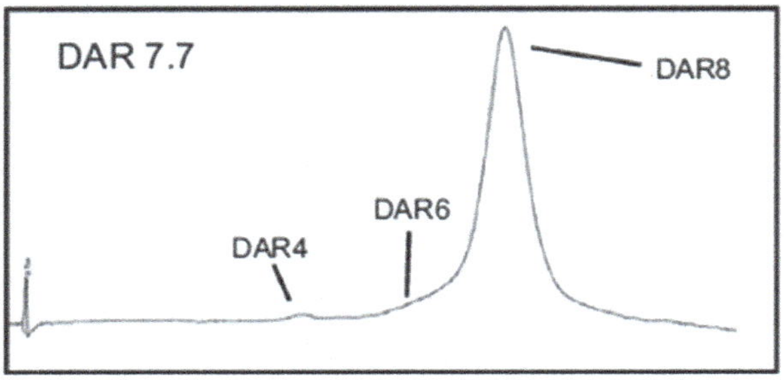

Fig. 8 The conjugated drug distribution by hydrophobic interaction chromatography. (Reproduced from Ogitani et al. [42])
The drug-to-antibody ratio (DAR) of DS-8201a was determined as approximately 8 by the reverse-phase chromatographic analysis. Homogeneous conjugation was observed in the hydrophobic interaction chromatography chart.

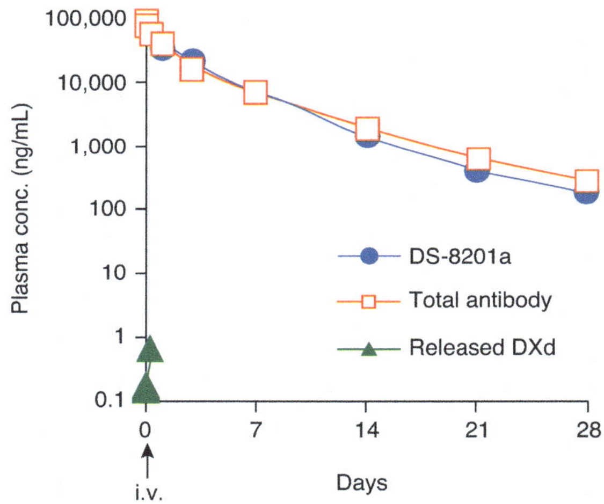

Fig. 9 Pharmacokinetics of DS-8201a in cynomolgus monkeys. (Reproduced from Ogitani et al. [42])
Mean plasma concentrations of DS-8201a, total antibody, and DXd after single intravenous administration of DS-8201a at 3 mg/kg to cynomolgus monkeys (n = 3)

only at the limited time points with much lower level compared to DS-8201a, indicating that the drug-linker of DS-8201a is stable in plasma (Fig. 9) [42].

Overexpression of HER2 [immunohistochemistry (IHC) 3+ or IHC 2+/ in-situ hybridization (ISH) +] is observed in 10–20% of breast cancer cases. In addition, low levels of HER2 (IHC 2+/ISH- or IHC1+) were reported to be around 50% in breast cancer [54]. T-DM1 is known to show antitumor efficacy against HER2-positive breast cancer but not against HER2 low expressing tumors [36, 42, 55]. On the other hand, DS-8201a showed preliminary evidence of activity in HER2 low expressing tumors while T-DM1 and a lower DAR HER2 ADC (DAR3.4) with the same drug-linker did not (Fig. 10) [42]. DS-8201a also showed antitumor efficacy in heterogeneous tumors while T-DM1 did not (Fig. 11) [43]. The potent antitumor efficacy of DS-8201a was confirmed in several patient-derived xenograft models

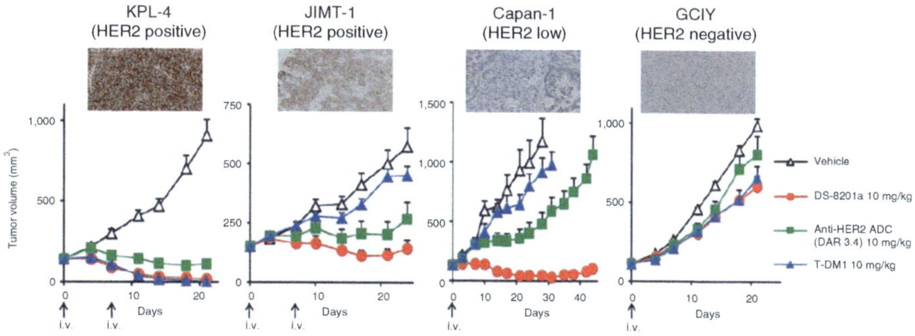

Fig. 10 Antitumor activity of DS-8201a against tumors with various HER2 expression levels. (Reproduced from Ogitani et al. [42])
Antitumor activity of DS-8201a was evaluated in mouse xenograft models in comparison with T-DM1 and DS-8201a with DAR 3.4 [HER2-ADC (DAR 3.4)]

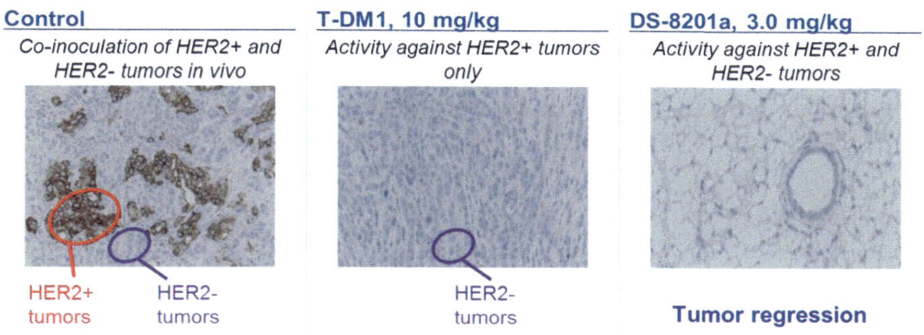

Fig. 11 Bystander cytotoxic activity of DS-8201a. (Reproduced from Ogitani et al. [43])
Mice were inoculated with a mixture of HER2 positive and HER2 negative cancer cells and received 10 mg/kg T-DM1 or 3 mg/kg DS-8201a 7 days after inoculation. The tumors were collected and HER2 status was assessed by immunohistochemisty using HercepTest II kit 14 days after treatment

including breast and gastric cancers as well [42]. DS-8201a was well tolerated in rats as the non-cross reactive species and cynomolgus monkeys as the cross reactive species. When DS-8201a was administered at doses from 20 to 197 mg/kg for rats and from 10 to 78.8 mg/kg for monkeys every 3 weeks for 3 times in total, the severely toxic dose of 10% in animals (STD10) was >197 mg/kg in rats and the highest non-severely toxic dose (HNSTD) was 30 mg/kg in monkeys (Table 2) [42].

In conclusion, DS-8201a showed potent antitumor efficacy in preclinical models with favorable pharmacokinetic and safety profiles. The preclinical safety profile was acceptable for entry into human trials.

Table 2 Summary of repeated dose toxicity studies in rats and monkeys

Species	Crl:CD(SD) rats	Cynomolgus monkeys
Doses	0, 20, 60 and 197 mg/kg	0, 10, 30 and 78.8 mg/kg
Regimens	Intravenous, every 3 weeks Days 1, 22, 43 (3 times in total)	Intravenous, every 3 weeks Days 1, 22, 43 (3 times in total)
No. of animals	10/sex/group (Main): All doses 5/sex/group (recovery): 60 and 197 mg/kg	3/sex/group (Main): All doses 2/sex/group (recovery): 30 and 78.8 mg/kg
Lethal dose	>197 mg/kg	78.8 mg/kg (1 female died)
Body weight	197 mg/kg: Low body weight gain	78.8 mg/kg: Decreased in 1 male and 1 female
Hematology	60 mg/kg: Decreased RBC and WBC parameters	78.8 mg/kg: Decreased RBC parameters
Target organs and tissues	20 mg/kg: Intestines, testis 60 mg/kg: Bone marrow, thymus, lymph node, skin, kidney, incisor	10 mg/kg: Intestines 30 mg/kg: Lung, skin, testis 78.8 mg/kg: Bone marrow, kidney
STD_{10}/HNSTD	STD_{10}: >197 mg/kg	HNSTD: 30 mg/kg

Reproduced from Ogitani Y et al. (2016) Clin Cancer Res 22: 5097
RBC red blood cell; *WBC* white blood cell; STD_{10} the severely toxic dose of 10% in animals; *HNSTD* the highest non-severely toxic dose

Clinical Studies

Phase I Study

In a dose-escalation phase I trial, 24 participants received initial intravenous doses of DS-8201a from 0.8 to 8.0 mg/kg (n = 3 for each of 0.8, 1.6, 3.2, and 8.0 mg/kg doses; n = 6 for each of 5.4 and 6.4 mg/kg) and dose-limiting toxicities were assessed over a 21-day cycle [10]. No dose-limiting toxicity was observed and the maximum tolerated dose was not reached. The most common adverse events were mild or moderate gastrointestinal and hematological events [10].

Analysis of pharmacokinetic properties showed a non-linear pharmacokinetic profile and increased $t_{1/2}$ at higher doses. Serum concentrations of DS-8201a, DXd, and total antibody versus time after treatment with 6.4 mg/kg of DS-8201a are shown in Fig. 12. Low concentrations of free DXd were observed and total antibody concentrations were similar to DS-8201a concentrations at all time points assessed after administration of 6.4 mg/kg, which suggests high stability of the drug-linker in the blood [10].

On the basis of the balance between activity and safety, the 5.4 mg/kg and 6.4 mg/kg DS-8201a doses were chosen for further investigation in the dose expansion phase of the ongoing study. In the dose expansion part of the phase I study, DS-8201a is given to patients with HER2-positive advanced or metastatic breast cancer or gastric cancer, HER2 low expressing breast cancer or other HER2 expressing or mutant solid tumors.

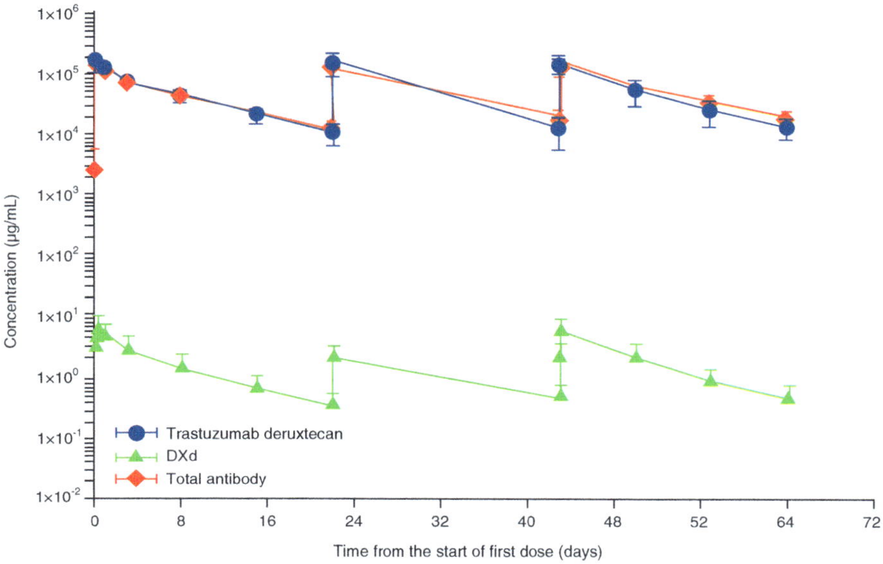

Fig. 12 Pharmacokinetic profile of DS-8201a (trastuzumab deruxtecan) (Reproduced from Doi et al. [10])
Mean serum concentration versus time curve of DS-8201a (trastuzumab deruxtecan), DXd, and total antibody over cycles 1–3 at a dose of 6.4 mg/kg. Error bars indicate standard deviation

Based on interim analysis reported at 2018 American Society of Clinical Oncology (ASCO) Annual Meeting [56], a 54.5% confirmed overall response rate (ORR) and a 93.9% disease control rate (DCR) was demonstrated by DS-8201a in updated subgroup analysis of 99 patients with HER2-positive metastatic breast cancer pretreated with trastuzumab and T-DM1 (as well as pertuzumab in the majority of cases). Furthermore, updated subgroup analysis in 34 heavily pretreated patients with HER2-low-expressing metastatic breast cancer, defined as IHC 2+/ISH- and IHC 1+, demonstrated a 50.0% confirmed ORR and an 85.3% DCR with DS-8201a. In the subgroup of 44 patients with HER2 positive gastric cancer or gastroesophageal junction adenocarcinoma previously treated with trastuzumab and chemotherapy, DS-8201a demonstrated a confirmed ORR of 43.2% (19/44 patients) and a DCR of 79.5% (35/44 patients). In addition, confirmed responses were observed across a range of other HER2-expressing or HER2-mutant solid tumors such as NSCLC, CRC, salivary gland carcinomas, and other tumor types.

Phase II Studies

Based on the phase I data, several clinical trials have been initiated to assess the efficacy and safety of DS-8201a in subjects with various HER2 expressing cancers. DS-8201a is currently in pivotal phase II clinical development for HER2 positive unresectable and/or metastatic breast cancer resistant or refractory to T-DM1 (DESTINY-Breast01, NCT03248492) in North America, Europe and Asia, and pivotal phase II development for HER2 positive advanced gastric cancer resistant or refractory to trastuzumab (DESTINY-Gastric01, NCT03329690) in Japan and South Korea. In addition to these pivotal phase II studies, global phase II studies in patients with advanced HER2 expressing colorectal cancer (NCT03384940), and advanced HER2 overexpressing or HER2 mutated non-small cell lung cancer are ongoing (NCT03505710) as well as a Phase Ib combination study with anti-PD1 antibody nivolumab.

DS-8201a has been granted Breakthrough Therapy designation for the treatment of patients with HER2-positive, locally advanced or metastatic breast cancer who have been treated with trastuzumab and pertuzumab and have disease progression after T-DM1, and Fast Track designation for the treatment of HER2-positive unresectable and/or metastatic breast cancer in patients who have progressed after prior treatment with HER2 targeted therapies including T-DM1 by the US FDA. DS-8201a has also received SAKIGAKE designation for the treatment of HER2 positive advanced gastric or gastroesophageal junction cancer by the Japan Ministry of Health, Labour and Welfare (MHLW).

Future Directions

In recent years, there have been many efforts to develop ADCs with adequate therapeutic window to establish a positive benefit/risk. Four ADCs have been approved and some promising ADC drug candidates are in late stage of clinical development. Selection of relevant target antigen, antibody with efficient drug delivery potential, potent payload and stable linker is essential for successful design to create therapeutic ADCs. Next generation of ADC research will also focus on antibody engineering to confer targeting with tumor specificity. The cumulative knowledge within the ADC scientific community and broader oncology drug development has led to fruitful advances in this field with the hope of further improving the outcomes of cancer patients using this promising class of drugs.

References

1. Lambert JM, Morris CQ (2017) Antibody-drug conjugates (ADCs) for personalized treatment of solid tumors: a review. Adv Ther 34:1015–1035
2. de Goeij BE, Lambert JM (2016) New developments for antibody-drug conjugate-based therapeutic approaches. Curr Opin Immunol 40:14–23
3. Trail PA, Dubowchik GM, Lowinger TB (2018) Antibody drug conjugates for treatment of breast cancer: novel targets and diverse approaches in ADC design. Pharmacol Ther 181:126–142

4. Uy N, Nadeau M, Stahl M, Zeidan AM (2018) Inotuzumab ozogamicin in the treatment of relapsed/refractory acute B cell lymphoblastic leukemia. J Blood Med 9:67–74
5. Ricart AD (2011) Antibody-drug conjugates of calicheamicin derivative: gemtuzumab ozogamicin and inotuzumab ozogamicin. Clin Cancer Res 17(20):6417–6427
6. Francisco JA, Cerveny CG, Meyer DL, Mixan BJ, Klussman K, Chase DF, Rejniak SX, Gordon KA, DeBlanc R, Toki BE, Law CL, Doronina SO, Spiegall CB, Senter PD, Wahl AF (2003) cAC10-vcMMAE, an anti-CD30-monomethyl auristatin E conjugate with potent and selective antitumor activity. Blood 102:1458–1465
7. Younes A, Bartlett NL, Leonard JP, Kennedy DA, Lynch CM, Sievers EL, Forero-Torres A (2010) Brentuximab vedotin (SGN-35) for relapsed CD30-positive lymphomas. N Engl J Med 363(19):1812–1821
8. LoRusso PM, Weiss D, Guardino E, Girish S, Sliwkowski MX (2011) Trastuzumab emtansine: a unique antibody-drug conjugate in development for human epidermal growth factor receptor 2-positive cancer. Clin Cancer Res 17(20):6437–6447
9. Bardia A, Mayer IA, Diamond JR, Moroose RL, Isakoff SJ, Starodub AN, Shah NC, O'Shaughnessy J, Kalinsky K, Guarino M, Abramson V, Juric D, Tolaney SM, Berlin J, Messersmith WA, Ocean AJ, Wegener WA, Maliakal P, Sharkey RM, Govindan SV, Goldenberg DM, Vahdat LT (2017) Efficacy and safety of anti-Trop-2 antibody drug conjugate Sacituzumab Govitecan (IMMU-132) in heavily pretreated patients with metastatic triple-negative breast cancer. J Clin Oncol 35(19):2141–2148
10. Doi T, Shitara K, Naito Y, Shimomura A, Fujiwara Y, Yonemori K, Shimizu C, Shimoi T, Kuboki Y, Matsubara N, Kitano A, Jikoh T, Lee C, Fujisaki Y, Ogitani Y, Yver A, Tamura K (2017) Safety, pharmacokinetics, and antitumour activity of trastuzumab deruxtecan (DS-8201), a HER2-targeting antibody-drug conjugate, in patients with advanced breast and gastric or gastro-oesophageal tumours: a phase 1 dose-escalation study. Lancet Oncol 18(11):1512–1522
11. Hamann PR, Hinman LM, Hollander I, Beyer CF, Lindh D, Holcomb R, Hallett W, Tsou HR, Upeslacis J, Shochat D, Mountain A, Flowers DA, Bernstein I (2002) Gemtuzumab ozogamicin, a potent and selective anti-CD33 antibody-calicheamicin conjugate for treatment of acute myeloid leukemia. Bioconjug Chem 13(1):47–58
12. Lee MD, Manning JK, Williams DR, Kuck NA, Testa RT, Borders DB (1989) Calicheamicins, a novel family of antitumor antibiotics. 3. Isolation, purification and characterization of calicheamicins beta 1Br, gamma 1Br, alpha 2I, alpha 3I, beta 1I, gamma 1I and delta 1I. J Antibiot (Tokyo) 42(7):1070–1087
13. Maiese WM, Lechevalier MP, Lechevalier HA, Korshalla J, Kuck N, Fantini A, Wildey MJ, Thomas J, Greenstein M (1989) Calicheamicins, a novel family of antitumor antibiotics: taxonomy, fermentation and biological properties. J Antibiot (Tokyo) 42(4):558–563
14. Ikemoto N, Kumar RA, Ling TT, Ellestad GA, Danishefsky SJ, Patel DJ (1995) Calicheamicin-DNA complexes: warhead alignment and saccharide recognition of the minor groove. Proc Natl Acad Sci USA 92(23):10506–10510
15. van der Velden VH, te Marvelde JG, Hoogeveen PG, Bernstein ID, Houtsmuller AB, Berger MS, van Dongen JJ (2001) Targeting of the CD33-calicheamicin immunoconjugate Mylotarg (CMA-676) in acute myeloid leukemia: in vivo and in vitro saturation and internalization by leukemic and normal myeloid cells. Blood 97(10):3197–3204
16. Godwin CD, Gale RP, Walter RB (2017) Gemtuzumab ozogamicin in acute myeloid leukemia. Leukemia 31(9):1855–1868
17. Bross PF, Beitz J, Chen G, Chen XH, Duffy E, Kieffer L, Roy S, Sridhara R, Rahman A, Williams G, Pazdur R (2001) Approval summary: gemtuzumab ozogamicin in relapsed acute myeloid leukemia. Clin Cancer Res 7(6):1490–1496
18. McKoy JM, Angelotta C, Bennett CL, Tallman MS, Wadleigh M, Evens AM, Kuzel TM, Trifilio SM, Raisch DW, Kell J, DeAngelo DJ, Giles FJ (2007) Gemtuzumab ozogamicin-associated sinusoidal obstructive syndrome (SOS): an overview from the research on adverse drug events and reports (RADAR) project. Leuk Res 31(5):599–604

19. Tanimoto T, Tsubokura M, Mori J, Pietrek M, Ono S, Kami M (2013) Differences in drug approval processes of 3 regulatory agencies: a case study of gemtuzumab ozogamicin. Investig New Drugs 31(2):473–478
20. Castaigne S, Pautas C, Terré C, Raffoux E, Bordessoule D, Bastie JN, Legrand O, Thomas X, Turlure P, Reman O, de Revel T, Gastaud L, de Gunzburg N, Contentin N, Henry E, Marolleau JP, Aljijakli A, Rousselot P, Fenaux P, Preudhomme C, Chevret S, Dombret H, Acute Leukemia French Association (2012) Effect of gemtuzumab ozogamicin on survival of adult patients with de-novo acute myeloid leukaemia (ALFA-0701): a randomised, open-label, phase 3 study. Lancet 379(9825):1508–1516
21. Sievers EL, Appelbaum FR, Spielberger RT, Forman SJ, Flowers D, Smith FO, Shannon-Dorcy K, Berger MS, Bernstein ID (1999) Selective ablation of acute myeloid leukemia using antibody-targeted chemotherapy: a phase I study of an anti-CD33 calicheamicin immunoconjugate. Blood 93(11):3678–3684
22. Linenberger ML (2005) CD33-directed therapy with gemtuzumab ozogamicin in acute myeloid leukemia: progress in understanding cytotoxicity and potential mechanisms of drug resistance. Leukemia 19(2):176–182
23. DiJoseph JF, Armellino DC, Boghaert ER, Khandke K, Dougher MM, Sridharan L, Kunz A, Hamann PR, Gorovits B, Udata C, Moran JK, Popplewell AG, Stephens S, Frost P, Damle NK (2004) Antibody-targeted chemotherapy with CMC-544: a CD22-targeted immunoconjugate of calicheamicin for the treatment of B-lymphoid malignancies. Blood 103(5):1807–1814
24. DiJoseph JF, Dougher MM, Armellino DC, Evans DY, Damle NK (2007) Therapeutic potential of CD22-specific antibody-targeted chemotherapy using inotuzumab ozogamicin (CMC-544) for the treatment of acute lymphoblastic leukemia. Leukemia 21(11):2240–2245
25. DiJoseph JF, Goad ME, Dougher MM, Boghaert ER, Kunz A, Hamann PR, Damle NK (2004) Potent and specific antitumor efficacy of CMC-544, a CD22-targeted immunoconjugate of calicheamicin, against systemically disseminated B-cell lymphoma. Clin Cancer Res 10(24):8620–8629
26. de Vries JF, Zwaan CM, De Bie M, Voerman JS, den Boer ML, van Dongen JJM, van der Velden VHJ (2012) The novel calicheamicin-conjugated CD22 antibody inotuzumab ozogamicin (CMC-544) effectively kills primary pediatric acute lymphoblastic leukemia cells. Leukemia 26(2):255–264
27. Kantarjian HM, DeAngelo DJ, Stelljes M, Martinelli G, Liedtke M, Stock W, Gökbuget N, O'Brien S, Wang K, Wang T, Paccagnella ML, Sleight B, Vandendries E, Advani AS (2016) Inotuzumab Ozogamicin versus standard therapy for acute lymphoblastic leukemia. N Engl J Med 375(8):740–753
28. van der Weyden CA, Pileri SA, Feldman AL, Whisstock J, Prince HM (2017) Understanding CD30 biology and therapeutic targeting: a historical perspective providing insight into future directions. Blood Cancer J 7(9):e603
29. Sanderson RJ, Hering MA, James SF, Sun MMC, Doronina SO, Siadak AW, Senter PD, Wahl AF (2005) In vivo drug-linker stability of an anti-CD30 dipeptide-linked auristatin immunoconjugate. Clin Cancer Res 11:843–852
30. Bai R, Pettit GR, Hamel E (1990) Dolastatin 10, a powerful cytostatic peptide derived from a marine animal. Inhibition of tubulin polymerization mediated through the vinca alkaloid binding domain. Biochem Pharmacol 39(12):1941–1949
31. Moskowitz CH, Nademanee A, Masszi T, Agura E, Holowiecki J, Abidi MH, Chen AI, Stiff P, Gianni AM, Carella A, Osmanov D, Bachanova V, Sweetenham J, Sureda A, Huebner D, Sievers EL, Chi A, Larsen EK, Hunder NN, Walewski J, AETHERA Study Group (2015) Brentuximab vedotin as consolidation therapy after autologous stem-cell transplantation in patients with Hodgkin's lymphoma at risk of relapse or progression (AETHERA): a randomised, double-blind, placebo-controlled, phase 3 trial. Lancet 385(9980):1853–1862
32. Connors JM, Jurczak W, Straus DJ, Ansell SM, Kim WS, Gallamini A, Younes A, Alekseev S, Illés Á, Picardi M, Lech-Maranda E, Oki Y, Feldman T, Smolewski P, Savage KJ, Bartlett NL, Walewski J, Chen R, Ramchandren R, Zinzani PL, Cunningham D, Rosta A, Josephson NC,

Song E, Sachs J, Liu R, Jolin HA, Huebner D1, Radford J, ECHELON-1 Study Group (2018) Brentuximab Vedotin with chemotherapy for stage III or IV Hodgkin's lymphoma. N Engl J Med 378(4):331–344
33. Masters JC, Nickens DJ, Xuan D, Shazer RL, Amantea M (2016) Clinical toxicity of antibody drug conjugates: a meta-analysis of payloads. Investig New Drugs 36(1):121–135
34. Lewis-Phillips GD, Dugger DL, Crocker LM, Parsons KL, Mai E, Blattler WA, Lambert JM, Chari RVJ, Lutz RJ, Wong WLT, Jacobson FS, Koeppen H, Schwall RH, Kenkare-Mitra SR, Spencer SD, Sliwkowski MX (2008) Targeting HER2-positive breast cancer with Trastuzumab-DM1, an antibody–cytotoxic drug conjugate. Cancer Res 68:9280–9290
35. Krop IE, Beeram M, Modi S, Jones SF, Holden SN, Yu W, Girish S, Tibbitts J, Yi JH, Sliwkowski MX, Jacobson F, Lutzker SG, Burris HA (2010) Phase I study of Trastuzumab-DM1, an HER2 antibody-drug conjugate, given every 3 weeks to patients with HER2-positive metastatic breast Cancer. J Clin Oncol 28:2698–2704
36. Burris HA 3rd, Hope S, Rugo HS, Vukelja SJ, Vogel CL, Borson RA, Limentani S, Tan-Chiu E, Krop IE, Michaelson RA, Girish S, Amler L, Zheng M, Chu YW, Klencke B, O'Shaughnessy JA (2011) Phase II study of the antibody drug conjugate Trastuzumab-DM1 for the treatment of human epidermal growth factor receptor 2 (HER2) –positive breast cancer after prior HER2-directed therapy. J Clin Oncol 29:398–405
37. Krop IE, LoRusso P, Miller KD, Modi S, Yardley D, Rodriguez G, Guardino E, Lu M, Zheng M, Girish S, Amler L, Winer EP, Rugo HS (2012) A phase II study of Trastuzumab Emtansine in patients with human epidermal growth factor receptor 2–positive metastatic breast cancer who were previously treated with Trastuzumab, Lapatinib, an Anthracycline, a Taxane, and Capecitabine. J Clin Oncol 30:3234–3241
38. Verma S, Miles D, Gianni L, Krop IE, Welslau M, Baselga J, Pegram M, Oh DY, Diéras V, Guardino E, Fang L, Lu MW, Olsen S, Blackwell K (2012) Trastuzumab Emtansine for HER2-positive advanced breast cancer. N Engl J Med 367:1783–1791
39. Ab O, Whiteman KR, Bartle LM, Sun X, Singh R, Tavares D, LaBelle A, Payne G, Lutz RJ, Pinkas J, Goldmacher VS, Chittenden T, Lambert JM (2015) IMGN853, a folate receptor-α (FRα)-targeting antibody-drug conjugate, exhibits potent targeted antitumor activity against FRα-expressing tumors. Mol Cancer Ther 14(7):1605–1613
40. Rudin CM, Pietanza MC, Bauer TM, Ready N, Morgensztern D, Glisson BS, Byers LA, Johnson ML, Burris HA 3rd, Robert F, Han TH, Bheddah S, Theiss N, Watson S, Mathur D, Vennapusa B, Zayed H, Lally S, Strickland DK, Govindan R, Dylla SJ, Peng SL, Spigel DR, SCRX16–001 investigators (2017) Rovalpituzumab tesirine, a DLL3-targeted antibody-drug conjugate, in recurrent small-cell lung cancer: a first-in-human, first-in-class, open-label, phase 1 study. Lancet Oncol 18(1):42–51
41. Moon SJ, Govindan SV, Cardillo TM, D'Souza CA, Hansen HJ, Goldenberg DM (2008) Antibody conjugates of 7-ethyl-10-hydroxycamptothecin (SN-38) for targeted cancer chemotherapy. J Med Chem 51(21):6916–6926
42. Ogitani Y, Aida T, Hagihara K, Yamaguchi J, Ishii C, Harada N, Soma M, Okamoto H, Oitate M, Arakawa S, Hirai T, Atsumi R, Nakada T, Hayakawa I, Abe Y, Agatsuma T (2016) DS-8201a, a novel HER2-targeting ADC with a novel DNA topoisomerase I inhibitor, demonstrates a promising antitumor efficacy with differentiation from T-DM1. Clin Cancer Res 22:5097–5108
43. Ogitani Y, Hagihara K, Oitate M, Naito H, Agatsuma T (2016) Bystander killing effect of DS-8201a, a novel anti-human epidermal growth factor receptor 2 antibody-drug conjugate, in tumors with human epidermal growth factor receptor 2 heterogeneity. Cancer Sci 107:1039–1046
44. Dokter W, Ubink R, van der Lee M, van der Vleuten M, van Achterberg T, Jacobs D, Loosveld E, van den Dobbelsteen D, Egging D, Mattaar E, Groothuis P, Beusker P, Coumans R, Elgersma R, Menge W, Joosten J, Spijker H, Huijbregts T, de Groot V, Eppink M, de Roo G, Verheijden G, Timmers M (2014) Preclinical profile of the HER2-targeting ADC SYD983/SYD985: introduction of a new duocarmycin-based linker-drug platform. Mol Cancer Ther 13(11):2618–2629

45. Rothenberg ML (2001) Irinotecan (CPT-11): recent developments and future directions--colorectal cancer and beyond. Oncologist 6(1):66–80
46. Kawato Y, Aonuma M, Hirota Y, Kuga H, Sato K (1991) Intracellular roles of SN-38, a metabolite of the camptothecin derivative CPT-11, in the antitumor effect of CPT-11. Cancer Res 51(16):4187–4191
47. Cardillo TM, Govindan SV, Sharkey RM, Trisal P, Arrojo R, Liu D, Rossi EA, Chang CH, Goldenberg DM (2015) Sacituzumab Govitecan (IMMU-132), an anti-Trop-2/SN-38 antibody–drug conjugate: characterization and efficacy in pancreatic, gastric, and other cancers. Bioconjug Chem 26(5):919–931
48. Cardillo TM, Govindan SV, Sharkey RM, Trisal P, Goldenberg DM (2011) Humanized anti-Trop-2 IgG-SN-38 conjugate for effective treatment of diverse epithelial cancers: preclinical studies in human Cancer Xenograft models and monkeys. Clin Cancer Res 17(10):3157–3169
49. Goldenberg DM, Cardillo TM, Govindan SV, Rossi EA, Sharkey RM (2015) Trop-2 is a novel target for solid cancer therapy with sacituzumab govitecan (IMMU-132), an antibody-drug conjugate (ADC). Oncotarget 6(26):22496–22512
50. Starodub AN, Ocean AJ, Manish A, Shah MA, Guarino MJ, Picozzi VJ Jr, Vahdat LT, Thomas SS, Govindan SV, Maliakal PP, Wegener WA, Hamburger SA, Sharkey RM, Goldenberg DM (2015) First-in-human trial of a novel anti-Trop-2 antibody-SN-38 conjugate, Sacituzumab Govitecan, for the treatment of diverse metastatic solid tumors. Clin Cancer Res 21(17):3870–3878
51. Bardia A, Mayer IA, Diamond JR, Moroose RL, Isakoff SJ, Starodub AN, Shah NC, O'Shaughnessy J, Kalinsky K, Guarino M, Abramson V, Juric D, Tolaney SM, Berlin J, Messersmith WA, Ocean AJ, Wegener WA, Maliakal P, Sharkey RM, Govindan SV, Goldenberg DM, Vahdat LT (2017) Efficacy and safety of anti-Trop-2 antibody drug conjugate Sacituzumab Govitecan (IMMU-132) in heavily pretreated patients with metastatic triple-negative breast cancer. J Clin Oncol 35(19):2141–2148
52. Gray JE, Heist RS, Starodub AN, Camidge DR, Kio E, Masters G, Purcell WT, Guarino MJ, Misleh J, Schneider CJ, Schneider BJ, Ocean AJ, Johnson T, Gandhi L, Kalinsky K, Scheff RJ, Messersmith WA, Govindan SV, Maliakal P, Mudenda B, Wegener WA, Sharkey RM, Goldenberg DM (2017) Therapy of small cell lung cancer (SCLC) with a topoisomerase-I-inhibiting antibody-drug conjugate (ADC) targeting Trop-2, Sacituzumab Govitecan. Clin Cancer Res 23(19):5711–5719
53. Heist RS, Guarino MJ, Masters G, Purcell WT, Starodub AN, Horn L, Scheff RJ, Bardia A, Messersmith WA, Berlin J, Ocean AJ, Govindan SV, Maliakal P, Mudenda B, Wegener WA, Sharkey RM, Goldenberg DM, Camidge DR (2017) Therapy of advanced non–small-cell lung cancer with an SN-38-anti-Trop-2 drug conjugate, Sacituzumab Govitecan. J Clin Oncol 35(24):2790–2797
54. Schalper KA, Kumar S, Hui P, Rimm DL, Gershkovich P (2014) A retrospective population-based comparison of HER2 immunohistochemistry and fluorescence in situ hybridization in breast carcinomas. Arch Pathol Lab Med 138:213–219
55. MMC L, Groothuis PG, Ubink R, MAJ V, Achterberg TA, Loosveld EM, Damming D, DCH J, Rouwette M, Egging DF, Dobbelsteen D, Beusker PH, Goedings P, GFM V, Lemmens JM, Timmers M, WHA D (2015) The preclinical profile of the Duocarmycin-based HER2-targeting ADC SYD985 predicts for clinical benefit in low HER2-expressing breast cancers. Mol Cancer Ther 14:692–703
56. Iwata H, Tamura K, Doi T, Tsurutani J, Modi S, Park H, Krop IE, Sagara Y, Redfern CH, Murthy RK, Redman RA, Shitara K, Fujisaki Y, Sugihara M, Zhang L, Shahidi J, Yver A, Takahashi S (2018) Trastuzumab deruxtecan (DS-8201a) in subjects with HER2-expressing advanced solid tumors: long-term efficacy and safety from a first-in-human phase 1 study with multiple expansion cohorts. J Clin Oncol 36(suppl; abstr 2501):2501

Part III
Hybrid Techniques of Active and Passive Targeting

Polymeric Micelles

Nobuhiro Nishiyama

Abstract The block copolymers composed of two kinds of polymers with different properties self-assemble into core-shell type nanoparticles, so-called polymer micelles. Polymeric micelles are formed by various driving forces such as hydrophobic interactions, electrostatic interactions, metal complex formation. Polymeric micelles are useful as a drug delivery system (DDS) because they can stably encapsulate a variety of compounds and can control release of these molecules. In addition, polymeric micelles are known to efficiently accumulate in solid cancers after systemic administration, thereby enhancing the efficacy and reducing the side effects of anticancer agents. Indeed, several micellar formulations of anticancer agents have progressed to clinical trials. This article outlines polymer micelle from the viewpoint of formulation and particle design.

Keywords Polymeric micelle

Introduction

Drugs categorized under the drug delivery system (DDS) are made primarily by utilizing nanotechnology. In various medical fields, DDS drugs have been produced and evaluated in preclinical or clinical trials, with some already approved for clinical use. In the field of oncology, DDS can be used for active or passive targeting of tumor tissues. Active targeting refers to the development of monoclonal antibodies directed against tumor-related molecules, allowing targeting of a tumor from the specific binding of antibodies with respective antigens. Passive targeting can be achieved by utilizing the enhanced permeability and retention (EPR) effect [1]. This effect is based on the pathophysiological characteristics of solid tumor tissues, namely, hypervascularity, incomplete vascular architecture, secretion of vascular permeability factors stimulating extravasation within cancer tissue, and absence of

N. Nishiyama (✉)
Laboratory for Chemistry and Life Science, Institute of Innovative Research, Tokyo Institute of Technology, Yokohama, Japan
e-mail: nishiyama.n.ad@m.titech.ac.jp

Y. Matsumura, D. Tarin (eds.), *Cancer Drug Delivery Systems Based on the Tumor Microenvironment*, https://doi.org/10.1007/978-4-431-56880-3_8

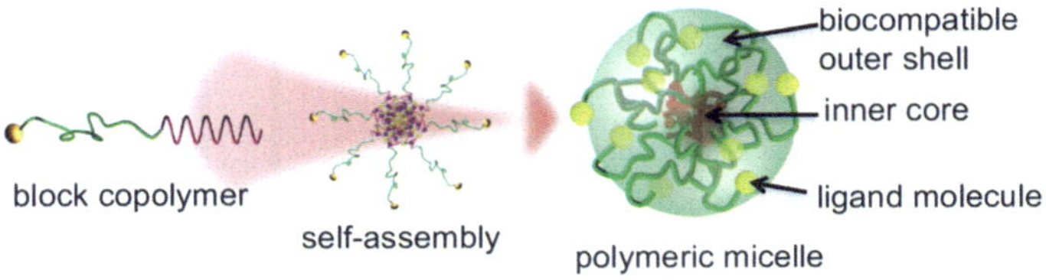

Fig. 1 Polymeric micelles formed through the self-assembly of block copolymers

effective lymphatic drainage from tumors that impedes the efficient clearance of macromolecules accumulated in solid tumor tissues.

Polymeric micelle-based anticancer drugs were originally developed by Kataoka et al. in the late 1980s or early 1990s [2–4]. Polymeric micelles were expected to increase the accumulation of drugs in tumor tissues by utilizing the EPR effect as well as to incorporate various kinds of drugs into their inner core with relatively high stability by chemical conjugation or physical entrapment (Fig. 1). Also, the size of micelles can be controlled within the diameter range of 10–100 nm to ensure that they do not penetrate normal vessel walls. With this development, it is expected that the incidence of drug-induced side effects may be decreased owing to the reduced drug distribution in normal tissues.

Characteristics and Particle Design of Polymer Micelles

Polymeric micelles have various features such as low critical micelle concentration (CMC), inner core of solid (glass) state, slow dissociation behavior unlike surfactant micelles. These properties largely depend on the composition and the chemical structure of the block copolymer. In addition, polymer micelles have the size of 10–100 nm, which can be controlled by the composition and chemical structure of the constituent block copolymers. In polymeric micelles from amphiphilic block copolymers, nanoparticles with a narrow particle size distribution are prepared by controlling the rate of solvent exchange from a good solvent (organic solvent) to a selective solvent (water) using the dialysis method [3]. Otherwise, the O/W emulsion method and sonication are used for the micelle preparation [3]. The equipment for manufacturing nanoparticles using shear stress (microfluidizer, etc.) is useful for mass preparation of polymeric micelles. On the other hand, since polymeric micelles utilizing electrostatic interactions and metal complex formation as driving force are formed by self-assembly in water, they can be prepared very reproducibly by simply mixing constituent molecules in aqueous media.

The method of encapsulating drugs into polymeric micelles can be roughly divided into "physical encapsulation" and "chemical bonding". For the physical encapsulation, the above-mentioned dialysis method and O/W emulsion method are used, and it is possible to load 10–20% (w/w) of drugs [3]. In addition, in the physical encapsulation, the chemical structure of the polymer constituting the micelle inner core is very important. First, in order to stably retain the drug in the micellar core, it is desirable that the inner core is in a glass state, and micelles with the mobile core are difficult to stably retain even if the drug can be solubilized [3]. Also, the affinity of the chemical structures between the polymer constituting inner core and the drug is also an important parameter. That is, by utilizing the π-π interactions or hydrogen bonding to the drug, it is possible to stably retain the drug in the micellar core (Fig. 2). On the other hand, the chemical conjugation of the drug to the polymer constituting inner core through the covalent bond allows the micelles to more stably retain the drug than the physical encapsulation [3]. In this case, the chemical bond needs to be cleaved at the target tissues and cells, ensuring the drug release.

Therefore, the chemical bonds that are cleaved in response to pH change, redox reaction, enzymatic reaction are utilized (Fig. 3). Although there is a restriction that

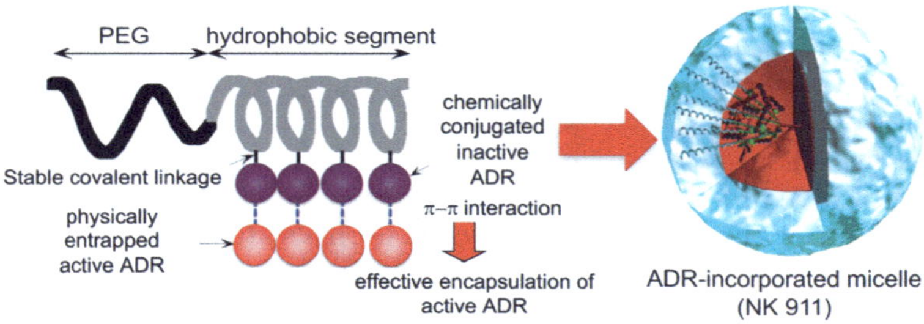

Fig. 2 Example of physical entrapment of drug molecules into polymeric micelles: In NK 911, Adriamycin(ADR) is chemically conjugated to the micellar core-forming polymer to physically entrap "active" ADR molecules through the π-π interactions. (Modified from Ref. 3)

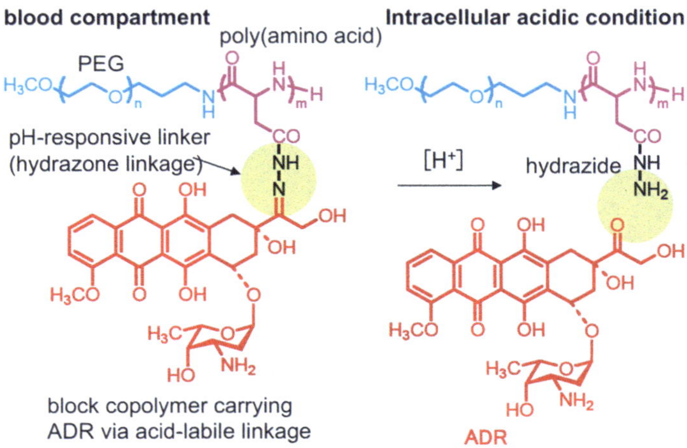

Fig. 3 Example of chemical conjugation of drug molecules into polymeric micelles: Adriamycin (ADR) is chemically conjugated to the micellar core-forming polymer though the acid-labile hydrazone linkage. (Modified from Ref. 4)

the drug needs to have functional groups available for the chemical conjugation, the use of the chemical conjugation allows precision control of drug action in the body.

Biodistribution of Polymeric Micelles and Their Optimization

Polymeric micelles can be applied to the treatment of various diseases, but the development of the micelles targeting solid tumors is greatly advanced. This is because solid tumors are characterized by hypervascular permeability and impaired lymphatic drainage in tumor tissues, allowing effective accumulation and prolonged retention of macromolecular carriers in tumors. This effect is called the Enhanced Permeability and Retention (EPR) effect [5], and it has been reported for hypervascular cancers such as breast and colon cancers.

However, the EPR effect might be limited in some intractable cancers. Among them, pancreatic cancers have the vascular structure in which the vascular endothelial cells are covered with the pericytes, and the exudation from the blood vessel to the cancer tissue is remarkably deteriorated. Also, pancreatic cancers are featured with the thick fibrosis. Accordingly, the tissue permeability of nanoparticles is largely restricted [6]. Indeed, adriamycin-incorporated PEG-coated liposomes (Doxil®) with the size of 100 nm remained in the interstitium and could not reach deep tumorous regions after the systemic administration to mice bearing subcutaneous human pancreatic cancer BxPC3 cells (Fig. 4a). We prepared polymeric micelles with the precisely controlled size in the range of 30 to 100 nm and studied the tissue permeability of these micelles using subcutaneous models of BxPC3 cells. As a result, the micelles smaller than 50 nm showed the enhanced tumor accumulation and homogeneous intratumoral distribution (Fig. 4b), whereas larger micelles localized at the perivascular region, resulting in the limited tumor accumulation [7]. Based on these results, we evaluated the efficacy of 30-nm micelles incorporating

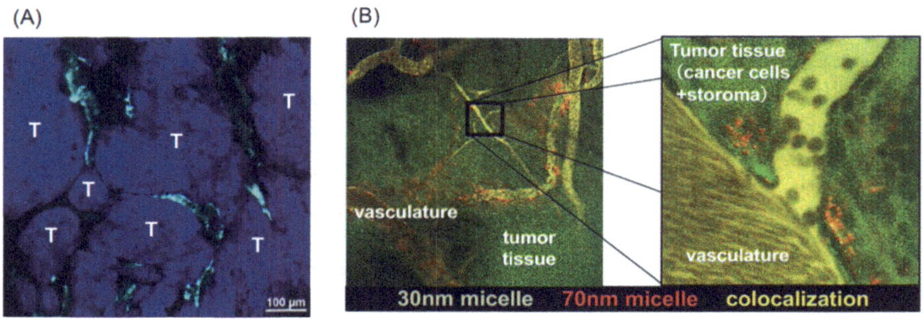

Fig. 4 (**a**) Intratumoral distribution of DOXIL (PEGylated liposome) in human pancreatic cancer model. (T: tumor nest, Blue: Hoechst, Black: stromal tissue, Cyan: DOXIL)
(**b**) Intratumoral distribution of 30 nm and 70 nm polymeric micelles in human pancreatic cancer model (30 nm micelles (Green) showed homogeneous distribution, whereas 70 nm micelles (Red) showed specific localization at perivascular region). (Modified from Ref. 7)

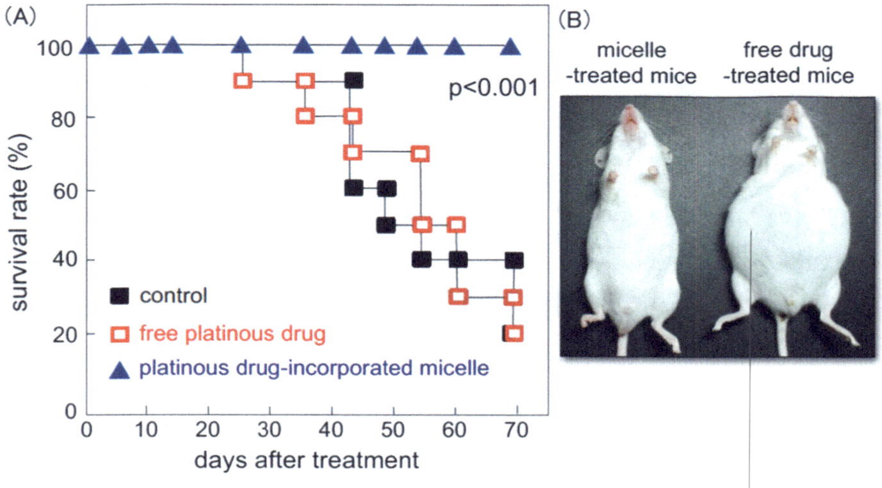

Fig. 5 Treatment of spontaneous pancreatic tumor model by polymeric micelles incorporating platinous anticancer agents. (Modified from Ref. 8)

platinum anticancer agents against spontaneous pancreatic cancer model by transgenic mouse (EL1-Luc/TAg). As a result, the mice treated with the micelles showed remarkably prolonged survival compared with non-treated or free platinum drug-treated mice (Fig. 5a). Also, the micelles inhibited cancerous ascites (Fig. 5b) [8]. Thus, it is likely that the size of polymeric micelles is important for targeting fibrotic tumors, and smaller micelles might be effective for the treatment of pancreatic cancers.

Glioblastoma (GBM) is another intractable tumor with the limited EPR effects. It is known that the vascular endothelium in GBM is characterized by reduced permeability similarly to the blood-brain barrier (BBB), which is known as the blood-brain tumor barrier (BBTB). Therefore, the accumulation of nanoparticles in GBM is significantly lower than that in other solid cancers. To circumvent BBTB, we introduced the cyclic RGD (cRGD) peptide that binds to $\alpha_V\beta_3$ integrin, which is highly expressed in tumor blood vessels, into the surface of polymeric micelles. Note that cRGD itself has been approved as an orphan drug for the antiangiogenic treatment of GBM. The cRGD-conjugated micelles showed greatly higher accumulation than the non-targeted micelles in a subcutaneous model of human GBM U87 cells (Fig. 6) [9]. Also, it is worth mentioning that the tumor accumulation of the cRGD-conjugated micelles is faster than that of non-targeted micelles based on the EPR effect, suggesting the active transport (e.g., transcytosis) of the cRGD-conjugated micelles beyond BBTB. Thus, the introduction of ligand molecules to the micelle surface is an effective approach to overcome biological barriers in intractable cancers.

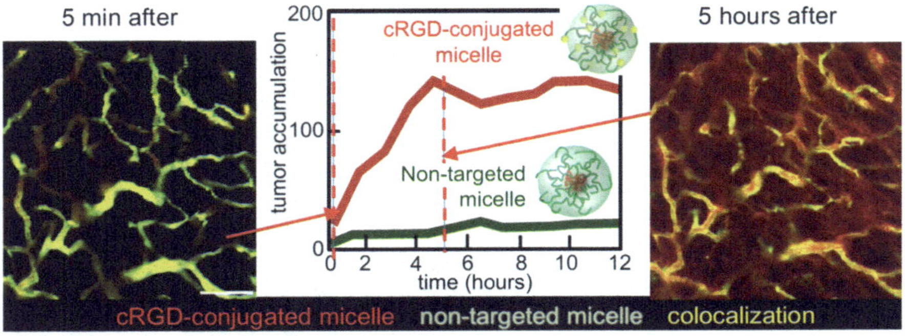

Fig. 6 Accumulation of cyclic RGD peptide(cRGD)-installed polymeric micelles in glioblastoma model (cRGD-installed micelles(Red) showed rapid tumor accumulation compared with non-targeted micelles(Green)). (Modified from Ref. 9)

Translation to Clinical Applications

DDS exerts the pharmacological effects by release of the incorporated drugs. Even if the drug release is too fast or too slow, sufficient therapeutic effect cannot be obtained. Therefore, it is necessary to optimize the stability of DDS and the drug release rate. Generally, mice are used to evaluate the therapeutic effects of DDS. When sufficient therapeutic efficacy is obtained in mice, the development of DDS will progress to nonclinical and clinical studies. However, the differences of biodistribution by animal species should be considered: it is known that the blood half-life of DDS in humans is approximately 4 times longer than that in mice. It is natural considering that a circulation of blood requires more time in humans than mice and DDS is excreted from the body by passing through clearance organs such as the liver and kidneys. Therefore, even though DDS shows sufficient blood retention in mice and releases drugs 24 hours after reaching the target tissues, the duration of 24 h stability might be insufficient in humans. From this point of view, DDS should be sufficiently stabilized aiming the final application to human beings, while showing the efficacy in mice. Alternatively, the use of the environment-sensitive linkers for the drug conjugation is a promising approach to solve the aforementioned differences by animal species. There is a similarity in the microenvironment of the diseased sites between mice and humans. For example, the acidic environment is a common feature of solid tumors among various animal species due to excessive glycolysis producing lactic acids, which is known as the Warburg effect [10]. So that the positron emission tomography using ^{18}F-fluorodeoxy glucose (FDG-PET) is widely used for cancer diagnosis. Hence, the acid-responsive cleavable linkers such as Schiff base and acetal linkage are used for specific drug activation in solid tumors. Indeed, such linkers for drug conjugation are used for the core design of polymeric micelles and those micelles are now undergoing clinical studies.

Table 1 summarizes the clinical development of polymeric micelles in Japan [1]. To date, five micellar formulations of anticancer agents have been progressed to

Table 1 the clinical development of polymeric micelles in Japan

code name	drug	drug loading	modality (linker)	clinical study
NC-6004	Cisplatin	Chemical conjugation	Metal complex (pH responsive)	Phase 3
NC-4016	Dachplatin (active complex of oxaliplatin)	Chemical conjugation	Metal complex (pH responsive)	Phase 1
NC-6300	Epirubicin	Chemical conjugation	Shiff base (pH responsive)	Phase 1
NK105	Paclitaxel	Physical entrapment	–	Phase 3
NK012	SN-38	Chemical conjugation	Ester linkage (hydrolysis)	Phase 2

clinical studies. Four of them utilize the chemical conjugation (including coordination bond of metal complex) for the drug encapsulation. Epirubicin(EPI)-incorporated micelles (NC-6300) bind epirubicin to the micellar core through the pH-responsive hydrazone linkage (Schiff base) (Fig. 3) and the drug is selectively released in tumor tissues. As a result, enhanced therapeutic efficacy and reduced cardiotoxicity have been demonstrated in the animal experiments, and the phase I study is currently underway in Japan [11].

The benefits by polymeric micelles include the enhanced therapeutic efficacy, reduction of side effects and improvement of dosing compliance. For example, cisplatin induces nephrotoxicity as a dose-limiting toxicity (DLT), so that 3-day infusion accompanied by hydration is performed to reduce it. By contrast, cisplatin-incorporated micelles (NC-6004) reduced nephrotoxicity due to slow glomerular filtration [12]. Therefore, a short-term administration without hydration may be applicable to the clinical use of NC-6004. In addition, cisplatin is known to exhibit irreversible ototoxicity, but NC-6004 appeared to avoid the adverse effect [13]. The animal experiments using guinea pigs demonstrated that the accumulation of cisplatin in the cochlear cortis of the inner ear was greatly reduced by NC-6004, leading to excellent preservation of hair cells [13]. Thus, the incorporation of toxic low molecular weight drugs into polymeric micelles greatly improves the quality of life (QOL) of cancer patients.

Application of Polymeric Micelles to Theranostic Nanomedinces

The usefulness of DDS for the tumor targeting has been demonstrated in cancer patients. Indeed, IgG with the molecular weight of 150 kDa has become the main stream of current cancer therapeutic agents. Nevertheless, recent publications revealed that the accumulation of PEG-coated liposomes in spontaneous tumors of diseased dogs was different among individuals and their intratumoral distributions

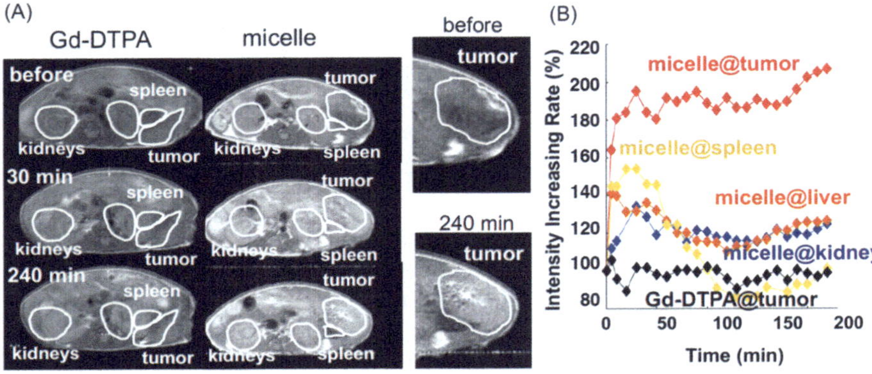

Fig. 7 (**a**) MR imaging of orthotopically inoculated pancreatic cancer cells after i.v. administration of polymeric micelles incorporating platinous anticancer drug and Gd-DTPA. (**b**) Intensity increasing ratio of the MR signals in each tissue. (Modified from Ref. 16)

were heterogeneous [14]. Likewise, such individual differences should occur in the tumor accumulation of DDS in cancer patients. Also, the accumulation of DDS may change depending on the past treatment history and process such as radiation therapy. Therefore, the concept of personalized medicine (precision medicine) may be applicable to DDS, i.e., the companion diagnosis is performed to predict the tumor accumulation and distribution of DDS in individual patients. In this regard, the clinical studies have already started to apply diagnostic DDS by PET or SPECT for confirmation of its tumor accumulation, followed by application of therapeutic DDS incorporating anticancer agents [15]. Polymeric micelles can carry both cisplatin and gadolinium-diethylenetriaminepentaacetic acid (Gd-DTPA), a clinically used MRI contrast agent known as Magnevist®, and they were applied to imaging and therapy of an orthotopic model of pancreatic cancers (Fig. 7) [16]. As a result, the orthotopic pancreatic cancer was successfully visualized by the micelles, while it was difficult to be detected by free Gd-DTPA. In addition, miniaturization of the tumor volume by delivery of cisplatin was monitored by MR imaging. Such real-time monitoring of tumor accumulation of DDS and its therapeutic efficacy may realize more reliable cancer treatment as well as personalized medicine.

Conclusion

The development of polymeric micelles is based on the design of the constituent block polymers. The polymer design need to be optimized depending on the drug to be incorporated, which is in contrast with the conventional encapsulation technology for nanoparticle formulations such as liposomes. The anticancer agent-incorporated polymeric micelles are undergoing clinical studies. Polymeric micelles currently load only approved drugs, but should be utilized for potent unapproved drugs, facilitating their clinical use by circumventing their side effects. Also, polymeric micelles will promote the practical application of emerging biopharmaceuticals such as nucleic acids and therapeutic proteins.

Integration of functional moieties to the polymers on demand allows to realize smart polymeric micelles. The ligand-installed micelles are useful for overcoming biological barriers such as BBTB. The theranostic micelles allow real-time monitoring of their tumor accumulation and therapeutic efficacy in a real-time manner, leading to more reliable cancer treatment. Thus, smart polymeric micelles should be a key technology for the future medicine.

References

1. Nishiyama N, Matsumura Y, Kataoka K (2016) Development of polymeric micelles for targeting intractable cancers. Cancer Sci 107(7):867–874
2. Cabral H, Miyata K, Osada K, Kataoka K (2018) Block copolymer micelles in nanomedicine applications. Chem Rev 118(14):6844–6892. https://doi.org/10.1021/acs.chemrev.8b00199
3. Nishiyama N, Kataoka K (2006) Nano-structured devices based on block copolymer assemblies for drug delivery: designing structures for enhanced drug function. Adv Polym Sci 193:67–101
4. Bae Y, Fukushima S, Harada A, Kataoka K (2003) Design of environment-sensitive supramolecular assemblies for intracellular drug delivery: polymeric micelles that are responsive to intracellular pH change. Angew Chem Int Ed 42(38):4640–4643
5. Matsumura Y, Maeda H (1986) A new concept for macromolecular therapeutics in cancer chemotherapy: mechanism of tumouritropic accumulation of proteins and the antitumour agent SMANCS. Cancer Res 46(12):6387–6392
6. Kano MR, Bae Y, Iwata C, Morishita Y, Yashiro M, Oka M, Fujii T, Komuro A, Kiyono K, Kaminishi M, Hirakawa K, Ouchi Y, Nishiyama N, Kataoka K, Miyazono K (2007) Improvement of cancer-targeting therapy, using nanocarriers for intractable solid tumors by inhibition of TGF-signaling. Proc Natl Acad Sci U S A 104(9):3460–3465
7. Cabral H, Matsumoto Y, Mizuno K, Chen Q, Murakami M, Kimura M, Terada Y, Kano MR, Miyazono K, Uesaka M, Nishiyama N, Kataoka K (2011) Accumulation of sub-100 nm polymeric micelles in poorly permeable tumours depends on size. Nat Nanotech 6(12):815–823
8. Cabral H, Murakami M, Hojo H, Terada Y, Kano MR, Chung U-I, Nishiyama N, Kataoka K (2013) Targeted therapy of spontaneous murine pancreatic tumors by polymeric micelles prolongs survival and prevents peritoneal metastasis. Proc Natl Acad Sci U S A 110(28):11397–11402
9. Miura Y, Takenaka T, Toh K, Wu S, Nishihara H, Kano MR, Ino Y, Nomoto T, Matsumoto Y, Koyama H, Cabral H, Nishiyama N, Kataoka K (2013) Cyclic RGD-linked polymeric micelles for targeted delivery of platinum anticancer drugs to glioblastoma through the blood-brain tumor barrier. ACS Nano 7(10):8583–8592
10. Vander Heiden MG, Cantley LC, Thompson CB (2009) Understanding the Warburg effect: the metabolic requirements of cell proliferation. Science 324(5930):1029–1033
11. Takahashi A, Yamamoto Y, Yasunaga M, Koga Y, Kuroda J, Takigahira M, Harada M, Saito H, Hayashi T, Kato Y, Kinoshita T, Ohkohchi N, Hyodo I, Matsumura Y (2013) NC-6300, an epirubicin-incorporating micelle, extends the antitumor effect and reduces the cardiotoxicity of epirubicin. Cancer Sci 104(7):920–925
12. Nishiyama N, Okazaki S, Cabral H, Miyamoto M, Kato Y, Sugiyama Y, Nishio K, Matsumura Y, Kataoka K (2003) Novel cisplatin-incorporated polymeric micelles can eradicate solid tumors in mice. Cancer Res 63(24):8977–8983

13. Baba M, Matsumoto Y, Kashio A, Cabral H, Nishiyama N, Kataoka K, Yamasoba T (2012) Micellization of cisplatin (NC-6004) reduces its ototoxicity in Guinea pigs. J Control Release 157(1):112–117
14. Hansen AE, Petersen AL, Henriksen JR, Boerresen B, Rasmussen P, Elema DR, af Rosenschöld PM, Kristensen AT, Kjær A, Andresen TL (2015) Positron emission tomography based elucidation of the enhanced rermeability and retention effect in dogs with cancer using copper-64 liposomes. ACS Nano 9(7):6985–6995
15. Lee H, Shields AF, Siegel BA, Miller KD, Krop I, Ma CX, LoRusso PM, Munster PN, Campbell K, Gaddy DF, Leonard SC, Geretti E, Blocker SJ, Kirpotin DB, Moyo V, Wickham TJ, Hendriks BS (2017) ^{64}Cu-MM-302 positron emission tomography quantifies variability of enhanced permeability and retention of nanoparticles in relation to treatment response in patients with metastatic breast cancer. Clin Cancer Res 23(15):4190–4202
16. Kaida S, Cabral H, Kumagai M, Kishimura A, Terada Y, Sekino M, Aoki I, Nishiyama N, Tani T, Kataoka K (2010) Visible-drug delivery by supra-molecular nanocarriers directing to single-platformed diagnosis and therapy of pancreatic tumor model. Cancer Res 70(18):7031–7041

Liposomes Conjugated with a Pilot Molecule

Kosuke Shimizu and Naoto Oku

Abstract The liposome is a lipid-based nanoparticle that is widely used as a drug encapsulation device and drug delivery system (DDS), because it enables the encapsulation of many drugs into it. Notably, the liposome is suitable as a targetable drug carrier, since it can be easily conjugated with functional pilot molecules such as antibody, peptide, sugar chain, etc., on its surface. These liposomes are referred to as active targeting liposomes, and they enable the encapsulated drugs or imaging agents to be delivered to specific sites for the treatment and diagnosis, respectively, of refractory diseases such as cancer. In this chapter, the advantages of the use of active-targeting liposomes as a tool of DDS and their application to cancer treatment are reviewed.

Keywords Angiogenic vessel · Antibody · Drug delivery system · Dual targeting · Liposome · Peptide · Pilot molecule · Sugar · Targeting · Tumor

---

K. Shimizu (✉)
Department of Molecular Imaging, Institute for Medical Photonics Research, Preeminent Medical Photonics Education & Research Center, Hamamatsu University School of Medicine, Hamamatsu, Shizuoka, Japan

Department of Medical Biochemistry, School of Pharmaceutical Sciences, University of Shizuoka, Shizuoka, Shizuoka, Japan
e-mail: kshimizu@hama-med.ac.jp

N. Oku
Department of Medical Biochemistry, School of Pharmaceutical Sciences, University of Shizuoka, Shizuoka, Shizuoka, Japan

Laboratory of Bioanalytical Chemistry, Faculty of Pharma-Science, Teikyo University, Tokyo, Japan

Y. Matsumura, D. Tarin (eds.), *Cancer Drug Delivery Systems Based on the Tumor Microenvironment*, https://doi.org/10.1007/978-4-431-56880-3_9

Abbreviations

APRPG	Ala-Pro-Arg-Pro-Gly
ASSHN	Ala-Ser-Ser-His-Asn
BBB	blood-brain barrier
BPA	*Bauhinia purprea agglutinin*
BNCT	boron neutron capture therapy
CT	computed tomography
DDS	drug delivery system
DCP	dicetylphosphate
DOX	doxorubicin
DSPE	distearoylphosphatidylethanolamine
ECM	extracellular matrix
EGF	epidermal growth factor
EGFP	enhanced green fluorescent protein
EPR	enhanced permeability and retention
FR	folate receptor
GPLPLR	Gly-Pro-Leu-Pro-Leu-Arg
HB-EGF	heparin-binding epidermal growth factor-like growth factor
HUVECs	human umbilical vein endothelial cells
Mal	maleimide
MRI	magnetic resonance imaging
MT1-MMP	membrane-type 1 matrix metalloproteinase
NGR	Asn-Gly-Arg
NHS	*N*-hydroxysuccinimide
OVA	ovalbumin
PC	phosphatidylcholine
PCL	polycation liposome
PDT	photodynamic therapy
PEG	polyethylene glycol
PEI	polyethylenimine
P-gp	P-glycoprotein
PS	phosphatidylserine
PTX	paclitaxel
RES	reticuloendothelial system
RGD	Arg-Gly-Asp
siRNA	small interfering RNA
sLe^X	sialyl LewisX
SM	sphingomyelin
SWKLPPS	Ser-Trp-Lys-Leu-Pro-Pro-Ser
TAMs	tumor-associated macrophages
TEPA	tetraethylenepentamine
Tf	transferrin
VEGFR-1	vascular endothelial growth factor receptor-1
WGA	*Wheat germ agglutinin*

Introduction

Liposomes and Tumor Targeting

A drug delivery system is a useful pharmaceutical technique for achieving effective and safe therapy for various diseases, since it enables the delivery of drugs to target sites such as tumor tissue after injection into the body. Furthermore, it is used to avoid the unwanted accumulation of drugs at normal organs, thus reducing side effects. There are many approaches to realize an effective drug delivery system (DDS), and one of the strategies is the use of lipid nanoparticles as drug carriers. Allen and Cullis previously reviewed the advantages of the usage of lipid-based or polymer-based nanoparticles from the view point of their pharmaceutical properties [4]: The use of them can improve the poor solubility of certain ingredients, reduce tissue damage on extravasation, avoid rapid clearance from the body, lessen unfavorable pharmacokinetics and poor distribution, and afford selectivity for target sites. The liposome is a lipid-based nanoparticle that is widely used for drug encapsulation and DDS devices, because it enables the encapsulation of many drugs into either its internal aqueous phase or lipid-bilayer phase, resulting in protection from degradation of the drugs by catabolic enzymes in the bloodstream after injection into the body. Another feature of a liposomal drug carrier, is that its physicochemical properties can be modified by changing the lipid composition: The surface charge of the liposomes can be altered by adding the desired charged lipids, and the membrane fluidity of the liposomes can be regulated by changing the length and saturation level of the fatty acyl chains in their phospholipid structure. The most important point to note is that liposomes can be engineered to provide pharmaceutically beneficial properties such as long-circulating retention in the bloodstream and ability to target to a desired organ by conjugation of functional molecules on the surface of the liposome membrane. Furthermore, since Matsumura and Maeda had advocated the enhanced permeability and retention (EPR) effect on tumor accumulation of macromolecules about 30 years ago [109], the use of long-circulating liposomes, in particular polyethylene glycol (PEG)-modified liposomes, has become a standard formulation of drug nanocarriers for tumor targeting of drugs. Thus, the liposome is a suitable DDS tool for effective delivery of anti-tumor drugs and nucleic acid medicines to tumor tissues. In fact, some liposomal anti-tumor drugs have been commercialized for worldwide use in the clinic as cancer chemotherapeutics [5]. Although liposomalization of anti-tumor drugs has successfully reduced severe side effects of the original drugs, such as the cardiotoxicity of doxorubicin, uncertainty of the main anti-tumor effect of the liposomal drugs is often argued. One of the reasons why the liposome-encapsulated anti-tumor drugs fail to show the expected anti-tumor effect in the body is that they are often stabilized by PEGylation, which results in low interaction with serum proteins such as complement components. It also enables them to avoid removal by the reticuloendothelial system ("stealth" properties) and gives them longevity in the bloodstream. But also results in poor uptake into targeted cells such as tumor cells. This phenomenon is

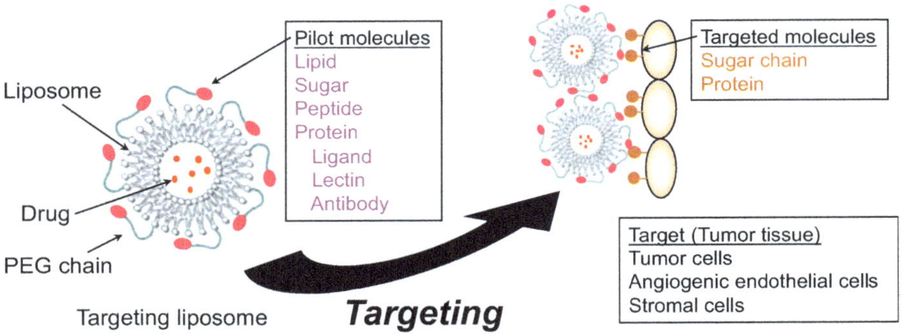

Fig. 1 Active targeting of liposomes to a target site with a pilot molecule
Liposomes conjugated with a pilot molecule such as a peptide and/or an antibody recognizes an actively targeted molecule such as a growth factor receptor or adhesion molecule via the strong and specific interaction among them. Consequently, effective delivery of a liposomal drug to the target site can be achieved

sometimes referred to as the "PEG dilemma" [55, 56]. To resolve this serious problem and to provide the tumor targetability of the liposome, modification of a pilot molecule on the liposome surface is a suitable approach (Fig. 1). In this chapter, we introduce various types of pilot molecules for tumor cell and tissue targeting by liposomes and their usefulness for cancer therapy.

Methodology of Surface Modification of a Pilot Molecule on the Liposome

For active delivery of liposomes to target cells, organs, and disease sites, conjugation of a pilot molecule on the liposome surface is quite commonly employed; and the methodology involved varies. The most commonly used method is the chemical synthesis of a lipid derivative conjugated to the desired pilot molecule and insertion of it into the liposomal membrane. Because some of the pilot molecules are peptides, proteins or antibodies, which contain either a carboxy or amine group in their structure, the amide reaction between the phospholipid and the pilot molecule is commonly used for their incorporation. We previously reported the chemical synthesis of a pilot peptide, Ala-Pro-Arg-Pro-Gly (APRPG), with the terminal group of the PEG chain in distearoylphosphatidylethanolamine (DSPE)-PEG phospholipid attached to the peptide and used this APRPG-PEG lipid derivative for modification of the liposomal surface [102]. On the other hand, utilizing a conjugation reaction between pilot molecules and linker phospholipids that contain reactive functional groups such as *N*-hydroxysuccinimide (NHS) or maleimide (Mal) in their chemical structure has become a mainstream method to obtain lipid derivatives. Since pilot molecules such as peptide and antibody contain either a primary amine group or thiol group in the amino acid residues in their structures, NHS and maleimide are often respectively used for the conjugation reaction. The NHS group easily reacts with the primary amine groups of amino acids such as lysine and arginine in the peptide sequence of targeting peptide, protein or antibody. We previously prepared ovalbumin (OVA)-conjugated liposomes by using a linker lipid containing an NHS group and obtained the conjugation of OVA with the liposomes: OVA containing a portion of [^{14}C]-labeled OVA was added to 3-(*N*-succinimicyloxyglutaryl) aminopropyl polyethyleneglycol-carbamyl distearoylphosphatidylethanolamine (DSPE-PEG-NHS) and incubated for the coupling reaction. Then, the obtained DSPE-PEG-OVA lipid derivative was mixed with liposomes and incubated for insertion of the lipid derivative into the liposomal membrane. For removal of the uncoupled OVA, gel filtration chromatography was performed, and OVA-conjugated liposomes could thus be prepared (Fig. 2) [63, 64, 151]. On the other hand, the maleimide group reacts easily with the thiol group derived from a cysteine residue in the amino acid sequence of a targeting peptide or protein [37], and with the enzymatically-digested fragment of whole antibody, such as Fab' [48, 96, 137, 147]. For preparation of antibody-conjugated liposomes, the Fab' fragment of the antibody is often used rather than the whole antibody, because Fab' fragment-conjugated liposomes have a longer circulation time than those conjugated with whole IgG antibody due to avoidance of recognition by Fc receptors on macrophages [107]. Herein, we simply introduce how to prepare the antibody-conjugated liposomes by using a linker lipid containing a maleimide group (Fig. 3): To expose the cysteine residues in the antibody structure, Fab' fragments must be prepared from the whole antibody such as IgG by both digesting with pepsin and reducing with cysteamine. Then, the purified Fab' fragment is reacted with a linker lipid such as *N*-[(3-maleimide-1-oxopropyl)aminopropyl polyethyleneglycol-carbamyl] distearoylphosphatidylethanolamine (DSPE-PEG-Mal) to obtain the DSPE-PEG-Fab' lipid derivative. Modification of liposomes with the DSPE-PEG-Fab' is done by the post-insertion method in a similar manner as described above; and antibody-conjugated liposomes, which are sometimes called as immunoliposomes, can thus be prepared. Because antibody has high and specific affinity for the antigen, the immunoliposome has a high-binding potential for the targeted molecule. In addition to this method, modification of a pilot molecule on liposomes can be achieved by utilizing strong interactions such as the biotin-streptavidin interaction to prepare the targeting liposomes.

Since 2001, when Sharpless and the colleagues advocated the concept of click chemistry, which is an easy and quick chemical reaction with high reaction efficiency to produce a covalent bond [82], click chemistry has been used in many fields to make new synthetic products. Indeed, this reaction has been applied to conjugate a pilot molecule to the surface of DDS carriers to prepare targeted nanoparticles. Said Hassane et al. applied click chemistry to the covalent coupling of ligand molecules to the surface of preformed liposomes and succeeded in preparing targeting liposomes [38, 141]; and later, other research groups utilized this methodology to prepare pilot molecule-conjugated lipid nanoparticles including targeted liposomes [38, 68, 93, 156]. Bostic et al. subsequently succeeded in the

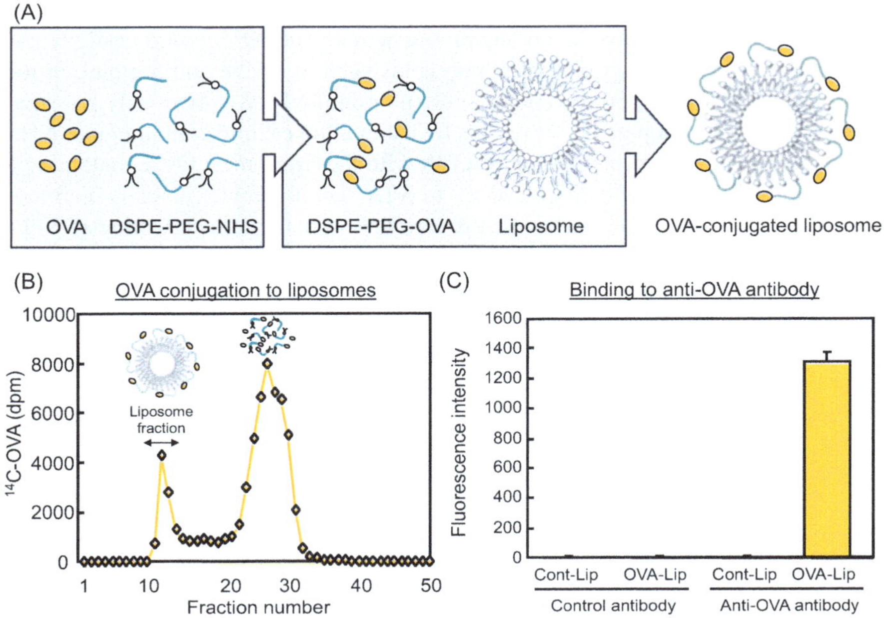

Fig. 2 Preparation of a pilot molecule-conjugated liposome
(**a**) Preparation of ovalbumin (OVA)-conjugated liposome
OVA was used as a model pilot molecule. An adequate amount of OVA was chemically conjugated to the end of the PEG chain of a linker phospholipid containing the reactive functional group, *N*-hydroxysuccinimide (NHS), and DSPE-PEG-NHS and the DSPE-PEG-OVA lipid derivative were obtained. Then, the synthesized DSPE-PEG-OVA was inserted into liposomes in a warm state, resulting in OVA-conjugated liposomes
(**b**) Confirmation of OVA conjugation to liposomes
[^{14}C]-labeled OVA-conjugated liposomes and uncoupled [^{14}C]OVA were separated by gel filtration chromatography. Radioactivity in each fraction was measured with a liquid scintillation counter
(**c**) Binding of OVA-conjugated liposomes to OVA-specific antibody
Nonspecific control antibody or anti-OVA antibody was coated onto the wells of a 96-well plate Then, fluorescently labeled non-conjugated liposomes (Cont-Lip) or OVA-conjugated liposomes (OVA-Lip) were added to the wells. The binding of OVA-Lip to anti-OVA antibody was confirmed by measuring the fluorescence intensity

preparation of biotinylated liposomes by using click chemistry under a copper-free condition and demonstrated high affinity of the biotinylated liposomes with a streptavidin-coated plate [17]. Furthermore, Kim and colleagues have developed a new tumor-targeting strategy using *in vivo* copper-free click chemistry, by which artificial sialic acid containing an azide group was metabolically expressed on the membrane surface of tumor cells orthogonally reacted with nanoparticles, and succeeded in achieving enhanced accumulation of the nanoparticles in tumor tissue [86]. These reports strongly indicate that click chemistry is becoming a mainstream method to prepare targeting liposomes, because of its ability to provide high efficiency for modification of liposomes with a pilot molecule [158].

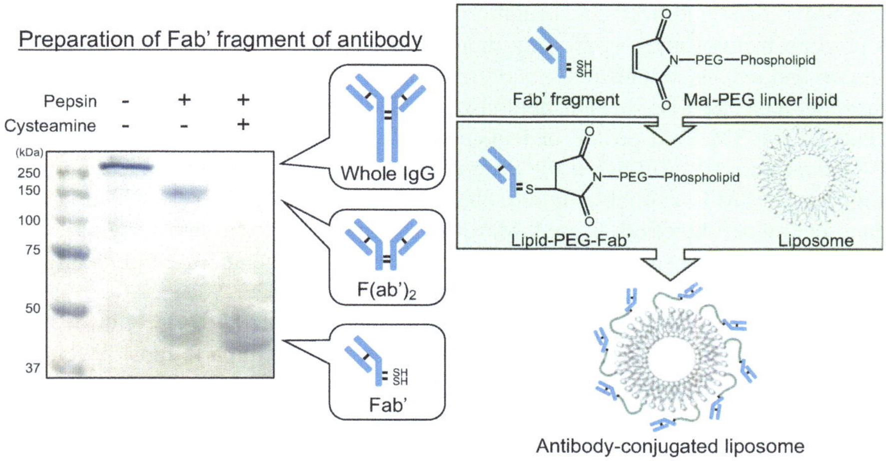

Fig. 3 Preparation of Fab' fragment of antibody-conjugated liposomes
Before antibody conjugation to liposomes, the Fab' fragment of the antibody was prepared by both digesting with pepsin and reducing with cysteamine. The Fab' fragment was reacted with a maleimide (Mal)-linker PEG lipid derivative such as DSPE-PEG-maleimide, resulting in the DSPE-PEG-Fab' lipid derivative. Then the DSPE-PEG-Fab' was inserted into liposomes in a warm state, and the Fab' fragment of the antibody-conjugated liposome (immunoliposome) was thus prepared. The gel image shows the results of Coomassie Brilliant Blue (CBB) staining of the antibodies incubated with pepsin and/or cysteamine following the SDS-PAGE

Pilot Molecules for Active Targeting

Lipids

Phosphatidylcholine (PC) is a most abundant phospholipid and universally distributed in biological membranes throughout the body. Because of its universality and biodegradability, PC is often used as a main component to prepare the basic liposome. Although PC is derived from natural biological components, most PC liposomes are recognized by the immune surveillance system in the body: PC liposomes are trapped in reticuloendothelial system (RES) organs such as liver and spleen once systemically injected into the body. By capitalizing on this trapping of PC liposomes by the RES, in which there is a large number of phagocytic macrophages, liposomes encapsulating clodronate are used to deplete systemic macrophages [88, 112] and to prepare immunodeficient model animals.

On the other hand, as described above, it is easy to change the physicochemical property of liposomes by adding functional lipids during preparation of the liposomes. Sphingomyelin (SM) is one of the sphingolipids found in the cell membrane, and it contributes to stabilization of membrane fluidity by forming lipid rafts in the cell membrane. Because of this function of SM, SM is often used as a component lipid for liposome preparation. Some researchers previously demonstrated that SM affords stabilized encapsulation of anti-tumor drugs such as vincristine and topotecan in liposomes, conferring them with long circulation time in the bloodstream and enhanced anti-tumor action in human xenograft models [161, 167]. On the other hand, Zembruski et al. demonstrated that liposomes containing SM directly changed the SM lipid profile of leukemia cells, but did not affect the activity of P-glycoprotein (P-gp) toward the drug-resistant tumor cells [183]. These findings suggest that SM is a suitable lipid to change the liposomal property without having an effect on cell function. Indeed, Marqibo is a liposome formulation mainly composed of SM, encapsulating the anti-tumor drug vincristine and used clinically for the treatment of acute lymphatic leukemia [154].

Synthetic cationic lipids such as *N*-[1-(2,3-dioleoyloxy)propyl]-*N,N,N*-trimethylammonium methyl-sulfate (DOTAP) are used to impart a positive charge on the surface of liposomes, and these cationic liposomes are suitable for use as non-viral gene delivery vectors. Because nucleic acid medicines such as plasmid DNA and small interfering RNA (siRNA) are negatively charged, cationic liposomes easily retain them in an electrostatic manner. Furthermore, the plasma membrane of cells is generally charged negatively; and thus, the electrostatic interaction between cationic liposomes and negatively-charged cell membrane is consequently accelerated, resulting in effective uptake of these liposomes into cells and delivery of nucleic acid medicines into them. Our previous studies demonstrated that polycation liposomes (PCLs), which were conjugated with polycationic cetyl polyethylenimine (PEI) polymers on the liposome membrane, were highly taken up into mammalian cells and induced the expression of exogenous genes in the cells [162]. Furthermore, *in vivo* gene transfection of mice with enhanced green fluorescent protein (EGFP) by cetyl-PEI-based PCLs was successful; and the expression of EGFP was observed in the lungs and liver, when they were injected respectively into the tail vein and portal vein. We also developed PCLs containing biological polyamine dicetyl phosphate-tetraethylenepentamine (DCP-TEPA) conjugate for siRNA and miRNA delivery and succeeded to knockdown target proteins *in vitro* and *in vivo* [7, 13].

Phosphatidylserine (PS) is a negative-charged phospholipid and is also a component of the cell membrane. PS is located on the cytosolic side of the lipid bilayer of the plasma membrane in living cells and plays a key role in cell-cycle signaling. Especially, PS is related to cell apoptosis and is transferred to the outer side of the membrane when cells undergo apoptosis. This location transfer of PS in apoptotic cells triggers a "find-me" or "eat-me" signal, and is important to display the message for phagocytes such as macrophages to recognize the cell for destruction, because macrophages highly express PS receptors on their cell surface. Thus, the PS-modified liposome is used to target phagocytes such as Kupffer cells in the liver [27, 71], cardiac macrophages [53], inflammatory macrophages involved in the development of obesity [168], *Mycobacterium tuberculosis* [49], and infiltrating macrophages in atherosclerotic vulnerable plaques [122]. Because macrophages play important roles in the progression of various diseases, the PS liposome has the possibility of being a useful tool for both diagnosis and therapy of such diseases.

Sugars

Sugar chains, which are composed of various monosaccharides and linked to lipids and proteins, are universally found in bacteria, viruses, fungi, and mammalians, and are involved variously in development, proliferation, mobility, memory formation, the immune system, and so on. Sugar chains are cleaved by glycolytic enzymes such as sialidases, and the function of them are locally controlled in the microenvironment of organs [111]. Sialic acid, an acidic monosaccharide, is frequently expressed at the end of glycans and gives a negative charge to the cell surface. This negative charge on the cell surface affords cells with a long circulation potential, with the erythrocyte, in which sialyl glycans are highly expressed, being a representative example. Thus, liposome formulations mimicking the erythrocyte membrane by conjugating gangliosides and sialic acid derivatives, such as monosialoganglioside (GM1) on the liposomal surface, were previously developed to equip these liposomes with a long circulating potential [6, 40]. We also developed glucuronate-conjugated liposomes, which showed a long circulation time in the bloodstream by avoiding the RES system and an enhanced anti-tumor action due to the subsequent EPR effect [118, 126]. On the other hand, it is known that some tumor-associated carbohydrate antigens, such as sialyl LewisX (sLe^{X}), are highly expressed in leukocytes and some types of metastatic tumor cells [159]. Because E-selectin is a receptor of sLe^{X} and is highly expressed on the angiogenic endothelial cells in tumor tissues, sLe^{X} is considered to be one of the key tumor-associated carbohydrate antigens strongly involved in hematogenous metastasis. Therefore, an anti-metastatic strategy based on the inhibition of the interaction between sLe^{X} and E-selectin has been proposed [120, 160]. Indeed, sLe^{X} is useful a targeting probe for liposomes to inhibit the interaction with E-selectin [2, 22, 142, 182]. Furthermore, these SLe^{X}-conjugated liposomes encapsulating cytotoxic drugs can suppress tumor growth and metastasis by delivering the encapsulated cytotoxic drugs to E-selectin-expressed endothelial cells in tumor tissue and subsequently disrupting the tumor vessels [60, 91].

Monosaccharide-conjugated liposomes are often used to deliver drugs ([51, 57],), nucleic acid medicines [73], and immunogens [94, 172] to the liver, because hepatic cells highly express monosaccharide receptors on their surface [94, 172]. Kawakami et al. previously compared the distribution of galactosylated (Gal), mannosylated (Man), and fucosylated (Fuc) liposomes after intravenous injection into mice, and demonstrated that all Gal, Man, and Fuc liposomes were transferred to the liver from the blood circulation immediately after systemic injection but that their intrahepatic distribution differed: Gal liposomes were mainly distributed to hepatic parenchyma cells, whereas Man and Fuc liposomes had accumulated on non-parenchyma cells in the liver [116]. They concluded that the different accumulation was dependent on the receptor expression in the liver: Gal liposomes were taken up into the parenchyma cells via asialoglycoprotein receptors; and Man and Fuc liposomes were trapped by mannose receptors and fucose receptors, respectively, in the non-parenchyma cells such as Kupffer cells [74]. Furthermore, the accumulation of Gal liposomes in parenchyma cells was highly dependent on the moiety of galactosyl lipids modified on the liposomes [106]. Utilizing the targeting potentials of galactosyl liposomes for hepatic parenchyma cells, a therapeutic approach for either hepatic carcinoma or liver metastatic tumors has been invented, resulting in a potent anti-tumor effect of anti-tumor drug-encapsulated galactosyl liposomes on hepatic tumors [185]. Meanwhile, the involvement of tumor-associated macrophages (TAMs) in tumor development has been recently a matter of focus; and the regulation of TAM activity by using DDS technology has become one of the therapeutic approaches for effective tumor suppression. This approach is reasonable, because TAMs, which are classified as anti-inflammatory M2 macrophages, express mannose receptors and support tumor cell growth by inhibiting inflammatory reaction in tumor tissue [3]. Therefore, active targeting liposomes such as Man liposomes can be used for TAM targeting to down-regulate the activity of them in tumor tissue [84, 88, 112].

The above lines of evidence point to the importance of liposome-surface sugar modification for targeting phagocytes, especially macrophages, which modification contributes to the effective delivery of encapsulated drugs for cancer treatment.

Lectins

Lectins are generally defined as proteins that have the potential to bind to the sugar chains of glycosylated proteins and glycolipids, and the expression of lectins is widely found in mammalian cells. Because the binding of lectins to sugar chains plays important roles in cell recognition and communication, lectins are thought to be one of key molecules that control signals for tumor development [28]. Therefore, it follows that a lectin can be a targeting probe for delivery of liposomes to a tumor [16, 18, 97]. *Wheat germ agglutinin* (WGA) is a lectin having a specific affinity for inflamed endothelial cells including angiogenic endothelial cells in which sugar chains terminated in *N*-acetyl-D-glucosamine or sialic acid were expressed by stimulation of inflammatory cytokines such as tumor necrosis factor alpha (TNF-α) [31, 52]. Because of the targetability of WGA and possibility of using it as a pilot molecule for tumor vasculature, WGA-conjugated liposomes have been developed. Geng et al. reported that the treatment of irradiated tumor-bearing mice with WGA-conjugated liposomes containing the anti-tumor drug cisplatin led to a significant delay of tumor growth by targeting cisplatin to irradiated tumor blood vessels [46]. Liu et al. also demonstrated the application of WGA-conjugated liposomes encapsulating doxorubicin to the treatment of drug-resistant tumors [100].

Bauhinia purprea agglutinin (BPA) is a well-known lectin that recognizes glycoproteins terminating in galactose, such as Thomsen–Friedenreich antigen, which is a tumor-associated molecule involved in prostate carcinogenesis; and its expression in prostate cancer cells increases with the development of tumor malignancy [180]. Pacis et al. previously studied the relationship between the expression of carbohydrates on the surface of prostate cancer cells and tumor malignancy, and they reported that an increase in the expression of galactosyl carbohydrates facilitates the bone metastasis of human prostate cancer cells [130]. On the other hand, we previously found that BPA strongly bound to human prostate cancer specimens, whereas it did not bind to normal prostate tissue [66]. Then, to demonstrate whether BPA could be applicable as a targetable liposomal probe for human prostate cancers, we prepared BPA-conjugated PEG-modified liposomes (BPA-PEG-LP) and investigated the association of the liposomes with two human prostate cancer cell lines, LNCaP and DU145. The results indicated that BPA-PEG-LP dominantly bound to both cell lines, was internalized into the cytosol, and associated with the cells to a significantly higher degree than did the unconjugated liposomes. Additionally, BPA-PEG-LP highly accumulated in DU145 solid tumors after systemic injection and became widely distributed in the tumor tissue. Furthermore, treatment with doxorubicin (DOX)-encapsulated BPA-PEG-LP significantly suppressed the solid tumor growth of DU145 prostate tumor cells. These results indicate that liposome-surface BPA molecules recognized prostate cell-surface galactocyl glycoproteins and that BPA-PEG-LP could thus be a potent DDS carrier for the treatment of human prostate cancers [66].

Ligand Proteins

In general, tumor growth is regulated by many kinds of cytokines and hormones; and receptor-mediated cell signaling plays crucial roles in the proliferation, differentiation, survival, and metastasis of tumor cells. Because tumor cells highly express the receptors for these ligand proteins for their maintenance, a ligand-receptor interaction is a key event during tumor development. Considering the specific interaction of ligand with receptor, ligand proteins involved in tumor development would be pilot molecules for effective tumor delivery of DDS carriers.

Folate receptor (FR), a glycosyl phosphatidylinositol (GPI)-anchored membrane protein, is known to be highly expressed in many kinds of tumor cells including ovarian cancer, but not in normal cells; and the ligand protein, folate, is often used as a pilot molecule for targeting FR-expressing tumor cells [41, 59, 89, 146]. Indeed, many researchers have developed folate-conjugated liposomes as tumor-targeting nanocarriers for cancer therapy with the liposomal anti-tumor drugs [23], such as daunorubicin [121, 174], DOX [42, 131, 139, 176], mitomycin C [136], zoledronic acid [152, 153], and molecular target drugs [58, 178]. These liposomes have also be used for cancer diagnosis using magnetic resonance imaging (MRI) [70], fluorescence imaging [32] and for cancer theranostics using microbubbles [128]. To further enhance the uptake of FR-targeted liposomes into tumor cells, cell-penetrating peptides such as TAT [186] and Pep-1 [72] have been used for modification of FR-targeted liposomes.

Iron is a mineral essential for cell growth and maintenance and transferrin (Tf) is a glycoprotein containing ferric iron in its structure which transports this ion throughout the whole body. It is well known that the receptor for Tf is highly expressed on the surface of many kinds of tumor cells and promotes their growth

[125]. In the process of incorporation of Tf into the cells, it is internalized by an endocytotic mechanism after binding to the cognate receptor. This process can be used as an important method to effectively deliver liposome-encapsulated drugs into cells and to have a higher anti-tumor effect. Thus, Tf-conjugated liposomes are expected to be suitable as drug delivery carriers for treatment of Tf receptor-expressing tumors [155]. Many researchers previously prepared Tf-conjugated liposomes by conjugation with Tf protein at the terminus of the PEG chains of PEG-modified liposomes via a linker lipid such as DSPE-PEG-NHS and examined their potential as a drug carrier for delivery to Tf receptor-expressing tumors [87, 184]. Maruyama et al. previously prepared TF-PEG-conjugated liposomes encapsulating platinum anti-tumor agents such as cisplatin and oxaliplatin, and demonstrated their anti-tumor effects on Tf receptor-expressing gastric [65] and colon [67] cancers. When they examined the uptake of Tf-PEG-conjugated liposomes into tumor cells, they found that these liposomes were highly taken up into Tf receptor-expressing colon cancer cells and that the uptake of these liposomes was blocked in the presence of an excess amount of transferrin, suggesting that the Tf-PEG-conjugated liposomes had been taken up into the cells via Tf receptor-mediated endocytosis [165]. They also demonstrated the potent therapeutic effects of Tf-PEG-conjugated liposomes encapsulating platinum anticancer drugs on solid tumor growth in gastric and colon cancer-bearing mice. Other researchers have similarly used Tf-conjugated liposomes for the delivery of nucleic acid medicines including plasmid DNA and siRNA to TF receptor-expressing tumors [110, 170, 175]. Kobayashi et al. focused on the endosomal uptake of Tf receptor-targeted liposomes into tumor cells by which the liposomes could avoid the efflux via a P-glycoprotein (P-gp) and demonstrated the application of Tf receptor-targeted liposomes encapsulating doxorubicin to cancer chemotherapy of multi-drug resistant tumors [80]. Gao et al. further found that dual modification of the liposome surface with both folate and Tf enabled the liposomes to pass through the blood-brain barrier (BBB) by avoiding efflux involved with P-gp in the BBB and demonstrated the usage of the targeting liposomes for glioma chemotherapy [45]. In light of the potential value of Tf-conjugated liposomes for the penetration of the BBB, brain delivery of nucleic acid medicines has been also been proposed [187]. Cardoso et al. reported that Tf-conjugated liposomes complexed with siRNA could mediate efficient gene silencing in neuronal cells and subsequently protect against neuronal cell death, suggesting that this approach could be useful for the treatment of neurodegenerative disorders [20, 21].

Boron neutron capture therapy (BNCT), which is based on the nuclear reaction between ^{10}B and thermal neutrons has been clinically applied to the treatment of malignant brain tumors and melanomas, since strong tumor cell eradication by BNCT can be achieved. However, the selective delivery of ^{10}B compounds to the tumor cells is required, because the range of α-rays produced by the boron neutron capture reaction is quite limited. Therefore, the application of Tf receptor-targeted liposomes to cancer BNCT is awaited, because Tf-PEG-conjugated liposomes are promisingly internalized into the tumor cells. Indeed, some researchers have tried to encapsulate ^{10}B compounds into Tf receptor-targeted liposomes and demonstrated strong tumor suppressive efficacy *in vivo* by BNCT [33, 108, 117]. Photodynamic therapy (PDT), using a photosensitizer and tissue-penetrating visible laser light, shows significant efficacy in the treatment of malignant tumors: Laser light promotes the photosensitizer to enter its excited state; and the photosensitizer, in turn, interacts with molecular oxygen and returns to its ground state, resulting in the generation of a real photodynamic agent, singlet oxygen, which affords tumor destruction [34]. To deliver the photosensitizers to tumors for PDT, Tf-mediated liposomal targeting of the photosensitizers is also a promising potential tool for cancer PDT [29, 47, 135]. On the other hand, the application of Tf-conjugated liposomes to molecular imaging, not only for early diagnosis of cancer, but also to monitor therapeutic results after anti-tumor treatment. has been reported; and imaging modalities such as MRI [148], computed tomography (CT) [114], and fluorescence imaging [157] are used for this application.

Peptides

Peptides are often used as tools both for active targeting of drugs and as DDS carriers, because they have high affinity for their target molecule, and are easy-to-use from the view point of chemical synthesis, manufacturing, and cost. To obtain a candidate peptide binding to the desired target molecule, biopanning with a phage-displayed peptide library is appropriate, because it is an easy and high throughput method to determine the targetable peptide that has high and specific affinity for the target molecule. Arap and colleagues injected a phage-displayed peptide library in the bloodstream of tumor-bearing mice to determine the peptides binding to tumor cells and thereby identified Asn-Gly-Arg (NGR) and Arg-Gly-Asp (RGD) peptide motifs as strong binders [8]. NGR selectively binds to CD13 and aminopeptidase N [132], whereas RGD binds to integrin $\alpha_v\beta_3$ and integrin $\alpha_v\beta_5$ [140]. These membrane proteins are highly expressed in not only tumor cells but also angiogenic endothelial cells in many kinds of tumor tissues, because they function in tumor cell growth, adhesion on the extracellular matrix (ECM), motility, invasion into ECM, and metastasis in distal organs. Therefore, these tumor-targetable peptides have been used as pilot molecules that enable drugs [24], biopharmaceuticals [25, 26, 188], and nanoparticles [36, 61, 173] to be effectively delivered to tumor sites for cancer therapy [164], metastasis suppression [90], and imaging [92, 95]. Among them, the RGD peptide, which was originally identified as a part of the amino acid sequence within the fibronectin ECM protein [62], is most widely used for tumor-targeting probe in DDS research. We previously developed an RGD-based siRNA delivery system using PCLs for treatment of malignant melanoma. In brief, we prepared PCLs containing DCP-TEPA, which were conjugated with RGD-peptide to target integrin-expressing lung-metastatic melanoma cells and angiogenic endothelial cells (RGD-PEG-PCL). The results revealed that the intravenous injection of RGD-PEG-PCL carrying siRNA against luciferase showed high knockdown efficiency against luciferase-expressing B16F10 tumors in the lungs of mice [179]; and the treatment with RGD-PEG-PCL carrying an siRNA cocktail for therapeutic treatment was able to significantly inhibit the growth of B16F10 melanoma in a lung-metastatic tumor model [75]. Harashima and colleagues also demonstrated the potential usage of a tumor-targeted siRNA delivery system using RGD peptide for cancer anti-angiogenic therapy [77, 143]. These findings suggest that targeted delivery of siRNAs by use of the RGD-based drug delivery system has considerable potential for cancer treatment.

Our colleagues previously identified a novel peptide that targets peritoneal tumors of human gastric cancer by *in vivo* intraperitoneal biopanning: Ser-Trp-Lys-Leu-Pro-Pro-Ser (SWKLPPS). A phage clone displaying the SWKLPPS peptide sequence showed higher binding to the peritoneal tumors than control phages. Furthermore, we prepared SWKLPPS-conjugated liposomes and demonstrated the enhanced accumulation of these liposomes in tumor tissue after intraperitoneal injection into a peritoneal tumor metastasis model mouse of human gastric cancer and also found a significant cytotoxic effect on the tumor cells [1], suggesting that SWKLPPS peptide would be a guide molecule for liposome-mediated drug delivery to peritoneal tumors.

Antibodies

Antibodies, also known as immunoglobulins, are large-molecular-weight proteins produced by immune cells to protect the body from exogenous pathogens by serving as part of an immune surveillance and protection system. Notably, an antibody recognizes a unique antigenic protein via the variable regions of the Fab sequence of its molecular structure; and it shows high affinity and specific recognition of the immunogen generating it. Therefore, an antibody is one of the most suitable pilot molecules for targeting liposome delivery to tumors; and antibody-conjugated liposomes strongly promise their specific delivery to tumor cells [169], since tumor cells often express characteristic molecules on their surface that are not found on normal cells. Among targeted molecules that are highly expressed in tumor cells, members of the epidermal growth factor (EGF) receptor family, such as EGF receptor and HER2 are commonly used for the targets of antibody-conjugated liposomes, because antibody drugs such as Cetuximab (anti-EGFR) and Trastuzumab (anti-HER2) have been commercialized and are used for the treatment of some types of tumors [104]. Many researchers have developed anti-EGFR antibody-conjugated liposomes encapsulating anti-tumor drugs [78, 105, 177, 181], anti-inflammatory drugs [98], and nucleic acid medicines [43, 44] and have demonstrated their potent therapeutic effect on EGFR-expressing tumors. In addition, anti-HER2 antibody-conjugated liposomes [19, 85, 166, 177] have also been developed for the cancer therapy of breast tumors.

Membrane-type 1 matrix metalloproteinase (MT1-MMP), which is identified as a member of the zinc-dependent matrix metalloproteinase family, is known to be required for the degradation of the ECM surrounding both tumor and endothelial cells. Since MT1-MMP is highly expressed in malignant tumor cells and angiogenic endothelial cells, it can be expected to be a useful target for cancer treatment. We developed the Fab' fragment of anti-MT1-MMP antibody-conjugated PEG liposomes to target both MT1-MMP-expressing human fibrosarcoma HT1080 and human umbilical endothelial cells (HUVECs) and found enhanced uptake of the liposome into both cells via the interaction with MT1-MMP molecule [14]. Hatakeyama et al. further demonstrated that the treatment with anti-MT1-MMP antibody-conjugated PEG liposomes encapsulating DOX significantly suppressed the tumor growth of HT1080 solid tumor-bearing mice and prolonged the survival time of the mice [54].

In a recent study, we focused on heparin-binding epidermal growth factor-like growth factor (HB-EGF), a membrane protein, as a target molecule of liposome delivery; because HB-EGF is known to be highly expressed on the cell surface of various cancers, such as breast cancer and gynecological cancer, including ovarian cancer [113, 129]. The precursor of HB-EGF is expressed on the cell surface as a membrane-anchored form (proHB-EGF) and then processed to a soluble form of HB-EGF, which mediates the intracellular signaling. Thus, we expected the membrane-binding form of HB-EGF to be a useful target molecule for delivering drugs to tumors. To deliver the drugs to the tumors by using the liposomal DDS approach, we previously prepared immunoliposomes conjugated with the Fab' fragment of anti-HB-EGF antibody and confirmed the therapeutic effect of the antibody-conjugated liposomes on human breast MDA-MB-231 tumor-bearing mice. The results indicated that the immunoliposomes highly bound to and were taken up into HB-EGF-overexpressing Vero cells and MDA-MB-231 cells compared with low-expressing wild-type Vero cells, whereas PEG-modified liposomes were little taken up into any cells. In addition, treatment with DOX-encapsulated immunoliposomes resulted in tumor regression in MDA-MB-231 tumor-bearing mice. Furthermore, we more recently demonstrated the capability of siRNA delivery to MDA-MB-231 tumors by using anti-HB-EGF antibody-conjugated lipid nanoparticles [123] resulting in a potent gene-silencing and subsequent suppression of solid tumor growth on the tumor-bearing mice [124]. These results indicate that anti-HB-EGF antibody-conjugated immunoliposomes could be a useful DDS carrier for the treatment of HB-EGF-expressing cancers such as breast and ovarian cancers.

Considering that the actual demand for antibodies in the development of molecular target drugs for cancer chemotherapy is rising, antibodies may be the premier pilot molecule for liposome delivery to tumors in the near future.

Targeting Strategy

Angiogenic Vessel Targeting

As describe above, liposomes conjugated with appropriate pilot molecules have been used for delivery of encapsulated or associated materials to specific organs or cells *in vivo* after injection of the liposomes into the bloodstream of model animals or humans. Liposomes or nanoparticles injected into animals firstly encounter blood cells and blood vessels after systemic injection; and, therefore, blood vessels are appropriate targets for DDS carriers. We previously selected tumor angiogenic vessels as a target of the drug carriers to achieve cancer treatment. Physiological angiogenesis observed in wound healing and female reproductive cycle is controlled by the combination of both pro-angiogenic factors and anti-angiogenic factors. In contrast, uncontrolled angiogenesis vigorously occurs in tumor tissues, and leaky angiogenic vessels are formed. In an earlier study, we prepared pilot molecules for targeting angiogenic vessels by using a phage-displayed random peptide library. The appropriate phage library was injected into the dorsal air sac model mice, and phages were recovered from the dorsum skin attached to an implanted chamber containing tumor cells. This biopanning was repeated several times for obtaining phage clones that specifically accumulated in angiogenic vessels. To analyze the phage clones expressed specific peptides, we determined the epitope sequence of the peptide showing the highest affinity for angiogenic vessels: APRPG [127]. We also clarified the target of APRPG peptide as being vascular endothelial growth factor receptor-1 (VEGFR-1), which is specifically present on angiogenic vessels [81]. Then, *in vivo* trafficking of liposomes modified with polyethylene glycol (PEG) was examined in tumor-bearing mice treated with APRPG peptide conjugated or not at the end of the liposomal PEG chains [103]. The results indicated that the APRPG-conjugated liposomes strongly accumulated in the tumor site and that this anti-neovascular therapy was appropriate for cancer therapy [11, 12]. In other research for exploring the angiogenic vessel-targetable peptide, our collaborators previously isolated the substrate peptides for MT1-MMP by using a phage-displayed hexamer substrate peptide library and determined the peptide sequence as Gly-Pro-Leu-Pro-Leu-Arg (GPLPLR). Because MT1-MMP is a key enzyme for formation of angiogenic vessels in tumor tissue, it is a suitable target molecule for delivery of anti-tumor agent to angiogenic vessels. We conjugated the GPLPLR peptide to the liposomes containing DPP-CNDAC, a hydrophobized derivative of the anti-tumor nucleoside CNDAC [9, 10] and demonstrated the utility of GPLPLR-conjugated liposomal DPP-CNDAC for anti-neovascular therapy [83]. We recently found another angiogenic vessel-targeting peptide, Ala-Ser-Ser-His-Asn (ASSHN), by *in vitro* phage biopanning with human endothelial progenitor cells. Then, PEGylated liposomes conjugated with the ASSHN peptide were designed and prepared for the delivery of anti-cancer agents. Treatment with ASSHN-PEG-conjugated liposomes encapsulating DOX strongly suppressed the tumor growth of mouse Colon 26 tumor in the tumor-bearing mice compared with PEGylated liposomes encapsulating DOX [39]. Similar strategies of angiogenic vessel targeting by using RGD-conjugated liposomes by Jassen et al. [69] and NGR-conjugated liposomes by Pastorino et al. [133] were reported, and these groups succeeded to deliver DOX [134, 145] and siRNA against VEGFR-2 [171] to angiogenic endothelial cells and obtained strong tumor suppression through the damage of the angiogenic vessels.

Since angiogenesis is critical for the growth of solid tumors and angiogenic endothelial cells are growing during this process, those cells are caused to degenerate by anti-cancer drugs. In the tumor-bearing mouse model, angiogenic vessel-targeted delivery of anti-cancer drugs eradicated the cancer cells, probably due to cutting off the oxygen and nutrient supplies to the cancer cells [149, 150]. Kerbel et al. reported that repeated low-dose administration of an anti-cancer drug, namely, metronomic-dosing chemotherapy, suppressed the growth of cancer cells without causing severe side effects [15, 76]. It is possible that the metronomic-dosing chemotherapy is similarly as effective as the anti-angiogenic therapy targeting angiogenic vessels for eradicating cancer cells without severe side effects. Luo et al. further demonstrated by applying the combination of the metronomic chemotherapy with angiogenic vessel-targeting paclitaxel (PTX) delivered by sterically-stabilized NGR-conjugated liposomes that low-dose metronomic injection of the NGR-PEG-conjugated liposome encapsulated PTX into HT1080 fibrosarcoma-bearing mice showed the most significant anti-tumor and anti-angiogenic activities compared with that of non-targeted PEGylated liposomes encapsulated with PTX and high-dose non-metronomic injection of NGR-PEG-conjugated liposomes encapsulated with PTX. [101].

Dual Targeting

In order to add more value to the targeted liposomes which are conjugated with one kind of pilot molecule, the strategy of dual ligand modification to the liposomes has been tried. As an example, we herein introduce brain tumor targeting by the liposomes conjugated with dual ligands. For the treatment of brain tumors by a liposome delivery system, BBB penetration of liposomes is most essential, since the BBB restricts the influx of not only drugs but also DDS carriers into brain tissue. To achieve the BBB penetration of liposomes, Tf is often used as a pilot molecule, because Tf has functional potentials not only to target Tf receptor-expressing tumors, but also to be transported into the endothelial cells. Therefore, dual modification of liposomes with both Tf and a tumor-targeting ligand is carried out for the treatment of brain tumors including malignant gliomas. Some researchers have reported the utility of such agents for the treatment of brain tumors: They used either folate [45] or RGD peptide [138] as the tumor-targeting probe. Du et al. used tamoxifen as the BBB-transportable molecule for the targeted liposome delivery across the BBB, because tamoxifen has a potential to inhibit efflux of multi-drug resistance proteins such as P-gp in the BBB. They additionally conjugated WGA lectin to facilitate endocytosis in both BBB and tumor cells and developed topotecan-incorporating liposome dually conjugated with tamoxifen and WGA for the treatment of glioma in the brain [35]. These pilot molecules work in an independent manner for delivery of the liposomes to tumor sites.

On the other hand, a synergistic increase in targeting potential of dual ligand modification of liposomes and the combination effect are not yet clear, although there is the possibility that pilot ligands on a liposomal membrane interact cooperatively with different target molecules on the plasma membrane of the target cells. In fact, as shown above, pilot peptides such as NGR [30, 101, 119], RGD [77, 99, 171], and APRPG were used individually to modify liposomes to ensure the targetability to angiogenic vessels and to prove the angiogenic vessel-targeting strategy for cancer therapy. Therefore, we sought to prove the "dual-targeting" of liposomes by using two different pilot molecules on the surface of the same liposomes for clarifying whether the association of the liposomes with the target cell membranes was cooperatively enhanced or not (Fig. 4a).

We first prepared dual-targeting liposomes conjugated with both APRPG and GNGRG, and investigated the usefulness of them for delivery of anticancer drugs to

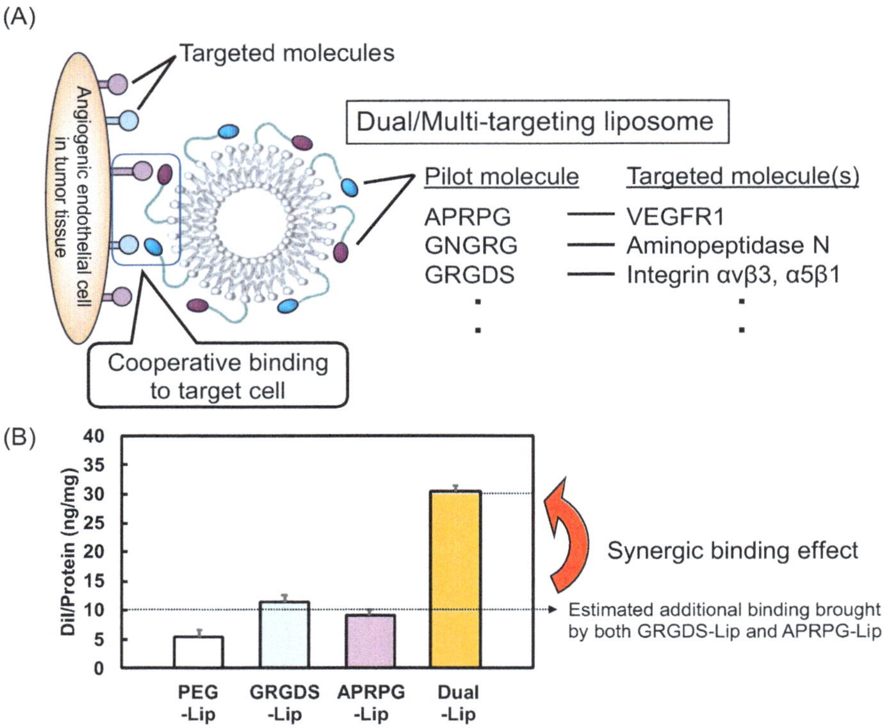

Fig. 4 Dual targeting of liposomes to angiogenic endothelial cells
(**a**) Concept of dual targeting strategy with a targeted liposome
Dual/Multi targeting liposome is prepared by conjugating two or more pilot molecules on the liposome surface. Enhanced targeting effect can be achieved via cooperative binding of the pilot molecules to targeted molecules
(**b**) Association of dual-targeting liposomes with angiogenic endothelial cells
$DiIC_{18}$-fluorescently labeled PEG-Lip (DSPC: Cholesterol (Chol): DSPE-PEG = 10: 5: 1), GRGDS-Lip (DSPC: Chol: DSPE-PEG-GRGDS = 10: 5: 1), APRPG-Lip (DSPC: Chol: DSPE-PEG-APRPG = 10: 5: 1), and Dual-Lip (DSPC: Chol: DSPE-PEG-GRGDS: DSPE-PEG-APRPG = 10: 5: 0.5: 0.5) were prepared. Then, the liposomes were incubated with human umbilical endothelial cells (HUVECs) for 4 h at 37 °C. After washing, the amount of liposomes associated with the HUVECs was determined fluorometrically. The association of liposomes is presented as the amount of $DiIC_{18}$ per cellular protein amount

tumor angiogenic vessels. Liposomes composed of DSPC and cholesterol with either DSPE-PEG, DSPE-PEG-APRPG or DSPE-PEG-GNGRG (10:5:1 as a molar ratio), or with both DSPE-PEG-APRPG and DSPE-PEG-GNGRG (10:5:0.5:0.5 as a molar ratio) were prepared by the hydration method and sized with 100-nm-pored filter. Then, the association of dual-targeting liposomes (Dual-PEG-Lip) to cytokine-stimulated human umbilical vein endothelial cells (HUVECs) was examined in comparison with that of single-targeting liposomes (APRPG-PEG-Lip or GNGRG-PEG-Lip). As a result, the association of Dual-PEG-Lip with HUVECs was significantly higher than that of either single-targeting liposomes [115]. Moreover, the anti-proliferative effect of DOX encapsulated in dual-targeting liposomes (Dual-PEG-DOX) on HUVECs was also significantly higher than that of DOX encapsulated in single-targeting liposomes (APRPG-PEG-DOX, GNGRG-PEG-DOX). Next, we determined the intratumoral localization of those liposomes in Colon26 NL-17 cancer-bearing mice. At 3 h after administration of 1,1′-dioctadecyl-3,3,3′,3′-tetramethyl-indocarbocyanine (DiIC$_{18}$)-labeled liposomes, the tumors were dissected and the tumor sections were observed under a confocal laser-scanning microscope, after which CD31 was visualized with Alexa Fluor 488. The results indicated that angiogenic endothelial cell-targeting liposomes were co-localized with CD31 and that the dual-targeting liposomes localized more intensely with CD31 than did the single-targeting liposomes [115].

An experiment similar to that described above was performed by using another couple of pilot molecules, namely, APRPG and GRGDS. Dual-targeting liposomes containing both DSPE-PEG-APRPG and DSPE-PEG-GRGDS synergistically associated with cytokine-stimulated HUVECs more so compared with single-targeting liposomes (APRPG- Lip or GRGDS-Lip) (Fig. 4b). Here, we have discussed about dual-targeting liposomes bearing different kinds of pilot molecules targeting angiogenic endothelial cells, although integrin $\alpha v\beta 3$ is known to be expressed on various tumor cells. In fact, the intratumoral distribution of dual-targeting liposomes composed of DSPE-PEG-APRPG and DSPE-PEG-GRGDS was monitored by confocal laser-scanning microscopy, and these liposomes found to be distributed around the blood vessels as well as to be co-localized with endothelial cells, suggesting that APRPG and GRGDS-conjugated liposomes targeted both angiogenic endothelial cells and tumor cells [163]. Kulza et al. similarly developed dual-targeting liposomes for angiogenic vessels, using both galectin-1-specific anginex [50] and RGD peptide, and they demonstrated the improved accumulation of MRI contrast agent at the tumor angiogenic vessels by using these targeting liposomes and the application of MRI for monitoring angiogenic activity [79]. To enhance the effective delivery to and selectivity of liposomes for tumors, this tumor-targeted dual ligand approach has been carried out. Saul et al. designed the tumor-targeted liposomes using both folate and a monoclonal antibody against EGF receptor as the tumor-targeting pilot molecules and found synergistic cytotoxicity of the liposomal DOX against human KB tumor cells expressing both folate and EGF receptors [144].

According to the above findings, dual-targeting and/or multi-targeting liposomes might be useful for not only enhancing the association to target cells, but also delivering the carrying materials to different kinds of cells, for example, such as endothelial cells, tumor cells, and other tumor-related cells in a tumor tissue.

Conclusions

In this chapter, we introduced the pilot molecules for tumor-targeted liposome delivery and the targeting strategies using them for effective cancer chemotherapy. At present, there are some liposomal anti-tumor agents that have been commercialized and used for cancer chemotherapy. In reality, however, they have limited applicability to certain tumors; and their anti-tumor mechanisms are poorly understood. Considering the advantages of tumor-targeting strategy with liposomes as described in this chapter, the next generation of liposomal formulations for cancer therapy will definitely come to possess a greater targeting potential by incorporating a pilot molecule. We expect that the usage of liposomes with pilot molecules will become one of the mainstream methods for tumor-targeted drug delivery in the near future.

References

1. Akita N, Maruta F, Seymour LW, Kerr DJ, Parker AL, Asai T, Oku N, Nakayama J, Miyagawa S (2006) Identification of oligopeptides binding to peritoneal tumors of gastric cancer. Cancer Sci 97:1075–1081
2. Alekseeva A, Kapkaeva M, Shcheglovitova O, Boldyrev I, Pazynina G, Bovin N, Vodovozova E (2015) Interactions of antitumour Sialyl Lewis X liposomes with vascular endothelial cells. Biochim Biophys Acta 1848:1099–1110
3. Allavena P, Mantovani A (2012) Immunology in the clinic review series; focus on cancer: tumour-associated macrophages: undisputed stars of the inflammatory tumour microenvironment. Clin Exp Immunol 167:195–205
4. Allen TM, Cullis PR (2004) Drug delivery systems: entering the mainstream. Science 303:1818–1822
5. Allen TM, Cullis PR (2013) Liposomal drug delivery systems: from concept to clinical applications. Adv Drug Deliv Rev 65:36–48
6. Allen TM, Hansen C, Rutledge J (1989) Liposomes with prolonged circulation times: factors affecting uptake by reticuloendothelial and other tissues. Biochim Biophys Acta 981:27–35
7. Ando H, Okamoto A, Yokota M, Asai T, Dewa T, Oku N (2013) Polycation liposomes as a vector for potential intracellular delivery of microRNA. J Gene Med 15:375–383
8. Arap W, Pasqualini R, Ruoslahti E (1998) Cancer treatment by targeted drug delivery to tumor vasculature in a mouse model. Science 279:377–380
9. Asai T, Kurohane K, Shuto S, Awano H, Matsuda A, Tsukada H, Namba Y, Okada S, Oku N (1998) Antitumor activity of 5′-O-dipalmitoylphosphatidyl 2′-C-cyano-2′-deoxy-1-beta-D-arabino-pentofuranosylcytosine is enhanced by long-circulating liposomalization. Biol Pharm Bull 21:766–771
10. Asai T, Shuto S, Matsuda A, Kakiuchi T, Ohba H, Tsukada H, Oku N (2001) Targeting and anti-tumor efficacy of liposomal 5′-O-dipalmitoylphosphatidyl 2′-C-cyano-2′-deoxy-1-beta-D-arabino-pentofuranosylcytosine in mice lung bearing B16BL6 melanoma. Cancer Lett 162:49–56

11. Asai T, Shimizu K, Kondo M, Kuromi K, Watanabe K, Ogino K, Taki T, Shuto S, Matsuda A, Oku N (2002) Anti-neovascular therapy by liposomal DPP-CNDAC targeted to angiogenic vessels. FEBS Lett 520:167–170
12. Asai T, Miyazawa S, Maeda N, Hatanaka K, Katanasaka Y, Shimizu K, Shuto S, Oku N (2008) Antineovascular therapy with angiogenic vessel-targeted polyethyleneglycol-shielded liposomal DPP-CNDAC. Cancer Sci 99:1029–1033
13. Asai T, Matsushita S, Kenjo E, Tsuzuku T, Yonenaga N, Koide H, Hatanaka K, Dewa T, Nango M, Maeda N, Kikuchi H, Oku N (2011) Dicetyl phosphate-tetraethylenepentamine-based liposomes for systemic siRNA delivery. Bioconjug Chem 22:429–435
14. Atobe K, Ishida T, Ishida E, Hashimoto K, Kobayashi H, Yasuda J, Aoki T, Obata K, Kikuchi H, Akita H, Asai T, Harashima H, Oku N, Kiwada H (2007) In vitro efficacy of a sterically stabilized immunoliposomes targeted to membrane type 1 matrix metalloproteinase (MT1-MMP). Biol Pharm Bull 30:972–978
15. Bocci G, Kerbel RS (2016) Pharmacokinetics of metronomic chemotherapy: a neglected but crucial aspect. Nat Rev Clin Oncol 13:659–673
16. Bogdanov AA Jr, Gordeeva LV, Torchilin VP, Margolis LB (1989) Lectin-bearing liposomes: differential binding to normal and to transformed mouse fibroblasts. Exp Cell Res 181:362–374
17. Bostic HE, Smith MD, Poloukhtine AA, Popik VV, Best MD (2012) Membrane labeling and immobilization via copper-free click chemistry. Chem Commun (Camb) 48:1431–1433
18. Bruck A, Abu-Dahab R, Borchard G, Schafer UF, Lehr CM (2001) Lectin-functionalized liposomes for pulmonary drug delivery: interaction with human alveolar epithelial cells. J Drug Target 9:241–251
19. Burks SR, Macedo LF, Barth ED, Tkaczuk KH, Martin SS, Rosen GM, Halpern HJ, Brodie AM, Kao JP (2010) Anti-HER2 immunoliposomes for selective delivery of electron paramagnetic resonance imaging probes to HER2-overexpressing breast tumor cells. Breast Cancer Res Treat 124:121–131
20. Cardoso AL, Simoes S, de Almeida LP, Plesnila N, Pedroso de Lima MC, Wagner E, Culmsee C (2008) Tf-lipoplexes for neuronal siRNA delivery: a promising system to mediate gene silencing in the CNS. J Control Release 132:113–123
21. Cardoso AL, Costa P, de Almeida LP, Simoes S, Plesnila N, Culmsee C, Wagner E, de Lima MC (2010) Tf-lipoplex-mediated c-Jun silencing improves neuronal survival following excitotoxic damage in vivo. J Control Release 142:392–403
22. Chantarasrivong C, Ueki A, Ohyama R, Unga J, Nakamura S, Nakanishi I, Higuchi Y, Kawakami S, Ando H, Imamura A, Ishida H, Yamashita F, Kiso M, Hashida M (2017) Synthesis and functional characterization of novel Sialyl LewisX mimic-decorated liposomes for E-selectin-mediated targeting to inflamed endothelial cells. Mol Pharm 14:1528–1537
23. Chaudhury A, Das S (2015) Folate receptor targeted liposomes encapsulating anti-cancer drugs. Curr Pharm Biotechnol 16:333–343
24. Chen X, Plasencia C, Hou Y, Neamati N (2005) Synthesis and biological evaluation of dimeric RGD peptide-paclitaxel conjugate as a model for integrin-targeted drug delivery. J Med Chem 48:1098–1106
25. Curnis F, Sacchi A, Borgna L, Magni F, Gasparri A, Corti A (2000) Enhancement of tumor necrosis factor alpha antitumor immunotherapeutic properties by targeted delivery to aminopeptidase N (CD13). Nat Biotechnol 18:1185–1190
26. Curnis F, Arrigoni G, Sacchi A, Fischetti L, Arap W, Pasqualini R, Corti A (2002) Differential binding of drugs containing the NGR motif to CD13 isoforms in tumor vessels, epithelia, and myeloid cells. Cancer Res 62:867–874
27. Daemen T, Regts J, Scherphof GL (1996) Liposomal phosphatidylserine inhibits tumor cytotoxicity of liver macrophages induced by muramyl dipeptide and lipopolysaccharide. Biochim Biophys Acta 1285:219–228
28. de Oliveira FE, Albuquerque da Cunha CR, Albuquerque PBS, de Paula RA, Aranda-Souza MA, Alves MS, Zagmignan A, Carneiro-da-Cunha MG, Nascimento da Silva LC, Dos Santos

Correia MT (2017) Lectin-carbohydrate interactions: implications for the development of new anticancer agents. Curr Med Chem 24:3667–3680
29. Derycke AS, Kamuhabwa A, Gijsens A, Roskams T, De Vos D, Kasran A, Huwyler J, Missiaen L, de Witte PA (2004) Transferrin-conjugated liposome targeting of photosensitizer AlPcS4 to rat bladder carcinoma cells. J Natl Cancer Inst 96:1620–1630
30. Di Paolo D, Pastorino F, Zuccari G, Caffa I, Loi M, Marimpietri D, Brignole C, Perri P, Cilli M, Nico B, Ribatti D, Pistoia V, Ponzoni M, Pagnan G (2013) Enhanced anti-tumor and anti-angiogenic efficacy of a novel liposomal fenretinide on human neuroblastoma. J Control Release 170:445–451
31. DiCorleto PE, de la Motte CA (1989) Role of cell surface carbohydrate moieties in monocytic cell adhesion to endothelium in vitro. J Immunol 143:3666–3672
32. Ding N, Lu Y, Lee RJ, Yang C, Huang L, Liu J, Xiang G (2011) Folate receptor-targeted fluorescent paramagnetic bimodal liposomes for tumor imaging. Int J Nanomedicine 6:2513–2520
33. Doi A, Kawabata S, Iida K, Yokoyama K, Kajimoto Y, Kuroiwa T, Shirakawa T, Kirihata M, Kasaoka S, Maruyama K, Kumada H, Sakurai Y, Masunaga S, Ono K, Miyatake S (2008) Tumor-specific targeting of sodium borocaptate (BSH) to malignant glioma by transferrin-PEG liposomes: a modality for boron neutron capture therapy. J Neuro-Oncol 87:287–294
34. Dougherty TJ, Gomer CJ, Henderson BW, Jori G, Kessel D, Korbelik M, Moan J, Peng Q (1998) Photodynamic therapy. J Natl Cancer Inst 90:889–905
35. Du J, Lu WL, Ying X, Liu Y, Du P, Tian W, Men Y, Guo J, Zhang Y, Li RJ, Zhou J, Lou JN, Wang JC, Zhang X, Zhang Q (2009) Dual-targeting topotecan liposomes modified with tamoxifen and wheat germ agglutinin significantly improve drug transport across the blood-brain barrier and survival of brain tumor-bearing animals. Mol Pharm 6:905–917
36. Dubey PK, Mishra V, Jain S, Mahor S, Vyas SP (2004) Liposomes modified with cyclic RGD peptide for tumor targeting. J Drug Target 12:257–264
37. Fleiner M, Benzinger P, Fichert T, Massing U (2001) Studies on protein-liposome coupling using novel thiol-reactive coupling lipids: influence of spacer length and polarity. Bioconjug Chem 12:470–475
38. Frisch B, Hassane FS, Schuber F (2010) Conjugation of ligands to the surface of preformed liposomes by click chemistry. Methods Mol Biol 605:267–277
39. Fukuta T, Asai T, Kiyokawa Y, Nakada T, Bessyo-Hirashima K, Fukaya N, Hyodo K, Takase K, Kikuchi H, Oku N (2017) Targeted delivery of anticancer drugs to tumor vessels by use of liposomes modified with a peptide identified by phage biopanning with human endothelial progenitor cells. Int J Pharm 524:364–372
40. Gabizon A, Papahadjopoulos D (1988) Liposome formulations with prolonged circulation time in blood and enhanced uptake by tumors. Proc Natl Acad Sci U S A 85:6949–6953
41. Gabizon A, Shmeeda H, Horowitz AT, Zalipsky S (2004) Tumor cell targeting of liposome-entrapped drugs with phospholipid-anchored folic acid-PEG conjugates. Adv Drug Deliv Rev 56:1177–1192
42. Gabizon A, Tzemach D, Gorin J, Mak L, Amitay Y, Shmeeda H, Zalipsky S (2010) Improved therapeutic activity of folate-targeted liposomal doxorubicin in folate receptor-expressing tumor models. Cancer Chemother Pharmacol 66:43–52
43. Gao J, Liu W, Xia Y, Li W, Sun J, Chen H, Li B, Zhang D, Qian W, Meng Y, Deng L, Wang H, Chen J, Guo Y (2011) The promotion of siRNA delivery to breast cancer overexpressing epidermal growth factor receptor through anti-EGFR antibody conjugation by immunoliposomes. Biomaterials 32:3459–3470
44. Gao J, Yu Y, Zhang Y, Song J, Chen H, Li W, Qian W, Deng L, Kou G, Chen J, Guo Y (2012) EGFR-specific PEGylated immunoliposomes for active siRNA delivery in hepatocellular carcinoma. Biomaterials 33:270–282
45. Gao JQ, Lv Q, Li LM, Tang XJ, Li FZ, Hu YL, Han M (2013) Glioma targeting and blood-brain barrier penetration by dual-targeting doxorubincin liposomes. Biomaterials 34:5628–5639

46. Geng L, Osusky K, Konjeti S, Fu A, Hallahan D (2004) Radiation-guided drug delivery to tumor blood vessels results in improved tumor growth delay. J Control Release 99:369–381
47. Gijsens A, Derycke A, Missiaen L, De Vos D, Huwyler J, Eberle A, de Witte P (2002) Targeting of the photocytotoxic compound AlPcS4 to Hela cells by transferrin conjugated PEG-liposomes. Int J Cancer 101:78–85
48. Golkar N, Tamaddon AM, Samani SM (2016) Effect of lipid composition on incorporation of trastuzumab-PEG-lipid into nanoliposomes by post-insertion method: physicochemical and cellular characterization. J Liposome Res 26:113–125
49. Greco E, Quintiliani G, Santucci MB, Serafino A, Ciccaglione AR, Marcantonio C, Papi M, Maulucci G, Delogu G, Martino A, Goletti D, Sarmati L, Andreoni M, Altieri A, Alma M, Caccamo N, Di Liberto D, De Spirito M, Savage ND, Nisini R, Dieli F, Ottenhoff TH, Fraziano M (2012) Janus-faced liposomes enhance antimicrobial innate immune response in Mycobacterium tuberculosis infection. Proc Natl Acad Sci U S A 109:E1360–E1368
50. Griffioen AW, van der Schaft DW, Barendsz-Janson AF, Cox A, Struijker Boudier HA, Hillen HF, Mayo KH (2001) Anginex, a designed peptide that inhibits angiogenesis. Biochem J 354:233–242
51. Guo B, Cheng Y, Li N, Li X, Jin M, Li T, Li J (2012) In vitro and in vivo studies of galactose-modified liver-targeting liposomes. J Drug Target 21(3):257–264
52. Hanasaki K, Varki A, Stamenkovic I, Bevilacqua MP (1994) Cytokine-induced beta-galactoside alpha-2,6-sialyltransferase in human endothelial cells mediates alpha 2,6-sialylation of adhesion molecules and CD22 ligands. J Biol Chem 269:10637–10643
53. Harel-Adar T, Ben Mordechai T, Amsalem Y, Feinberg MS, Leor J, Cohen S (2011) Modulation of cardiac macrophages by phosphatidylserine-presenting liposomes improves infarct repair. Proc Natl Acad Sci U S A 108:1827–1832
54. Hatakeyama H, Akita H, Ishida E, Hashimoto K, Kobayashi H, Aoki T, Yasuda J, Obata K, Kikuchi H, Ishida T, Kiwada H, Harashima H (2007) Tumor targeting of doxorubicin by anti-MT1-MMP antibody-modified PEG liposomes. Int J Pharm 342:194–200
55. Hatakeyama H, Akita H, Harashima H (2011) A multifunctional envelope type nano device (MEND) for gene delivery to tumours based on the EPR effect: a strategy for overcoming the PEG dilemma. Adv Drug Deliv Rev 63:152–160
56. Hatakeyama H, Akita H, Harashima H (2013) The polyethyleneglycol dilemma: advantage and disadvantage of PEGylation of liposomes for systemic genes and nucleic acids delivery to tumors. Biol Pharm Bull 36:892–899
57. Hattori Y, Kawakami S, Yamashita F, Hashida M (2000) Controlled biodistribution of galactosylated liposomes and incorporated probucol in hepatocyte-selective drug targeting. J Control Release 69:369–377
58. Heger Z, Polanska H, Merlos Rodrigo MA, Guran R, Kulich P, Kopel P, Masarik M, Eckschlager T, Stiborova M, Kizek R, Adam V (2016) Prostate tumor attenuation in the nu/nu murine model due to anti-sarcosine antibodies in folate-targeted liposomes. Sci Rep 6:33379
59. Hilgenbrink AR, Low PS (2005) Folate receptor-mediated drug targeting: from therapeutics to diagnostics. J Pharm Sci 94:2135–2146
60. Hirai M, Minematsu H, Hiramatsu Y, Kitagawa H, Otani T, Iwashita S, Kudoh T, Chen L, Li Y, Okada M, Salomon DS, Igarashi K, Chikuma M, Seno M (2010) Novel and simple loading procedure of cisplatin into liposomes and targeting tumor endothelial cells. Int J Pharm 391:274–283
61. Huang P, Wu J, Li X, Liu X, Li Y, Cui G (2015) RGD peptide-pegylated PLLA nanoparticles containing epirubicin hydrochloride exhibit receptor-dependent tumor trafficking in vitro and in vivo. Pharm Res 32:2328–2343
62. Humphries MJ, Olden K, Yamada KM (1986) A synthetic peptide from fibronectin inhibits experimental metastasis of murine melanoma cells. Science 233:467–470
63. Ichikawa K, Urakami T, Yonezawa S, Miyauchi H, Shimizu K, Asai T, Oku N (2007) Enhanced desensitization efficacy by liposomal conjugation of a specific antigen. Int J Pharm 336:391–395

64. Ichikawa K, Asai T, Shimizu K, Yonezawa S, Urakami T, Miyauchi H, Kawashima H, Ishida T, Kiwada H, Oku N (2013) Suppression of immune response by antigen-modified liposomes encapsulating model agents: a novel strategy for the treatment of allergy. J Control Release 167:284–289
65. Iinuma H, Maruyama K, Okinaga K, Sasaki K, Sekine T, Ishida O, Ogiwara N, Johkura K, Yonemura Y (2002) Intracellular targeting therapy of cisplatin-encapsulated transferrin-polyethylene glycol liposome on peritoneal dissemination of gastric cancer. Int J Cancer 99:130–137
66. Ikemoto K, Shimizu K, Ohashi K, Takeuchi Y, Shimizu M, Oku N (2016) Bauhinia purprea agglutinin-modified liposomes for human prostate cancer treatment. Cancer Sci 107:53–59
67. Ishida O, Maruyama K, Tanahashi H, Iwatsuru M, Sasaki K, Eriguchi M, Yanagie H (2001) Liposomes bearing polyethyleneglycol-coupled transferrin with intracellular targeting property to the solid tumors in vivo. Pharm Res 18:1042–1048
68. Ito H, Kamachi T, Yashima E (2012) Specific surface modification of the acetylene-linked glycolipid vesicle by click chemistry. Chem Commun (Camb) 48:5650–5652
69. Janssen AP, Schiffelers RM, ten Hagen TL, Koning GA, Schraa AJ, Kok RJ, Storm G, Molema G (2003) Peptide-targeted PEG-liposomes in anti-angiogenic therapy. Int J Pharm 254:55–58
70. Kamaly N, Kalber T, Thanou M, Bell JD, Miller AD (2009) Folate receptor targeted bimodal liposomes for tumor magnetic resonance imaging. Bioconjug Chem 20:648–655
71. Kamps JA, Morselt HW, Scherphof GL (1999) Uptake of liposomes containing phosphatidylserine by liver cells in vivo and by sinusoidal liver cells in primary culture: in vivo-in vitro differences. Biochem Biophys Res Commun 256:57–62
72. Kang MH, Yoo HJ, Kwon YH, Yoon HY, Lee SG, Kim SR, Yeom DW, Kang MJ, Choi YW (2015) Design of multifunctional liposomal nanocarriers for folate receptor-specific intracellular drug delivery. Mol Pharm 12:4200–4213
73. Kawakami S, Yamashita F, Nishikawa M, Takakura Y, Hashida M (1998) Asialoglycoprotein receptor-mediated gene transfer using novel galactosylated cationic liposomes. Biochem Biophys Res Commun 252:78–83
74. Kawakami S, Wong J, Sato A, Hattori Y, Yamashita F, Hashida M (2000) Biodistribution characteristics of mannosylated, fucosylated, and galactosylated liposomes in mice. Biochim Biophys Acta 1524:258–265
75. Kenjo E, Asai T, Yonenaga N, Ando H, Ishii T, Hatanaka K, Shimizu K, Urita Y, Dewa T, Nango M, Tsukada H, Oku N (2013) Systemic delivery of small interfering RNA by use of targeted polycation liposomes for cancer therapy. Biol Pharm Bull 36:287–291
76. Kerbel RS, Kamen BA (2004) The anti-angiogenic basis of metronomic chemotherapy. Nat Rev Cancer 4:423–436
77. Kibria G, Hatakeyama H, Ohga N, Hida K, Harashima H (2013) The effect of liposomal size on the targeted delivery of doxorubicin to integrin alphavbeta3-expressing tumor endothelial cells. Biomaterials 34:5617–5627
78. Kim IY, Kang YS, Lee DS, Park HJ, Choi EK, Oh YK, Son HJ, Kim JS (2009) Antitumor activity of EGFR targeted pH-sensitive immunoliposomes encapsulating gemcitabine in A549 xenograft nude mice. J Control Release 140:55–60
79. Kluza E, Jacobs I, Hectors SJ, Mayo KH, Griffioen AW, Strijkers GJ, Nicolay K (2012) Dual-targeting of alphavbeta3 and galectin-1 improves the specificity of paramagnetic/fluorescent liposomes to tumor endothelium in vivo. J Control Release 158:207–214
80. Kobayashi T, Ishida T, Okada Y, Ise S, Harashima H, Kiwada H (2007) Effect of transferrin receptor-targeted liposomal doxorubicin in P-glycoprotein-mediated drug resistant tumor cells. Int J Pharm 329:94–102
81. Koide H, Asai T, Furuya K, Tsuzuku T, Kato H, Dewa T, Nango M, Maeda N, Oku N (2011) Inhibition of Akt (ser473) phosphorylation and rapamycin-resistant cell growth by knockdown of mammalian target of rapamycin with small interfering RNA in vascular endothelial growth factor receptor-1-targeting vector. Biol Pharm Bull 34:602–608

82. Kolb HC, Finn MG, Sharpless KB (2001) Click chemistry: diverse chemical function from a few good reactions. Angew Chem Int Ed Eng 40:2004–2021
83. Kondo M, Asai T, Katanasaka Y, Sadzuka Y, Tsukada H, Ogino K, Taki T, Baba K, Oku N (2004) Anti-neovascular therapy by liposomal drug targeted to membrane type-1 matrix metalloproteinase. Int J Cancer 108:301–306
84. Kono Y, Kawakami S, Higuchi Y, Maruyama K, Yamashita F, Hashida M (2014) Tumour-associated macrophages targeted transfection with NF-kappaB decoy/mannose-modified bubble lipoplexes inhibits tumour growth in tumour-bearing mice. J Drug Target 22:439–449
85. Kono K, Takashima M, Yuba E, Harada A, Hiramatsu Y, Kitagawa H, Otani T, Maruyama K, Aoshima S (2015) Multifunctional liposomes having target specificity, temperature-triggered release, and near-infrared fluorescence imaging for tumor-specific chemotherapy. J Control Release 216:69–77
86. Koo H, Lee S, Na JH, Kim SH, Hahn SK, Choi K, Kwon IC, Jeong SY, Kim K (2012) Bioorthogonal copper-free click chemistry in vivo for tumor-targeted delivery of nanoparticles. Angew Chem Int Ed Eng 51:11836–11840
87. Koshkaryev A, Piroyan A, Torchilin VP (2012) Increased apoptosis in cancer cells in vitro and in vivo by ceramides in transferrin-modified liposomes. Cancer Biol Ther 13:50–60
88. Krug S, Abbassi R, Griesmann H, Sipos B, Wiese D, Rexin P, Blank A, Perren A, Haybaeck J, Huttelmaier S, Rinke A, Gress TM, Michl P (2018) Therapeutic targeting of tumor-associated macrophages in pancreatic neuroendocrine tumors. Int J Cancer 143(7):1806–1816
89. Kularatne SA, Low PS (2010) Targeting of nanoparticles: folate receptor. Methods Mol Biol 624:249–265
90. Kurohane K, Namba Y, Oku N (2000) Liposomes modified with a synthetic Arg-Gly-Asp mimetic inhibit lung metastasis of B16BL6 melanoma cells. Life Sci 68:273–281
91. Kuznetsova NR, Stepanova EV, Peretolchina NM, Khochenkov DA, Boldyrev IA, Bovin NV, Vodovozova EL (2013) Targeting liposomes loaded with melphalan prodrug to tumour vasculature via the Sialyl Lewis X selectin ligand. J Drug Target 22(3):242–250
92. Kwon S, Ke S, Houston JP, Wang W, Wu Q, Li C, Sevick-Muraca EM (2005) Imaging dose-dependent pharmacokinetics of an RGD-fluorescent dye conjugate targeted to alpha v beta 3 receptor expressed in Kaposi's sarcoma. Mol Imaging 4:75–87
93. Lallana E, Sousa-Herves A, Fernandez-Trillo F, Riguera R, Fernandez-Megia E (2012) Click chemistry for drug delivery nanosystems. Pharm Res 29:1–34
94. Latif N, Bachhawat BK (1984) The effect of surface sugars on liposomes in immunopotentiation. Immunol Lett 8:75–78
95. Lee MH, Kim JY, Han JH, Bhuniya S, Sessler JL, Kang C, Kim JS (2012) Direct fluorescence monitoring of the delivery and cellular uptake of a cancer-targeted RGD peptide-appended naphthalimide theragnostic prodrug. J Am Chem Soc 134:12668–12674
96. Leung SL, Zha Z, Cohn C, Dai Z, Wu X (2014) Anti-EGFR antibody conjugated organic-inorganic hybrid lipid nanovesicles selectively target tumor cells. Colloids Surf B: Biointerfaces 121:141–149
97. Li XT, He ML, Zhou ZY, Jiang Y, Cheng L (2015) The antitumor activity of PNA modified vinblastine cationic liposomes on Lewis lung tumor cells: in vitro and in vivo evaluation. Int J Pharm 487:223–233
98. Limasale YD, Tezcaner A, Ozen C, Keskin D, Banerjee S (2015) Epidermal growth factor receptor-targeted immunoliposomes for delivery of celecoxib to cancer cells. Int J Pharm 479:364–373
99. Liu XQ, Song WJ, Sun TM, Zhang PZ, Wang J (2011) Targeted delivery of antisense inhibitor of miRNA for antiangiogenesis therapy using cRGD-functionalized nanoparticles. Mol Pharm 8:250–259
100. Liu S, Song XL, Wang YH, Wang XM, Xiao Y, Wang X, Cheng L, Li XT (2017) The efficacy of WGA modified daunorubicin anti-resistant liposomes in treatment of drug-resistant MCF-7 breast cancer. J Drug Target 25:541–553

101. Luo LM, Huang Y, Zhao BX, Zhao X, Duan Y, Du R, Yu KF, Song P, Zhao Y, Zhang X, Zhang Q (2013) Anti-tumor and anti-angiogenic effect of metronomic cyclic NGR-modified liposomes containing paclitaxel. Biomaterials 34:1102–1114
102. Maeda N, Takeuchi Y, Takada M, Namba Y, Oku N (2004a) Synthesis of angiogenesis-targeted peptide and hydrophobized polyethylene glycol conjugate. Bioorg Med Chem Lett 14:1015–1017
103. Maeda N, Takeuchi Y, Takada M, Sadzuka Y, Namba Y, Oku N (2004b) Anti-neovascular therapy by use of tumor neovasculature-targeted long-circulating liposome. J Control Release 100:41–52
104. Mamot C, Drummond DC, Greiser U, Hong K, Kirpotin DB, Marks JD, Park JW (2003) Epidermal growth factor receptor (EGFR)-targeted immunoliposomes mediate specific and efficient drug delivery to EGFR- and EGFRvIII-overexpressing tumor cells. Cancer Res 63:3154–3161
105. Mamot C, Drummond DC, Noble CO, Kallab V, Guo Z, Hong K, Kirpotin DB, Park JW (2005) Epidermal growth factor receptor-targeted immunoliposomes significantly enhance the efficacy of multiple anticancer drugs in vivo. Cancer Res 65:11631–11638
106. Managit C, Kawakami S, Yamashita F, Hashida M (2005) Effect of galactose density on asialoglycoprotein receptor-mediated uptake of galactosylated liposomes. J Pharm Sci 94:2266–2275
107. Maruyama K, Takahashi N, Tagawa T, Nagaike K, Iwatsuru M (1997) Immunoliposomes bearing polyethyleneglycol-coupled Fab' fragment show prolonged circulation time and high extravasation into targeted solid tumors in vivo. FEBS Lett 413:177–180
108. Maruyama K, Ishida O, Kasaoka S, Takizawa T, Utoguchi N, Shinohara A, Chiba M, Kobayashi H, Eriguchi M, Yanagie H (2004) Intracellular targeting of sodium mercaptoundecahydrododecaborate (BSH) to solid tumors by transferrin-PEG liposomes, for boron neutron-capture therapy (BNCT). J Control Release 98:195–207
109. Matsumura Y, Maeda H (1986) A new concept for macromolecular therapeutics in cancer chemotherapy: mechanism of tumoritropic accumulation of proteins and the antitumor agent smancs. Cancer Res 46:6387–6392
110. Mendonca LS, Firmino F, Moreira JN, Pedroso de Lima MC, Simoes S (2010) Transferrin receptor-targeted liposomes encapsulating anti-BCR-ABL siRNA or asODN for chronic myeloid leukemia treatment. Bioconjug Chem 21:157–168
111. Minami A, Otsubo T, Ieno D, Ikeda K, Kanazawa H, Shimizu K, Ohata K, Yokochi T, Horii Y, Fukumoto H, Taguchi R, Takahashi T, Oku N, Suzuki T (2014) Visualization of sialidase activity in mammalian tissues and cancer detection with a novel fluorescent sialidase substrate. PLoS One 9:e81941
112. Miselis NR, Wu ZJ, Van Rooijen N, Kane AB (2008) Targeting tumor-associated macrophages in an orthotopic murine model of diffuse malignant mesothelioma. Mol Cancer Ther 7:788–799
113. Miyamoto S, Yagi H, Yotsumoto F, Kawarabayashi T, Mekada E (2006) Heparin-binding epidermal growth factor-like growth factor as a novel targeting molecule for cancer therapy. Cancer Sci 97:341–347
114. Miyata S, Kawabata S, Hiramatsu R, Doi A, Ikeda N, Yamashita T, Kuroiwa T, Kasaoka S, Maruyama K, Miyatake S (2011) Computed tomography imaging of transferrin targeting liposomes encapsulating both boron and iodine contrast agents by convection-enhanced delivery to F98 rat glioma for boron neutron capture therapy. Neurosurgery 68:1380–1387. discussion 1387
115. Murase Y, Asai T, Katanasaka Y, Sugiyama T, Shimizu K, Maeda N, Oku N (2010) A novel DDS strategy, "dual-targeting", and its application for antineovascular therapy. Cancer Lett 287:165–171
116. Nag A, Ghosh PC (1999) Assessment of targeting potential of galactosylated and mannosylated sterically stabilized liposomes to different cell types of mouse liver. J Drug Target 6:427–438

117. Nakamura H (2009) Liposomal boron delivery for neutron capture therapy. Methods Enzymol 465:179–208
118. Namba Y, Sakakibara T, Masada M, Ito F, Oku N (1990) Glucuronate-modified liposomes with prolonged circulation time. Chem Pharm Bull (Tokyo) 38:1663–1666
119. Negussie AH, Miller JL, Reddy G, Drake SK, Wood BJ, Dreher MR (2010) Synthesis and in vitro evaluation of cyclic NGR peptide targeted thermally sensitive liposome. J Control Release 143:265–273
120. Ng KK, Weis WI (1997) Structure of a selectin-like mutant of mannose-binding protein complexed with sialylated and sulfated Lewis(x) oligosaccharides. Biochemistry 36:979–988
121. Ni S, Stephenson SM, Lee RJ (2002) Folate receptor targeted delivery of liposomal daunorubicin into tumor cells. Anticancer Res 22:2131–2135
122. Ogawa M, Umeda IO, Kosugi M, Kawai A, Hamaya Y, Takashima M, Yin H, Kudoh T, Seno M, Magata Y (2014) Development of 111In-labeled liposomes for vulnerable atherosclerotic plaque imaging. J Nucl Med 55:115–120
123. Okamoto A, Asai T, Kato H, Ando H, Minamino T, Mekada E, Oku N (2014) Antibody-modified lipid nanoparticles for selective delivery of siRNA to tumors expressing membrane-anchored form of HB-EGF. Biochem Biophys Res Commun 449:460–465
124. Okamoto A, Asai T, Hirai Y, Shimizu K, Koide H, Minamino T, Oku N (2018) Systemic administration of siRNA with Anti-HB-EGF antibody-modified lipid nanoparticles for the treatment of triple-negative breast cancer. Mol Pharm 15:1495–1504
125. Okazaki F, Matsunaga N, Okazaki H, Utoguchi N, Suzuki R, Maruyama K, Koyanagi S, Ohdo S (2010) Circadian rhythm of transferrin receptor 1 gene expression controlled by c-Myc in colon cancer-bearing mice. Cancer Res 70:6238–6246
126. Oku N (1999) Anticancer therapy using glucuronate modified long-circulating liposomes. Adv Drug Deliv Rev 40:63–73
127. Oku N, Asai T, Watanabe K, Kuromi K, Nagatsuka M, Kurohane K, Kikkawa H, Ogino K, Tanaka M, Ishikawa D, Tsukada H, Momose M, Nakayama J, Taki T (2002) Anti-neovascular therapy using novel peptides homing to angiogenic vessels. Oncogene 21:2662–2669
128. Omata D, Negishi Y, Hagiwara S, Yamamura S, Endo-Takahashi Y, Suzuki R, Maruyama K, Aramaki Y (2012) Enhanced gene delivery using bubble liposomes and ultrasound for folate-PEG liposomes. J Drug Target 20:355–363
129. Ongusaha PP, Kwak JC, Zwible AJ, Macip S, Higashiyama S, Taniguchi N, Fang L, Lee SW (2004) HB-EGF is a potent inducer of tumor growth and angiogenesis. Cancer Res 64:5283–5290
130. Pacis R, Pilat M, Yamazaki K, Pienta K (1995) Differential carbohydrate expression in tumorigenic vs nontumorigenic prostate cell-lines. Int J Oncol 7:1349–1354
131. Pan XQ, Wang H, Lee RJ (2003) Antitumor activity of folate receptor-targeted liposomal doxorubicin in a KB oral carcinoma murine xenograft model. Pharm Res 20:417–422
132. Pasqualini R, Koivunen E, Kain R, Lahdenranta J, Sakamoto M, Stryhn A, Ashmun RA, Shapiro LH, Arap W, Ruoslahti E (2000) Aminopeptidase N is a receptor for tumor-homing peptides and a target for inhibiting angiogenesis. Cancer Res 60:722–727
133. Pastorino F, Brignole C, Marimpietri D, Cilli M, Gambini C, Ribatti D, Longhi R, Allen TM, Corti A, Ponzoni M (2003) Vascular damage and anti-angiogenic effects of tumor vessel-targeted liposomal chemotherapy. Cancer Res 63:7400–7409
134. Pastorino F, Brignole C, Di Paolo D, Nico B, Pezzolo A, Marimpietri D, Pagnan G, Piccardi F, Cilli M, Longhi R, Ribatti D, Corti A, Allen TM, Ponzoni M (2006) Targeting liposomal chemotherapy via both tumor cell-specific and tumor vasculature-specific ligands potentiates therapeutic efficacy. Cancer Res 66:10073–10082
135. Paszko E, Vaz GM, Ehrhardt C, Senge MO (2013) Transferrin conjugation does not increase the efficiency of liposomal Foscan during in vitro photodynamic therapy of oesophageal cancer. Eur J Pharm Sci 48:202–210

136. Patil Y, Amitay Y, Ohana P, Shmeeda H, Gabizon A (2016) Targeting of pegylated liposomal mitomycin-C prodrug to the folate receptor of cancer cells: intracellular activation and enhanced cytotoxicity. J Control Release 225:87–95
137. Petrilli R, Eloy JO, Lee RJ, Lopez RFV (2018) Preparation of immunoliposomes by direct coupling of antibodies based on a thioether bond. Methods Mol Biol 1674:229–237
138. Qin L, Wang CZ, Fan HJ, Zhang CJ, Zhang HW, Lv MH, Cui SD (2014) A dual-targeting liposome conjugated with transferrin and arginine-glycine-aspartic acid peptide for glioma-targeting therapy. Oncol Lett 8:2000–2006
139. Riviere K, Huang Z, Jerger K, Macaraeg N, Szoka FC Jr (2011) Antitumor effect of folate-targeted liposomal doxorubicin in KB tumor-bearing mice after intravenous administration. J Drug Target 19:14–24
140. Ruoslahti E (1996) RGD and other recognition sequences for integrins. Annu Rev Cell Dev Biol 12:697–715
141. Said Hassane F, Frisch B, Schuber F (2006) Targeted liposomes: convenient coupling of ligands to preformed vesicles using "click chemistry". Bioconjug Chem 17:849–854
142. Saiki I, Koike C, Obata A, Fujii H, Murata J, Kiso M, Hasegawa A, Komazawa H, Tsukada H, Azuma I, Okada S, Oku N (1996) Functional role of sialyl Lewis X and fibronectin-derived RGDS peptide analogue on tumor-cell arrest in lungs followed by extravasation. Int J Cancer 65:833–839
143. Sakurai Y, Hatakeyama H, Sato Y, Hyodo M, Akita H, Ohga N, Hida K, Harashima H (2014) RNAi-mediated gene knockdown and anti-angiogenic therapy of RCCs using a cyclic RGD-modified liposomal-siRNA system. J Control Release 173:110–118
144. Saul JM, Annapragada AV, Bellamkonda RV (2006) A dual-ligand approach for enhancing targeting selectivity of therapeutic nanocarriers. J Control Release 114:277–287
145. Schiffelers RM, Koning GA, ten Hagen TL, Fens MH, Schraa AJ, Janssen AP, Kok RJ, Molema G, Storm G (2003) Anti-tumor efficacy of tumor vasculature-targeted liposomal doxorubicin. J Control Release 91:115–122
146. Scomparin A, Salmaso S, Eldar-Boock A, Ben-Shushan D, Ferber S, Tiram G, Shmeeda H, Landa-Rouben N, Leor J, Caliceti P, Gabizon A, Satchi-Fainaro R (2015) A comparative study of folate receptor-targeted doxorubicin delivery systems: dosing regimens and therapeutic index. J Control Release 208:106–120
147. Shahinian S, Silvius JR (2004) High-yield coupling of antibody Fab' fragments to liposomes containing maleimide-functionalized lipids. Methods Enzymol 387:3–15
148. Shan L, Wang S, Sridhar R, Bhujwalla ZM, Wang PC (2007) Dual probe with fluorescent and magnetic properties for imaging solid tumor xenografts. Mol Imaging 6:85–95
149. Shimizu K, Oku N (2004) Cancer anti-angiogenic therapy. Biol Pharm Bull 27:599–605
150. Shimizu K, Asai T, Oku N (2005) Antineovascular therapy, a novel antiangiogenic approach. Expert Opin Ther Targets 9:63–76
151. Shimizu K, Miyauchi H, Urakami T, Yamamura-Ichikawa K, Yonezawa S, Asai T, Oku N (2016) Specific delivery of an immunosuppressive drug to splenic B cells by antigen-modified liposomes and its antiallergic effect. J Drug Target 24:890–895
152. Shmeeda H, Amitay Y, Gorin J, Tzemach D, Mak L, Ogorka J, Kumar S, Zhang JA, Gabizon A (2010) Delivery of zoledronic acid encapsulated in folate-targeted liposome results in potent in vitro cytotoxic activity on tumor cells. J Control Release 146:76–83
153. Shmeeda H, Amitay Y, Tzemach D, Gorin J, Gabizon A (2013) Liposome encapsulation of zoledronic acid results in major changes in tissue distribution and increase in toxicity. J Control Release 167:265–275
154. Silverman JA, Deitcher SR (2013) Marqibo(R) (vincristine sulfate liposome injection) improves the pharmacokinetics and pharmacodynamics of vincristine. Cancer Chemother Pharmacol 71:555–564
155. Singh M (1999) Transferrin as a targeting ligand for liposomes and anticancer drugs. Curr Pharm Des 5:443–451

156. Smyth T, Petrova K, Payton NM, Persaud I, Redzic JS, Graner MW, Smith-Jones P, Anchordoquy TJ (2014) Surface functionalization of exosomes using click chemistry. Bioconjug Chem 25:1777–1784
157. Sonali SRP, Singh N, Sharma G, Vijayakumar MR, Koch B, Singh S, Singh U, Dash D, Pandey BL, Muthu MS (2016) Transferrin liposomes of docetaxel for brain-targeted cancer applications: formulation and brain theranostics. Drug Deliv 23:1261–1271
158. Spanedda MV, De Giorgi M, Hassane FS, Schuber F, Bourel-Bonnet L, Frisch B (2017) Coupling of ligands to the liposome surface by click chemistry. Methods Mol Biol 1522:93–106
159. Springer TA, Lasky LA (1991) Cell adhesion. Sticky sugars for selectins. Nature 349:196–197
160. Stahn R, Schafer H, Kernchen F, Schreiber J (1998) Multivalent sialyl Lewis x ligands of definite structures as inhibitors of E-selectin mediated cell adhesion. Glycobiology 8:311–319
161. Storm G, Roerdink FH, Steerenberg PA, de Jong WH, Crommelin DJ (1987) Influence of lipid composition on the antitumor activity exerted by doxorubicin-containing liposomes in a rat solid tumor model. Cancer Res 47:3366–3372
162. Sugiyama M, Matsuura M, Takeuchi Y, Kosaka J, Nango M, Oku N (2004) Possible mechanism of polycation liposome (PCL)-mediated gene transfer. Biochim Biophys Acta 1660:24–30
163. Sugiyama T, Asai T, Nedachi YM, Katanasaka Y, Shimizu K, Maeda N, Oku N (2013) Enhanced active targeting via cooperative binding of ligands on liposomes to target receptors. PLoS One 8:e67550
164. Sundaram S, Trivedi R, Durairaj C, Ramesh R, Ambati BK, Kompella UB (2009) Targeted drug and gene delivery systems for lung cancer therapy. Clin Cancer Res 15:7299–7308
165. Suzuki R, Takizawa T, Kuwata Y, Mutoh M, Ishiguro N, Utoguchi N, Shinohara A, Eriguchi M, Yanagie H, Maruyama K (2008) Effective anti-tumor activity of oxaliplatin encapsulated in transferrin-PEG-liposome. Int J Pharm 346:143–150
166. Tang Y, Soroush F, Tong Z, Kiani MF, Wang B (2017) Targeted multidrug delivery system to overcome chemoresistance in breast cancer. Int J Nanomedicine 12:671–681
167. Tardi P, Choice E, Masin D, Redelmeier T, Bally M, Madden TD (2000) Liposomal encapsulation of topotecan enhances anticancer efficacy in murine and human xenograft models. Cancer Res 60:3389–3393
168. Toita R, Kawano T, Murata M, Kang JH (2016) Anti-obesity and anti-inflammatory effects of macrophage-targeted interleukin-10-conjugated liposomes in obese mice. Biomaterials 110:81–88
169. Torchilin V (2008) Antibody-modified liposomes for cancer chemotherapy. Expert Opin Drug Deliv 5:1003–1025
170. Tros de Ilarduya C, Arangoa MA, Moreno-Aliaga MJ, Duzgunes N (2002) Enhanced gene delivery in vitro and in vivo by improved transferrin-lipoplexes. Biochim Biophys Acta 1561:209–221
171. Vader P, Crielaard BJ, van Dommelen SM, van der Meel R, Storm G, Schiffelers RM (2012) Targeted delivery of small interfering RNA to angiogenic endothelial cells with liposome-polycation-DNA particles. J Control Release 160:211–216
172. Wang HW, Jiang PL, Lin SF, Lin HJ, Ou KL, Deng WP, Lee LW, Huang YY, Liang PH, Liu DZ (2013) Application of galactose-modified liposomes as a potent antigen presenting cell targeted carrier for intranasal immunization. Acta Biomater 9:5681–5688
173. Wang F, Chen L, Zhang R, Chen Z, Zhu L (2014) RGD peptide conjugated liposomal drug delivery system for enhance therapeutic efficacy in treating bone metastasis from prostate cancer. J Control Release 196:222–233
174. Xiong S, Yu B, Wu J, Li H, Lee RJ (2011) Preparation, therapeutic efficacy and intratumoral localization of targeted daunorubicin liposomes conjugating folate-PEG-CHEMS. Biomed Pharmacother 65:2–8
175. Xu L, Pirollo KF, Chang EH (2001) Tumor-targeted p53-gene therapy enhances the efficacy of conventional chemo/radiotherapy. J Control Release 74:115–128

176. Yamada A, Taniguchi Y, Kawano K, Honda T, Hattori Y, Maitani Y (2008) Design of folate-linked liposomal doxorubicin to its antitumor effect in mice. Clin Cancer Res 14:8161–8168
177. Yang T, Choi MK, Cui FD, Kim JS, Chung SJ, Shim CK, Kim DD (2007) Preparation and evaluation of paclitaxel-loaded PEGylated immunoliposome. J Control Release 120:169–177
178. Ye P, Zhang W, Yang T, Lu Y, Lu M, Gai Y, Ma X, Xiang G (2014) Folate receptor-targeted liposomes enhanced the antitumor potency of imatinib through the combination of active targeting and molecular targeting. Int J Nanomedicine 9:2167–2178
179. Yonenaga N, Kenjo E, Asai T, Tsuruta A, Shimizu K, Dewa T, Nango M, Oku N (2012) RGD-based active targeting of novel polycation liposomes bearing siRNA for cancer treatment. J Control Release 160:177–181
180. Yu LG (2007) The oncofetal Thomsen-Friedenreich carbohydrate antigen in cancer progression. Glycoconj J 24:411–420
181. Zalba S, Contreras AM, Haeri A, Ten Hagen TL, Navarro I, Koning G, Garrido MJ (2015) Cetuximab-oxaliplatin-liposomes for epidermal growth factor receptor targeted chemotherapy of colorectal cancer. J Control Release 210:26–38
182. Zeisig R, Stahn R, Wenzel K, Behrens D, Fichtner I (2004) Effect of sialyl Lewis X-glycoliposomes on the inhibition of E-selectin-mediated tumour cell adhesion in vitro. Biochim Biophys Acta 1660:31–40
183. Zembruski NC, Nguyen CD, Theile D, Ali RM, Herzog M, Hofhaus G, Heintz U, Burhenne J, Haefeli WE, Weiss J (2013) Liposomal sphingomyelin influences the cellular lipid profile of human lymphoblastic leukemia cells without effect on P-glycoprotein activity. Mol Pharm 10:1020–1034
184. Zhai G, Wu J, Yu B, Guo C, Yang X, Lee RJ (2010) A transferrin receptor-targeted liposomal formulation for docetaxel. J Nanosci Nanotechnol 10:5129–5136
185. Zhao C, Feng Q, Dou Z, Yuan W, Sui C, Zhang X, Xia G, Sun H, Ma J (2013) Local targeted therapy of liver metastasis from colon cancer by galactosylated liposome encapsulated with doxorubicin. PLoS One 8:e73860
186. Zhu Y, Cheng L, Cheng L, Huang F, Hu Q, Li L, Tian C, Wei L, Chen D (2014) Folate and TAT peptide co-modified liposomes exhibit receptor-dependent highly efficient intracellular transport of payload in vitro and in vivo. Pharm Res 31:3289–3303
187. Zhuo H, Peng Y, Yao Q, Zhou N, Zhou S, He J, Fang Y, Li X, Jin H, Lu X, Zhao Y (2013) Tumor imaging and interferon-gamma-inducible protein-10 gene transfer using a highly efficient transferrin-conjugated liposome system in mice. Clin Cancer Res 19:4206–4217
188. Zou M, Zhang L, Xie Y, Xu W (2012) NGR-based strategies for targeting delivery of chemotherapeutics to tumor vasculature. Anti Cancer Agents Med Chem 12:239–246

A Multifunctional Envelope-Type Nano Device for Cancer Therapy

Ikramy A. Khalil, Hiroto Hatakeyama, Takashi Nakamura, and Hideyoshi Harashima

Abstract Regarding cancer, innovative therapeutic approaches are needed for achieving a complete cure, since the conventional approaches are not satisfactory. The use of nucleic acids for modifying gene and protein levels in specific cells is one of the most promising approaches for cancer therapy, since it corrects the underlying cause of the disease. However, delivering various nucleic acids to their intracellular target sites is a difficult task, since they need to traverse the plasma membrane in order to be effective. Unlike conventional low molecular weight drugs, nucleic acids are large, hydrophilic and susceptible to degradation. Therefore, a sophisticated gene delivery system is required to protect and efficiently deliver these molecules to their target sites. We have been developing a series of innovative gene delivery systems based on our original design of a multifunctional envelope-type nano device (MEND). A MEND could be optimized at various levels to efficiently and safely deliver different nucleic acids to their target sites in specific cells in the body. We focused mainly on improving biodistribution after systemic administration as well as improving intracellular trafficking so as to achieve the maximum effect. Here, we summarize our efforts to develop various versions of MENDs for efficient gene delivery in vitro and in vivo. The focus is on the use of MENDs for the treatment of cancer through applying the concept of active targeting as well as passive targeting. We also summarize the applicability of a MEND system for use in cancer immunotherapy. The MEND system described here is expected to extend

I. A. Khalil
Faculty of Pharmaceutical Sciences, Hokkaido University, Sapporo, Japan

Faculty of Pharmacy, Assiut University, Assiut, Egypt

H. Hatakeyama
Clinical Pharmacology and Pharmacometrics, Graduate School of Pharmaceutical Sciences, Chiba University, Chiba, Japan

T. Nakamura · H. Harashima (✉)
Laboratory for Molecular Design of Pharmaceutics, Laboratory of Innovative Nanomedicine, Faculty of Pharmaceutical Sciences, Hokkaido University, Sapporo, Japan
e-mail: harasima@pharm.hokudai.ac.jp

Y. Matsumura, D. Tarin (eds.), *Cancer Drug Delivery Systems Based on the Tumor Microenvironment*, https://doi.org/10.1007/978-4-431-56880-3_10

the therapeutic applications of nucleic acids for the treatment of various currently incurable diseases, including cancer.

Keywords Non-viral vectors · MEND · pDNA · siRNA · R8 · YSK · GALA · Active targeting

Introduction

Despite the great progress in the field of medicine in the past few years, some diseases remain incurable. Cancer, a malignancy that shows a high mortality rate, is one of these diseases. Around 100 types of cancer affect humans where it is responsible for ~13% of all deaths in the world. Treatment strategies that are currently in use include chemotherapy using conventional low molecular weight drugs, radiation and surgery. These methods are still not satisfactory weapons in our fight against cancer. Efforts are ongoing to develop more efficient innovative medicines for use in the treatment of cancer and other incurable diseases. These efforts focus mainly on finding new drugs and developing more efficient drug delivery systems (DDSs). Recent advances in drug development rely on the use of macromolecular drugs such as peptides, proteins or nucleic acids to replace conventional low molecular weight drugs. Gene therapy or treatment with nucleic acids currently represents one of the most promising innovative technologies for the treatment of numerous incurable diseases. The strategy involves introducing a therapeutic nucleic acid such as plasmid DNA (pDNA) to increase the production of a certain protein. Alternatively, it may introduce antisense oligonucleotides (ASONs) or small interference RNA (siRNA) to silence the production of another protein. In both cases, efficient gene delivery systems are required to effectively deliver pDNA, ASONs or siRNA to their target sites, namely the nucleus or the cytosol.

The delivery of nucleic acids to their target sites is a much more complicated process compared to conventional low molecular weight drugs. Nucleic acids are large, negatively charged hydrophilic macromolecules, which limit their efficient cellular uptake since they have almost no ability to penetrate the negatively charged, lipophilic plasma membranes. Moreover, because nucleic acids are biologically active compounds, they are susceptible to the action of enzymes in the circulation. Therefore, specific gene delivery systems are required to neutralize the high negative charge of nucleic acids, to condense them in smaller nanoparticles (NPs) and to protect them from enzymes when they are in the circulation. Gene delivery systems based on the use of viral vectors are the most commonly used systems for efficient gene delivery to target cells. While they are much less efficient, non-viral vectors are safer and easier to prepare compared to viral vectors. Therefore, developing more efficient non-viral gene delivery systems appears to be the most straightforward approach in moving safe gene therapy from the laboratory bench to the clinic and in attempting to find innovative cures for various currently incurable diseases.

To increase the efficiency of non-viral gene delivery, various systems have been designed to overcome different extracellular and intracellular barriers [18, 89, 104]. We developed a novel multifunctional envelope-type nano device (MEND) to achieve efficient gene delivery based on a strategy that focuses on overcoming the barriers that it faces. The MEND was designed to mimic viral vectors, which have the inherent ability to efficiently transfect cells. The MEND is composed of a core of nucleic acids condensed with a polycation which is then coated by a lipid film to simulate envelope-type viruses. The lipid envelope can be decorated with a variety of different functional devices for controlling biodistribution, tissue targeting and cellular uptake. Furthermore, the MEND can be charged with different components to control intracellular trafficking such as enhancing endosomal escape, increasing its stability in the cytosol and enhancing nuclear delivery. Since the first report on a MEND in 2004, various versions were developed through optimization of the biomaterials used in its construction as well as through controlling the topology of these biomaterials to maximize their effect. Our continuous efforts directed at the efficient gene delivery after systemic administration eventually paid off. Some recently developed MEND versions produce high gene expression levels (in the case of pDNA) or high gene silencing (in the case of siRNA), which compete with the most efficient gene delivery systems that are currently known. The level of gene delivery achieved by different MENDs was sufficient to produce measurable therapeutic effects in the treatment of different diseases. This prompted us to develop even more efficient and targeted versions of MENDs in the future, which will permit us to successfully treat different currently incurable diseases.

In this chapter, we summarize the MEND system and its application in the field of cancer therapy. We overview our efforts to prepare an effective MEND system since its early development stage to achieve efficient gene delivery in vitro. We describe the shift in MEND use from in vitro to in vivo applications, with emphasis on efficient gene delivery to the liver after intravenous administration (IV) for the treatment of different liver diseases including liver cancer. We also focus on developing MENDs that are specific for use in cancer therapy. We describe novel strategies for enhancing tumor targeting and the efficient gene delivery to tumor cells. In addition, we review our efforts to use the MEND system for cancer immunotherapy, as one of the most promising strategies for cancer therapy.

The MEND as an Advanced Drug Delivery Technology

Overview of MENDs and Their In Vitro Applications

The Rationale for Producing a MEND

We developed a MEND as an original gene delivery system based on the concept of programmed packaging of nucleic acids. The MEND was prepared using a novel lipid film hydration method and was designed to mimic viral vectors, which have an inherent ability to efficiently transfect cells [47–49, 60, 80, 102]. The MEND is composed of a core-shell structure in which a nucleic acid is first condensed with a suitable polycation such as poly-l-lysine (PLL), polyethylene imine (PEI) or protamine to form a condensed core. This condensation protects nucleic acids during preparation and against degradation in the circulation. The cores are further coated with a lipid envelope, which was optimized to fuse with endosomal membranes under the influence of the acidic medium in the endosome. The lipid envelope also protects the nucleic acids further and serves as a platform for attaching other functional devices, which may be added to enhance cellular uptake and endosomal escape, or to improve the pharmacokinetics and biodistribution of the system. A number of different types of MENDs have been developed so far, the most important of which are discussed below and summarized in Fig. 1 and Table 1. Other MEND versions are not discussed here due to space limitations.

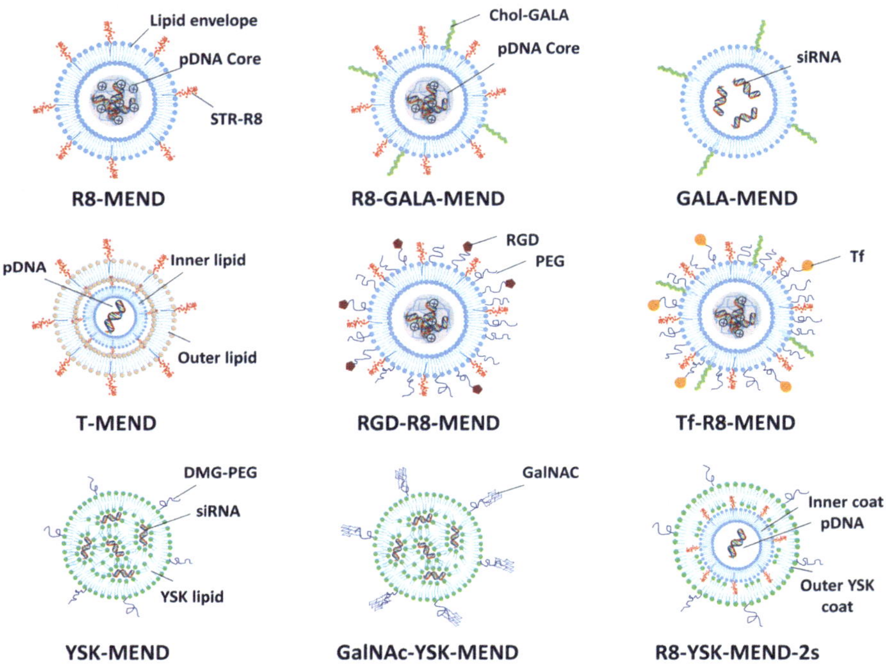

Fig. 1 A schematic diagram of some versions of MENDs
The original R8-MEND was composed of a pDNA core covered with a lipid envelope that had been modified with STR-R8. The efficiency was increased by adding Chol-GALA (R8-GALA-MEND). In the absence of STR-R8, the GALA MEND is used for targeting lung endothelial cells. A tetra-lamellar MEND (T-MEND) is particularly useful for transfecting non-dividing cells. Specificity is improved by using dual ligand MENDs such as the RGD-R8-MEND and the Tf-R8-MEND. A remarkable gene silencing in hepatocytes could be achieved after the IV administration of the YSK-MEND. The specificity and efficiency were further increased by using a pH-labile hepatocyte-targeted GalNAc-YSK-MEND. A two-step coating design (R8-YSK-MEND-2s) produced an efficient system through synergism between R8 and YSK. Several other MEND versions were developed but are not shown here

Table 1 Different Versions of MEND Systems

MEND	Description	Diameter (nm)	Zeta Potential (mV)	Application	References
R8-MEND	Positively charged pDNA/PLL condensed core covered with a negative lipid envelope (DOPE/CHEMS) modified with STR-R8 peptide.	185–330	38–59	Efficient gene expression in vitro in HeLa and A549 cells and in vivo in the skin after local application.	[41]
R8-GALA-MEND	Negatively charged pDNA or siRNA/PEI cores covered with a positive lipid (DOPE/Chol/STR-R8) modified with Chol-GALA peptide.	162–185	35–38	Efficient pDNA and siRNA delivery to the liver after IV administration.	([43]; [28]; [29])
GALA-MEND	Negatively charged siRNA/PEI core covered with a positive lipid (DOTMA/EPC/Chol) modified with Chol-GALA peptide.	100–150	16–22	Efficient siRNA delivery to lung endothelial cells after IV administration	[51]
T-MEND	A multi-layer system composed of pDNA/protamine core covered with an inner lipid for nuclear delivery (DOPE/CL/STR-R8) and an outer lipid for endosomal escape (DOPE/PA/STR-R8).	163	54	Efficient pDNA delivery to JAW II (non-dividing cells).	[3]
RGD-R8-MEND	Negatively charged pDNA/protamine core covered with a positive lipid (DOTAP/DOPE/Chol) modified with STR-R8 and PEG-RGD ligands.	139	26	Efficient and specific pDNA delivery to endothelial cells (HUVEC).	[45]
Tf-R8-MEND	Negatively charged pDNA/PEI core covered with a positive lipid (DOPE/Chol/STR-R8/Chol-GALA). The system is coated with a dense PEG coat and further modified with Tf.	205	(−) 17	Highly efficient and specific gene delivery to HeLa cells even in the presence of serum.	[38]

(continued)

Table 1 (continued)

MEND	Description	Diameter (nm)	Zeta Potential (mV)	Application	References
YSK-MEND	siRNA is condensed in a lipid envelope composed of YSK05/POPE/Chol/DMG-PEG. YSK05 was replaced with more potent lipids (YSK-13 and YSK-15).	116	4.5	Highly efficient gene silencing in the liver after IV administration. Used successfully for treatment of HBV and HCV infections.	([81]; [82])
YSK-R8-MEND-2s	A double-coated system in which pDNA is first coated with DOPE/STR-R8 to form negative inner lipid NPs. These NPs were further coated with YSK/Chol/DMG-PEG lipid under acidic conditions followed by proper neutralization.	162	(−) 27	Highly efficient gene delivery to HeLa cells even in the presence of serum.	[44]
GalNAc-YSK-MEND	siRNA is condensed in a lipid envelope composed of YSK05/Chol/PEG. A pH-labile maleic anhydride PEG was used and its tips were further modified with GalNAc ligand.	60–70	2.5	Highly efficient and specific gene silencing in liver hepatocytes after IV administration.	[84]

The first proto-type MEND appeared in 2004, was composed of a pDNA/PLL condensed core carrying a net positive charge coated with a negative lipid envelope consisting of dioleoyl-phosphoethanolamine (DOPE) and cholesteryl hemisuccinate (CHEMS) [47]. The lipid envelope was further modified with the cationic stearylated octaarginine peptide (STR-R8), which functions as a cell penetrating peptide (CPP), to enhance cellular uptake [47]. Coating the positive pDNA/PLL core with a lipid envelope resulted in a ten-fold increase in transfection activity. Furthermore, modifying the surface of MEND with STR-R8 resulted in a 1000-fold increase in transfection activities compared to the condensed core. This result encouraged us to further investigate the role of the octaarginine (R8) peptide in enhancing transfection activities. We found that R8 significantly increased the cellular uptake of a MEND [40]. Using a high density of R8 stimulated uptake via macropinocytosis, which resulted in an improved intracellular trafficking through avoiding the excessive lysosomal degradation associated with conventional clathrin-mediated endocytosis [39]. The presence of R8 also improved the migration of the pDNA to the nucleus [42]. The activity of R8 is not only related to its positive charge, as evidenced by the fact that a similar positively charged peptide (octalysine (K8)) did not improve transfection activities to the same extent [16]. We confirmed that presence of R8 in the MEND stimulated endosomal escape at both neutral and acidic pH, compared to K8, which mediates endosomal escape only at a neutral pH. Therefore, the R8 peptide is an essential component for realizing high transfection activities through enhanced cellular uptake as well as controlled intracellular trafficking [42]. The R8-MEND was further optimized for efficient siRNA and ASONs delivery in vitro, mainly through using cores produced by the condensation of nucleic acids using protamine or STR-R8 [63, 64]. Novel lipid-containing sorbitol-based molecular transporters carrying guanidine moieties to mimic the R8 peptide also showed improved gene delivery in vitro [30].

Optimizing the R8-MEND for Efficient Gene Delivery

We further optimized the MEND system containing pDNA by changing the lipid composition and the amount of polycation used in core formation. The transfection activity of the optimized MEND in HeLa and A549 cells was comparable to that for Adenovirus, one of the most efficient viral vectors known to date [41]. The high transfection activity of the R8-MEND was accompanied by remarkably lower cytotoxicities compared to Adenovirus. In addition, using the optimized R8-MEND, it was possible to successfully transfect hair follicles after a single topical application to the dorsal skin of ICR mice. An R8-MEND encapsulating pDNA *encoding LacZ* or green fluorescence protein (GFP) efficiently transfected hair follicles at the anagen phase while similar pDNAs condensed with Lipofectamine failed to transfect. An R8-MEND encapsulating the constitutively active bone morphogenetic protein receptor type IA (caBmpr1a) was used to investigate the role of bone morphogenetic protein (BMP) in hair follicle cycling. Hair follicles treated with an R8-MEND encapsulating caBmpr1a were observed in or close to the subcutis space. In contrast, hair follicles treated with a control R8-MEND were restricted to more superficial areas of the dermis. Alkaline phosphatase activity and a morphological examination of skin samples showed that hair follicles treated with the R8-MEND-caBmpr1a were in the late anagen or early catagen phase, whereas hair follicles that were treated with the control MEND were in the late catagen or telogen phase. Proliferating cells, indicative of an anagen phase, were detected in hair follicles that had been treated with the R8-MEND-caBmpr1a but not in follicles that had been treated with the control MEND. These results indicate that the R8-MEND successfully delivered constitutively active Bmpr1a, which resulted in the prolongation of the anagen phase and an extension of hair cycling. This was the first demonstration to show that the R8-MEND has the ability to produce a significant biological effect in vivo.

Increasing the Efficiency of the R8-MEND

In an attempt to improve its efficiency, we investigated the inclusion of different components in the R8-MEND system. The introduction of the pH-sensitive fusiogenic peptide GALA into the system has improved transfection activity in vitro and in vivo [24, 73]. GALA is a 30 amino acid synthetic peptide that changes from a random coil to α helix conformation in the acidic pH of endosomes to induce fusion between the liposomal and endosomal membranes [54]. This fusion significantly enhances the endosomal escape of nucleic acids and eventually improves transfection activities. In the absence of R8, GALA improved transfection activities by ~ten fold. In the presence of R8, the inclusion of GALA increased transfection activities by ~400 fold [38]. We hypothesized that the presence of R8 improved interactions between the liposomal and endosomal membranes and allowed the GALA peptide to more efficiently induce fusion and endosomal escape. A novel stearylated INF7 peptide derivative showed a similar improvement in transfection activities by virtue of its ability to enhance endosomal escape, even in the presence of a hydrophilic polyethylene glycol (PEG) coating, which is a known inhibitor of endosomal escape [17]. In addition, the incorporation of the KALA peptide, which functions as a pH-independent fusion inducer, further increased the transfection activities, especially in difficult-to-transfect cells such as dendritic cells [86].

The nuclear delivery of pDNA through the nuclear envelope is a critical rate-limiting step in transfection, especially in non-dividing cells. We concluded that the molecular design of a gene carrier plays an important role in enhancing the nuclear delivery of pDNA. Therefore, we rationally designed a tetra-lamellar MEND (T-MEND) to improve both endosomal escape and the nuclear delivery of pDNA through step-wise membrane fusion [3]. T-MEND is a multi-layered system composed of a pDNA-condensed core covered with an inner and an outer lipid envelope. The outer lipid envelope was optimized to cause efficient endosomal escape under conditions of an acidic pH while the inner lipid envelope was designed to cause efficient fusion with the nuclear envelope. The optimized T-MEND increased the level of gene expression in non-dividing cells by two-order of magnitude compared to the conventional MEND, thus greatly expanding the applicability of our MEND system for delivering genes to various cells.

MEND for Targeted Drug and Gene Delivery

The original R8-MEND carries a high positive charge, which makes it non-specific and not stable in serum. The transfection activities of such positive systems are severely impaired in the presence of serum due to interactions with several negative serum proteins. Therefore, we attempted to further optimize the MEND system to enhance its activity in the presence of serum and to increase its targeting ability. A common method for enhancing targeting to a specific tissue involves attaching a specific targeting ligand to the system to mediate efficient internalization through receptor-mediated endocytosis. The ligands that are used to accomplish this are frequently attached to the tip of PEG spacers to enhance flexibility for interactions with target receptors. MENDs that were modified with different specific ligands were prepared to circumvent the non-specificity associated with the R8-MEND. Transferrin (Tf)-modified MENDs were used for targeting tumor cells, which over-express Tf-receptors [36]. We then focused on targeting endothelial cells in different organs. A MEND modified with cyclic RGD (cRGD) was used to target tumor endothelial cells (TECs), which over-express $\alpha_v\beta_3$ integrins [77]. The MEND system could also be used to target TECs using our original aptamer ligand that was discovered from a diverse nucleic acid library using the cell-based systemic evolution of ligands by exponential enrichment (SELEX) [6, 7]. The IRQ ligand and the GALA peptides were both used to permit the MEND to specifically target lung endothelial cells [35, 51]. A peptide ligand (KGGRAKD) was similarly used to target endothelial cells of adipose tissue [31].

Dual Ligand MENDs

Despite the high promise of actively targeted MENDs, as discussed above, the efficiency of transfection of targeted cells continued to remain low and needed to be improved. Uptake that is mediated by specific ligands occurs mainly through receptor-mediated endocytosis, which is a saturable process. In addition, the presence of PEG inhibits the interaction between liposomal and endosomal membranes leading to reduced cellular uptake and reduced endosomal escape. In contrast, the use of CPPs results in a more efficient unsaturable cellular uptake, compared to specific targeting ligands. The use of CPPs also enhances the interaction between liposomal and cellular membranes to enhance endosomal escape. Therefore, we designed a rational strategy in which a specific targeting ligand was combined with a CPP to produce more specific and yet efficient PEGylated liposomal systems. We developed a dual ligand system in which the R8 peptide is masked with PEG chains, the tips of which carry a specific targeting ligand. The R8 is not expressed on the surface and hence would not be expected to bind to negative serum components. Cellular uptake occurs mainly through the targeting ligand on the tip of the PEG chain, while the presence of R8 would enhance both cellular uptake and endosomal escape.

We initially applied the dual ligand liposomal system for targeting TECs in vitro and in vivo. This is discussed in detail later in this chapter. Following the successful use of dual ligand MENDs to target TECs, we attempted to use a triple ligand concept for targeting tumor cells, rather than TECs [38]. We first coated the R8-MEND with a dense PEG coat to prevent nonspecific interactions and to confer long circulation properties, a prerequisite for tumor accumulation via the enhanced permeability and retention (EPR) effect. This PEG coating decreased transfection activities to less than 1% in a dose-dependent matter. We identified the mechanism responsible for this reduced transfection where we found that cellular uptake and endosomal escape were both reduced substantially. We attached Tf ligands on the tip of PEG chains to improve cellular uptake and to add a targeting property to the system.

However, the presence of this ligand improved the transfection only by three-fold, which was not enough for achieving efficient gene expression. The inclusion of the GALA peptide to enhance the endosomal escape enhanced the transfection activities by 400-fold. The triple R8-Tf-GALA-MEND (RTG-MEND) had a remarkably high transfection activity in cancerous HeLa cells, even in the presence of serum. The activity was comparable to the parent non-specific R8-MEND as well as the Lipofectamine transfection system. In addition, the activity was significantly lower in normal non-cancerous cells than in HeLa cells. The system was found to be internalized mainly through Tf receptors, which are over-expressed in rapidly dividing cells including most cancerous cells. No high transfection activity was observed in the absence of R8, which confirms the importance of the dual ligand concept, since the presence of R8 improved the cellular uptake mediated by Tf as well as endosomal escape mediated by GALA.

Synergism Between R8 and pH-Sensitive Ionizable Lipids

The discovery of our original pH-sensitive ionizable lipid, termed YSK05, represented a huge step forward in terms of improving the efficiency of transfection. The YSK05 lipid significantly enhances the endosomal escape of its nucleic acid cargos. However, the YSK-MEND has a neutral charge at physiological pH, which limits its cellular uptake in the absence of endogenous ligands that mediate the cellular uptake of neutral particles. The addition of R8 peptide in a dose sufficient to form a positively charged MEND enhanced the transfection activities of YSK-MEND encapsulating pDNA [44]. To avoid producing a positively charged MEND so as to avoid nonspecific interactions, we modified YSK-MEND with a low amount of R8. The non-positively charged YSK-MEND modified with low levels of R8 improved transfection activities compared to an unmodified YSK-MEND. However, the activity was still lower than the positively charged YSK-MEND that was modified with high levels of R8. To enhance transfection activities in the presence of low levels of R8, we designed a double-coated system that was intended to improve endosomal escape and nuclear delivery. In this design, pDNA is first coated with a lipid film containing a low level of R8 to form a negatively charged lipid NP. This lipid NP was further coated with YSK lipid under acidic conditions followed by neutralization to form a non-positively charged YSK-MEND. The double-coated R8-YSK-MEND had high transfection activities that were comparable to a positively charged YSK-MEND containing high levels of R8 but was even more serum resistant. Surprisingly, we found that the cellular uptake was higher in the case of the double-coated system compared to the original system. Furthermore, the presence of R8 improved endosomal escape as evidenced by confocal microscopy images. We hypothesized that the double-coated design allows the YSK lipid to be located in the outer layer, not bound to pDNA, which may enhance its fusiogenic activity with the endosomal membrane resulting in improved endosomal escape. In addition, the pDNA released from the endosomes would be bound to the inner lipid layer, which would be expected to further protect DNA and may improve nuclear delivery due to the presence of R8. This novel double-coated R8-miodified YSK-MEND represents a promising approach for improving transfection efficiency without imparting a positive charge to the final system.

In Vivo Applications of MEND

R8-MEND for In Vivo Liver Gene Delivery

Although the R8-MEND was used successfully for in vivo gene expression after topical administration, as explained earlier, the system, when administered intravenously, failed to produce a significant gene expression in any tissue. Early biodistribution studies of R8-liposomes showed that liposomes modified with a high density of R8 largely and rapidly accumulated in the liver [61]. Based on this study, we hypothesized that the R8-MEND could be used to deliver a gene to the liver for the treatment of liver-related diseases, particularly liver cancer. However, no detectable gene expression was observed in the liver despite this high localization in that organ. We attempted to increase the efficiency of gene expression in the liver by the inclusion of the GALA peptide to enhance endosomal escape. The transfection activity was only slightly increased in the liver when GALA was used. This R8-GALA-MEND was composed of a positively charged PEI-condensed core and a negatively charged lipid envelope, since this was the design that showed the highest gene expression in vitro [41]. Surprisingly, when the design of the MEND was changed to the use of a negatively charged PEI-condensed core and a positively charged lipid envelope (DOPE/R8), the transfection activity was increased by ~500 fold in the liver [43]. The optimized negative core R8-GALA-MEND resulted in the gene expression level in the liver to be increased by more than 6000-fold compared to the original R8-MEND. Both R8 and GALA are indispensible for producing efficient gene expression. We investigated the mechanism responsible for this change by shifting from a positively charged core to a negatively charged core by quantifying the pDNA delivered to cells and nuclei in each case using the quantitative polymerase chain reaction (PCR). We found that the amount of pDNA delivered to cells or to nuclei was comparable for MENDs with negatively charged or positively charged cores. We hypothesized that a negatively charged core and the presence of GALA may facilitate the decondensation of DNA in the nucleus, thus leading to more efficient gene expression. Our results indicated that the role of GALA may not be restricted only to enhancing endosomal escape but it may have multiple roles in gene delivery. Interestingly, the gene expression of the R8-GALA-MEND was higher in the liver than other organs, which indicates that this generally non-specific system may have the ability to target liver cells if systemically administered. This targeting activity was mainly related to R8 since a MEND prepared with the cationic lipid dimyristoyl trimethylammonium propane (DOTAP) had a lower gene expression and showed no specific liver delivery.

The successful delivery of a gene to the liver using the R8-GALA-MEND encouraged us to examine its ability to protect liver cells against toxicity induced by the lipopolysaccharide/D-galactosamine (LPS/D-GalN). The administration of R8-GALA-MEND containing a condensed pDNA encoding hepatocyte growth factor (HGF) to mice prior to the administration of LPS/D-GalN protected the liver against acute injury, as evidenced by the reduced levels of alanine aminotransferase (ALT) and aspartate aminotransferase (AST), which are indicators of liver injury [29]. The activity of the R8-GALA-MEND was comparable to that for a commercially available in vivo liver transfection reagent (jet-PEI-Gal). However, the survival rate was significantly increased in the case of the MEND compared to jet-PEI-Gal, which indicates that the MEND system has a superior safety profile. The R8-GALA-MEND was also used to deliver siRNA to the liver. A MEND containing siRNA against a scavenger receptor class B, member 1 (SR-BI) caused a significant reduction in SR-BI mRNA levels in primary hepatocytes (in vitro) and in the liver (after IV administration) [28]. The siRNA-loaded MEND caused no change in the level of markers for liver toxicity, i.e., AST and ALT or inflammatory cytokines. Collectively, these results indicate that the R8-MEND system successfully delivered pDNA and siRNA specifically to hepatocytes, despite its permanent positive charge, which would be expected to shift the system towards macrophages in major clearance organs such as the liver and the spleen.

YSK05-MEND for In Vivo Liver Gene Delivery

Our results for using a MEND in vivo clearly showed that improving endosomal escape has a remarkable effect in terms of enhancing the efficiency of gene delivery. Meanwhile, avoiding a permanent positive charge of the system is clearly needed to avoid non-specific interactions and uptake by organs of the reticulo endothelial system (RES). Therefore, we attempted to develop an efficient lipid system for use in preparing an optimized MEND with improved endosomal escape, albeit without a high positive charge. We developed our original pH-sensitive ionizable lipid, referred to as the YSK lipid. The first version of YSK was termed YSK05 and contained a pH-sensitive group (N-methylpiperidine) and two unsaturated carbon chains [81]. YSK05 has a neutral charge at physiological pH and adopts a positive charge in an acidic medium. It was used to condense siRNA or pDNA under acidic conditions to form NPs that turn neutral upon adjusting the pH of the medium to neutrality. These neutral NPs are suitable for systemic administration due to a lack of non-specific interactions. They are recognized by Apo-E proteins in the circulation, which mediate their efficient and selective cellular uptake by hepatocytes through low density lipoprotein (LDL) receptors. After internalization into cells, the NPs become positively charged in the acidic endosomes which mediates efficient interaction with negatively charged endosomal membranes. The presence of two long unsaturated carbon chains in YSK05 confirms the phase transition from a lamellar phase to an inverted hexagonal phase when it interacts with anionic membranes. This phase transition is a strong inducer of membrane fusion resulting in an improved endosomal escape. We prepared YSK05-MEND containing siRNA and compared it with MENDs prepared using a cationic lipid (DOTAP). The cellular uptake of the cationic DOTAP-MEND was higher than that of the YSK05-MEND. This can be explained by the fact that YSK05-MEND has a pKa value of 6.6, indicating that it would have a neutral charge under physiological conditions. Despite having a lower cellular uptake, the YSK05-MEND resulted in a significantly improved gene silencing. The improved gene silencing of the YSK05-MEND was well correlated with enhanced endosomal escape as a result of its fusiogenic activity, as evidenced by a hemolytic assay. The YSK05-MEND also showed improved gene silencing after intratumoral injection compared to the DOTAP-MEND.

We further optimized the YSK05-MEND through optimization of the lipid composition. After IV administration to mice, high levels of the optimized YSK05-MEND accumulated in the liver and resulted in a high and durable F7 gene silencing in hepatocytes with a median effective dose (ED50) as low as 0.06 mg/kg [99]. The therapeutic value of the optimized YSK05-MEND was confirmed in mice with humanized livers with persistent hepatitis C virus (HCV) infections. A MEND encapsulating siRNA against HCV RNA was successfully used to significantly reduce the level of HCV RNA (~25-fold lower than control siRNA), resulting in the normalization of chronically inflamed liver tissues as judged by reduced HCV-induced liver necrosis and inflammation in inducible-HCV transgenic mouse model (HCV-Tg mice). The YSK05-MEND was also used for efficient pDNA delivery and expression in hepatocytes in vivo [78]. Optimization of the pH, ionic strength and the presence of sodium chloride in the medium was employed in order to efficiently package pDNA in the YSK05-MEND. Improving the encapsulation efficiency of pDNA caused a remarkable effect on gene expression in hepatocytes. Confocal microscopic images of liver tissue showed that the YSK05-MEND caused a more homogenous gene expression in the liver, compared to hydrodynamic injection.

Second Generation YSK-MEND for Gene Delivery to the Liver

We developed a second-generation YSK lipid for enhancing endosomal escape efficiency and for confirming hepatocyte specific gene delivery. The YSK-13 and YSK-15 series contain two long unsaturated carbon chains linked to one sp2 carbon atom to produce a larger bond angle for emphasizing phase transition and fusiogenic activity under conditions of an acidic pH. In addition, the YSK-13 and YSK-15 series showed different pKa values ranging from 5.7 to 7.25 depending on the structures of the carbon chains that were used. MENDs prepared with the YSK-13 and YSK-15 series caused a higher gene silencing compared to YSK05-MEND [82]. Surprisingly, controlling the pKa value of each MEND affected the intrahepatic distribution of siRNA. Maximum gene silencing in hepatocytes was observed at a pKa value of 6.4, where maximum gene silencing in liver sinusoidal endothelial cells (LSECs) was observed at pKa values of 7 or higher. Moreover, we were able to enhance hepatocyte specificity by using an ester bond in the YSK13 series. This bond can be cleaved by the phospholipase A1 found in LSECs but not in hepatocytes. The YSK13-C3-MEND showed a strong gene silencing in the liver (ED50 of 0.015 mg/Kg) with a higher gene silencing in hepatocytes compared to LSECs (>67 fold). It is noteworthy that the activity of YSK13-C3 was comparable to that for MC3, one of the most efficient pH-sensitive lipids being currently tested in several clinical trials [92]. The YSK13-C3-MEND encapsulating siRNA against the hepatitis B virus (HBV) RNA was used for the efficient treatment of HBV infections [103]. It resulted in a strong and durable suppression of HBV antigens in mice with persistent HBV infections.

We further investigated the effect of the diameter of the YSK-MEND on gene silencing activity in vitro and in vivo in the liver. Using microfluidic mixing, we prepared different siRNA containing YSK-MENDs with different diameters (ranging from 32 to 67 nm) and evaluated their efficiencies [83]. In the liver, the activities of the smaller MENDs were lower than the larger MENDs. This could be explained by poor packaging as well as an increased instability of the siRNA in the presence of serum in the case of smaller MENDs. Moreover, the endosomal escape of the smaller MENDs was significantly lower than that of larger MENDs in the presence of serum. Taken together, we concluded that the activity of small sized MENDs could be enhanced by improving siRNA encapsulation and by improving serum resistance.

A MEND for Hepatocyte Active Targeting

We attempted to elucidate the mechanism responsible for the liver toxicity that is associated with a high dose of lipid NPs. We found that the accumulation of lipid NPs in LSECs caused the induction of various cytokines that lead to the inflammation of liver tissue [84]. Therefore, we attempted to decrease the delivery to LSECs and increase the delivery to hepatocytes through modification of the MEND with N-acetyl-d-galactosamine (GalNAc), an efficient active targeting ligand for hepatocytes. The GalNAc-MEND showed improved efficiency as well as a reduced toxicity. The use of PEG linked to the MEND through a pH-labile linkage (maleic anhydride) further reduced the toxicity of the GalNAc-MEND [22]. Under acidic conditions, PEG is rapidly desorbed from the MEND and fusiogenic activity with endosomal membranes is restored. The optimized GalNAc-MEND was successfully used for the treatment of HBV infections without any signs of toxicity in mice with humanized livers infected with HBV. Hepatotoxicity could be also reduced significantly by reducing the amount of pH-sensitive lipid used. Neutralizing the negative charges of siRNA by protamine allowed the use of a lower amount of pH-sensitive lipid without affecting gene silencing activity in vitro and in vivo [85]. The use of MENDs with a low lipid protamine core represented an important step for increasing the therapeutic window of YSK-MENDs, which is a highly desirable property for clinical applications.

MEND for In Vivo Lung Gene Delivery

Targeting lung endothelial cells (LECs) is a promising approach for treating various lung diseases, especially lung cancer. Targeting tissues other than the liver is a challenging task, since the liver represents a major clearance organ for injected NPs. To develop a lung targeted MEND, we needed a specific ligand that has the ability to specifically bind to LECs. We previously identified an IRQ peptide (IRQRRRR), which has the ability to be taken up by cells through a unique caveolar endocytosis pathway, which is involved in cellular uptake by endothelial cells [62]. Liposomes modified with PEG and the IRQ peptide (IRQ-PEG-Lip) showed about a 20-fold increase in lung accumulation compared to liposomes that were modified with PEG (PEG-Lip) [35]. Therefore, we hypothesized that an IRQ-modified MEND could be developed for efficient lung gene delivery. Despite its relatively high lung accumulation, the IRQ-MEND encapsulating luciferase encoding pDNA failed to produce a significant luciferase expression after IV administration. The activity in the lung was significantly improved by decreasing the length of the PEG spacer and the hydrophobic part of the PEG lipid which is attached to the MEND surface. Optimization of the polycation used for condensing pDNA as well as the cationic lipid used resulted in a further enhancement in gene expression in the lung. Eventually, an optimized IRQ-MEND was developed that resulted in gene expression in the lung that was 5-orders of magnitude higher than the unmodified first version.

Our attempts to find a more lung specific ligand continued. By accident, we discovered that MENDs containing the GALA peptide accumulated at higher levels in the lung than in other tissues. GALA-modified liposomes were delivered to the lung 24-fold higher than unmodified liposomes [51]. We proved that the GALA peptide binds to sialic acid terminated sugars in LECs since the uptake was blocked by lectins, which bind to sialic acid terminated sugars. Using intravital real time confocal microscopy in living mice confirmed that GALA-modified NPs were delivered to the lung in the form of single NPs, without aggregation, which is expected for conventional cationic systems. Furthermore, the uptake by LECs was blocked by the presence of an excess of free GALA peptide, which confirms the unique targeting ability of GALA. A GALA-MEND encapsulating siRNA against the endothelial marker CD31 caused strong gene silencing (~90% using a dose of 1 mg siRNA/kg) in lung endothelial cells in vivo [51]. In this system, the GALA peptide plays a dual role as a targeting ligand to LECs and as an endosomal escape device. The GALA-MEND was used successfully for the prevention of pulmonary metastasis of melanoma without significant toxic effects, as judged by measuring various biochemical and toxicity markers [51]. This unique system for lung targeting has great promise for use in treatment of various lung diseases.

We continued our efforts to enhance the activity of the GALA-MEND for gene silencing in the lung endothelium. Changing the preparation method from the lipid film hydration method to the ethanol dilution method significantly enhanced gene silencing in the lung [52]. The enhanced silencing effect was attributed mainly to the enhanced endosomal escape ability of the GALA-MEND that was prepared using ethanol dilution, since biodistribution and lung accumulation was not affected by the preparation method. The activity of the GALA-MEND in the lung was further improved by the addition of a PEG spacer between the GALA peptide and the MEND surface [79]. The presence of a PEG spacer improved cellular uptake by LECs in vitro as well as improved lung accumulation in vivo. The optimized GALA-PEG-MEND system showed a strong gene silencing in the lung with a low non-specific gene silencing in the liver endothelium. Collectively, this functional improvement in the GALA-MEND further increased its potential for use for the treatment of lung cancer.

Overcoming the PEG Dilemma

Delivery to Cancer Tissue and the PEG Dilemma

Doxil (Caelyx), one of the monumental achievements in the history of DDS, with a diameter of 100 nm, is a particle that contains liposomal doxorubicin (DOX) and is modified with PEG, a hydrophilic polymer [32]. Systemically injected PEGylated liposomes accumulate in tumor tissue because PEGylated liposomes have a prolonged circulation time which allowed them to extravasate through blood vessels and accumulate in the interstitial tumor space via the EPR effect [56, 57]. This is due to the presence of a discontinuous endothelium and the absence of lymphatic vessels in tumor tissue. Systemically injected PEGylated liposomes are retained in the circulation, permitting them to become distributed in off-target organs, for example, the heart in the case of DOX. Doxil has more moderate side effects compared with free DOX [32]. Doxil is currently used in the treatment of acquired immunodeficiency syndrome (AIDS)-related Kaposi's sarcoma, ovarian cancer and breast cancer [15].

However, despite the great success of delivering chemotherapeutic drugs to tumors using PEGylated NPs via passive targeting based on the EPR effect, PEGylation hampers in vivo applications of NPs as gene carriers of pDNA and siRNA to tumor delivery. A PEGylated MEND has a long circulation time in the blood and accumulates in tumors via the EPR effect, as expected. However, the surface aqueous phase formed by the PEG moiety inhibits the interaction of the MEND with the target cell surface. As a result, cellular uptake is nil or minimal. Furthermore, PEGylation improves the stability of the lipid envelope of a MEND, which results in poor endosomal escape via membrane fusion and in the degradation of cargos in lysosomes, the digestive compartment [58, 76]. These serious issues associated with the use of PEG in delivering genes and nucleic acids to cancer sites are referred to as the "PEG dilemma" [25, 27]. Therefore, a successful gene and nucleic acid delivery for cancer treatment requires a rational strategy and the preparation of carrier systems that are designed to overcome the issues associated with the use of PEG.

Overcoming the PEG Dilemma in the Tumor Microenvironment

Enhancement of Cellular Uptake by Targeting Ligands

One reason for issues associated with the use of PEG is the steric hindrance conferred by the PEG which inhibits interactions between the positively charged surface of the MEND and anionic molecules on the cell surface and the subsequent cellular uptake of the MEND. The first method for overcoming this problem involved displaying ligands for receptors on the surface of targeted cells on PEGylated carriers. This would be expected to improve the selectivity, binding and uptake of the carriers by targeted cells and is known as "active targeting". Target molecules are carefully selected based on their characteristics such as specificity, expression level on tumor cells, and ability to internalize ligand-modified PEGylated NPs.

However, as explained above, the route for the uptake of targeted NPs by cells is mainly governed by receptor-meditated endocytosis. This might restrict the amount of MENDs taken up, due to the limited number of available receptors and the recycling of endocytosis. On the other hand, CPPs are widely utilized for the delivery of drugs and genes as summarized above. Taking both specific ligand-mediated active targeting and CCPs-mediated efficient intracellular delivery into consideration, we proposed a rational strategy designed to take advantage of a combination of both specific ligands and CPPs which would allow PEGylated liposomes to be used as a more selective and efficient system for in vivo systemic applications. To accomplish this, we developed a dual-ligand system, in which specific ligand-modified PEGylated liposomes are combined with cationic ligands, such as CPPs (Fig. 2) [45, 90].

We utilized a dual-ligand MEND for delivering drugs to TECs. Endothelial cells in angiogenic vessels express several proteins that are absent or barely detectable in established blood vessels, including integrins, vascular endothelial growth factor receptors (VEGFRs) and other types of membrane molecules, such as aminopeptidase N (CD13). Peptides containing a specific motif in their sequence that recognize specific molecules have been identified via the use of phage-displayed peptide libraries. The arginine-glycine-aspartic acid (RGD) sequence is the most typical motif and has been used for the targeted delivery of drugs and genes because of its ability to recognize integrins that are expressed on both tumor cells and neovascular endothelial cells. Another peptide ligand, the asparagine-glycine-arginine (NGR) motif peptide was identified as a ligand for CD13, which is over expressed on TECs and tumor cells. Therefore, we attached RGD or NGR peptides on the ends of PEG chains [45, 46, 90, 91]. The RGD peptide enhanced the cellular uptake of PEG-liposomes in human umbilical vein endothelial (HUVEC) cells by 1.4 fold. The use of a combination of RGD and the R8 peptide synergistically enhanced cellular uptake by 5.1 fold. Liposomes modified with PEG-RGD and R8 showed a ~300 fold increase in gene expression compared to PEG-modified liposomes. The synergistic activity of dual-ligand RGD-R8-MENDs was observed only in HUVEC cells, and

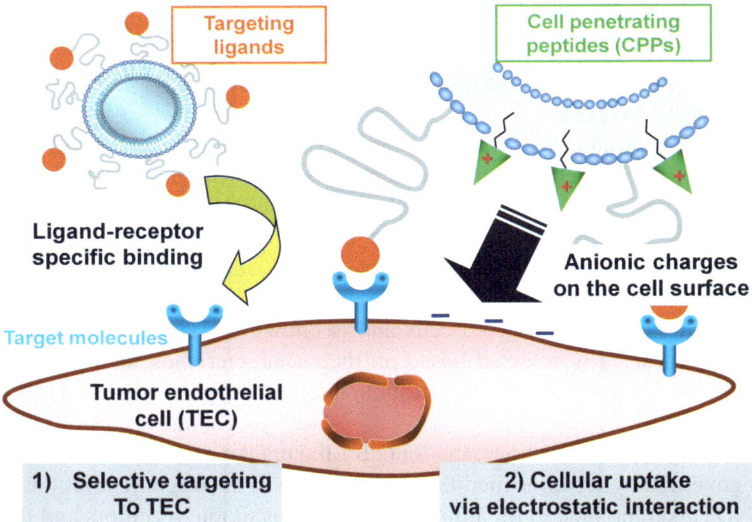

Fig. 2 A schematic diagram of the dual-ligand system
Specific ligands and CPPs can be attached to the end of the PEG chain and on the surface of liposomes, respectively. CPPs should not be functional as surface ligands and should be free from opsonins due to steric hindrance by the PEG layer in the blood circulation. While after arriving at the targeted tumor endothelial cells, cellular association via the specific ligands (1) allows CPPs to exert their powerful ability to allow the liposomes to be internalized into cells due to the proximity of the liposomes to the surface of target cells via electrostatic interactions (2)

no synergistic effect was observed in skin endothelial cells. Similarly, we tested the effect of combining oligoarginines and the NGR motif (68, 32). Because cationic CPPs must not be functional and should be opsonin-free in the systemic circulation, the surface of liposomes was modified with tetra arginine (R4), where R4 could be masked by PEG [90]. Among the oligoarginines, R4 produced a synergistic effect with the NGR motif on the cellular uptake of liposomes in CD13-positive cells (68).

It was previously reported that particle diameters >200 nm appear to be more effective in adhering firmly to the margins of vascular walls under conditions of flow than particles with diameters of <200 nm [19]. For the targeting of dual-ligand liposomes to tumor vessels, we employed a liposome formulation with a diameter of 300 nm, and not the 100 nm diameter that is generally used for targeting a drug to a tumor. The large size minimized the extravasation of a dual-ligand liposome into tumor tissue via the EPR effect in chemoresistant renal cell carcinoma (RCC) tumors [91]. Small liposomes loaded with doxorubicin accumulated in RCC tumors at higher levels than larger liposomes. However, tumor growth suppression was observed only with larger dual ligand liposomes targeting ECs. The large sized dual-ligand liposomes composed of NGR and R4 efficiently delivered cargos to TECs that are sensitive to chemodrugs, induced the death of TECs and an antiangiogenic effect, which resulted in the suppression of tumor growth. The dual ligand liposomes specifically disrupted the angiogenic blood vessels in drug resistant tumors and had a minor effect on normal ECs in the liver or the spleen, the major clearance organs of liposomal systems. Dual-ligand systems could also be applied for delivering pDNA to TECs [45].

De-PEGylation Systems

The endosomal release of cargos such as pDNA or siRNA into the cytoplasm is regarded as a critical step in achieving successful transgene expression and knock-down, because endosomes are fused with lysosomes, digestive compartments. Endosomal escape is caused by membrane fusion or disruption of the endosome. PEGylation confers steric stability to complexes and inhibits their disruption and fusion with the endosomal membrane. Therefore, valid approaches involve removing the PEG moiety from MENDs in the target site in response to the local environment, which would be expected to improve cellular uptake and the subsequent endosomal escape of MENDs. Intracellular environments with a low pH in endosomes/lysosomes, an intracellular reducing environment and enzymes are employed as triggers of PEG cleavage [34, 55, 87, 105]. We developed an SS-cleavable proton-activated lipid-like material (ssPalm), which contains two tertiary amines that are positively charged at acidic pH, and with a disulfide bond that can be cleaved in a reducing environment [4, 5, 93, 97].

To achieve tumor-specific de-PEGylation, we focused on enzymes that are specifically expressed in tumors such as matrix metalloproteinase (MMP). The MMP substrate peptide was inserted between the PEG and DOPE as a linker, and the resulting conjugated PEG-peptide-DOPE ternary conjugate is referred to as PPD (Fig. 3) [23]. A MEND modified with PPD (PPD-MEND) encapsulating pDNA or siRNA with a diameter of 100–200 nm exhibited a higher in vitro activity than a conventional (non-cleavable) PEG-lipid modified MEND (PEG-MEND) depending on the level of expression of MMP. The PPD-MEND was stable in the systemic circulation, and accumulated in tumors via the EPR effect when the PPD-MEND was intravenously injected into tumor bearing mice [23, 26]. The PPD-MEND could be used to deliver cargos, such as pDNA and siRNA, to tumor cells more efficiently than a PEG-MEND and exhibited significantly increased gene expression or gene silencing in xenografted tumors as a result of the enhanced cellular uptake and endosomal escape. Hashida et al. and other groups also reported on MMP based delivery systems [21, 50, 94, 106]. Other proteases and enzymes, such as phospholipases [2] and glucose-oxidase [74] have been utilized as stimulus triggers in drug delivery systems, as well.

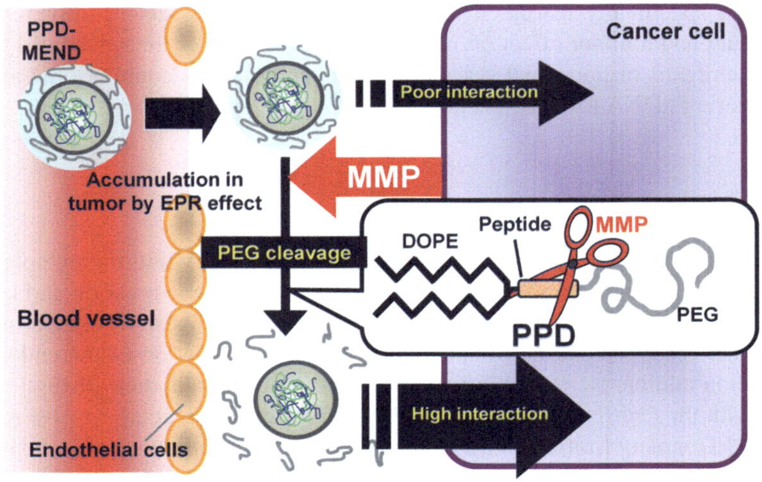

Fig. 3 Strategies for de-PEGylation in response to a tumor environment
A MEND modified with a cleavable PEG-lipid (PPD) is stable in the systemic circulation and accumulates in a tumor via the EPR effect. The peptide linker in PPD is degraded in response to MMP secreted from cancer cells, which results in de-PEGylation. The bare MEND can associate with cancer cells via electrostatic interactions between the cationic surface of the MEND and the anionic surface of cancer cells

MENDs in Cancer Immunotherapy

The success of immunotherapy in clinic has revolutionized the field of cancer therapy and immunotherapy has now joined the impregnable ranks as a pillar. To induce effective anti-tumor immune responses, enhancing the activity of immune cells and blocking immune suppression by tumor cells are necessary. Carrier type DDS can be a useful technology for achieving both an enhanced immune response and blocking immune suppression. In this section, the applications of MENDs in cancer immunotherapy are reviewed.

Dawn of Cancer Immunotherapy

Cancer Immunotherapy in Clinic

The use of immunotherapy in the field of cancer therapy was received with skepticism for many years. The impregnable ranks have been surgery, radiation, chemotherapy, and targeted therapy. However, the appearance of immune-checkpoint inhibitors and chimeric antigen receptor T-cell (CAR-T) therapies marked a turning point in cancer therapy [14]. The success of immune-checkpoint therapies and CAR-T therapies have revolutionized the field of cancer therapy and immunotherapy has joined the above impregnable ranks as a pillar. Immune-checkpoint inhibitors function by blocking receptor and/or ligand interactions, such as the cytotoxic T-lymphocyte-associated protein 4 (CTLA-4), programmed death-1 (PD-1), PD ligand 1 (PD-L1), that are involved in dulling T cell activation or function in lymphoid tissue and tumor tissue. PD-1, an inhibitory receptor, is expressed on the surface of effector T cells. Since PD-L1 is expressed on the surface of cancer cells, the functions of effector T cells are inhibited via PD-1/PD-L1 interactions. As of this writing, six types of immune checkpoint inhibitors have been approved for use; anti-CTLA4 antibody (Ipilimumab), anti-PD-1 antibodies (Nivolumab and Pembrolizumab), and anti-PD-L1 antibodies (Atezolizumab, Avelumab and Durmalumab). Anti-PD-1 is making particularly rapid progress and the scope of its applications has been expanded. For example, pembrolizumab has been approved for the treatment of melanoma, non-small cell lung cancer (first-line), head and neck squamous cell cancer, classical Hodgkin lymphoma, primary mediastinal large B-cell lymphoma, urothelial carcinoma and microsatellite instability-high cancer, gastric cancer and cervical cancer (FDA, June 2018). CAR-T cells are autologous T cells that are genetically engineered to express a CAR specific for tumor antigens and thus can specifically target a tumor independent of major histocompatibility complex (MHC) restriction [101]. Immunotherapy using CAR-T cells targeting CD19, a surface molecule expressed by B cells, showed a high response (90%) rate in B cell acute lymphoblastic leukemia (ALL) [101], and was approved by the FDA in 2017. Unfortunately, these immunotherapies are not effective for all types of cancer and patients, but for only a minor fraction of cancers and patients. The response rate of anti-PD-1 antibodies was 20–30%, and the treatment efficacy of solid tumors utilizing CAR-T cell is unsatisfactory. Therefore, the emerging focus is on overcoming these problems, and the use of a combination therapies with other immunotherapies and cancer therapies, or new immunotherapies are currently being developed.

Understanding the Spatial-Temporal Dynamics of Cancer Immune Responses

In our body, the immune system protects us from cancer invasion. A series of step-wise events are associated with the protection and the spatial-temporal events are summarized in the concept of the Cancer-Immunity Cycle (Fig. 4) [11]. For an anticancer immunity to lead to the effective elimination of cancer cells, it is necessary to iteratively proceed and expand the Cancer-Immunity Cycle. First, antigens/neoantigens derived from the death of cancer cells are captured by dendritic cells (DCs), followed by antigen processing and antigen presentation. DCs simultaneously receive maturation stimuli from dead cancer cells. Mature DCs move to lymph nodes and prime T cells via MHC molecules presenting antigen peptides. The activated T cells, namely effector T cells, traffic to and infiltrate the tumor tissues. Finally, the effector T cells specifically recognize cancer cells via T cell receptors (TCR) and kill the cancer cells. Additional tumor-associated antigens are released

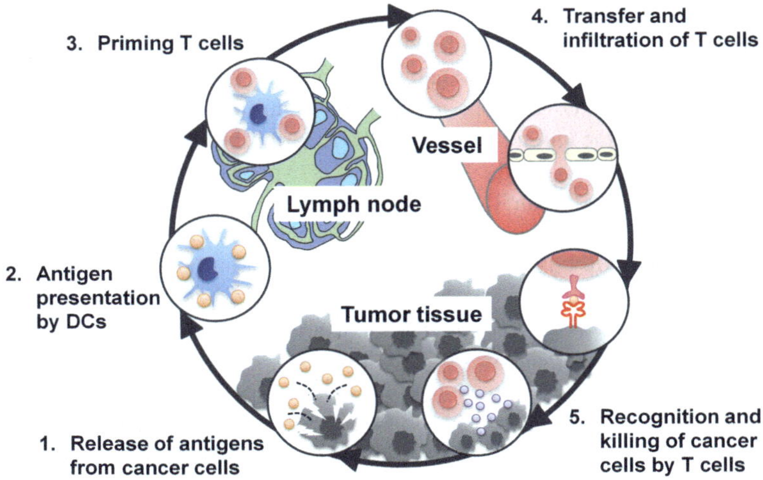

Fig. 4 Image of the Cancer-Immunity Cycle
Antigens produced as a result of the death of cancer cells are captured by dendritic cells (DCs), followed by antigen processing and antigen presentation. DCs simultaneously receive maturation stimuli from dead cancer cells. The mature DCs move to lymph nodes and prime T cells via MHC molecules presenting antigen peptides. The activated T cells, namely effector T cells, traffic to and infiltrate the tumor tissues. Finally, the effector T cells specifically recognize cancer cells via a T cell receptor and kill the cancer cells. The image was reproduced from a part of previous report [11], copyright 2013 Elsevier

from the killed cancer cells, leading to an increase in the breadth and depth of the response in subsequent revolutions of the cycle. Although the Cancer-Immunity Cycle functions correctly in healthy people, it does not perform optimally in cancer patients. Some or a plurality of parts are inhibited. If the main cause of the inhibition is PD-1/PD-L1 interactions, the blocking of PD-1 or PD-L1 results in a much-improved therapeutic effect. However, more than 70% of cancer patients have other inhibitory mechanisms in the Cancer-Immunity Cycle. Therefore, in order to apply appropriate immunotherapy to all patients, it is necessary to understand the immune status of the tumor immune microenvironment (TIME) [9, 12].

Based on our current knowledge, TIME can be segregated into three main classes; immune-desert TIME, immune-excluded TIME and inflamed TIME [12]. Although detailed information has been reviewed elsewhere [12], we briefly summarize these classifications as follows. In immune-desert TIME, it is thought that no immune responses against cancer cells occur due to a lack of appropriate T cell priming and activation, immunological ignorance, or tolerance. To overcome immune-desert TIME, it is necessary to activate anticancer immunity by increasing immunogenicity and using adjuvants. In immune-excluded TIME, a lack of an appropriate chemokine state, barriers by vascular factors and stroma likely inhibit the infiltration of effector immune cells. Although the detailed mechanism remains unknown, re-modeling tumor vessels and stroma appears to be efficient. In the case of inflamed TIME, several subtypes of immune cells, including both effector cells and suppressor cells, infiltrate the tumor. Effector cells such as tumor-specific cytotoxic T lymphocytes (CTLs) are maintained in a dysfunctional state by inhibitory factors that are expressed in cancer cells, including regulatory T (Treg) cells and myeloid-derived suppressor cells (MDSCs). In this case, therapies designed to eliminate inhibitory factors and cells by antibodies or chemical drugs are efficient. To provide an appropriate therapy against complicated TIME, the development of combination therapies is needed.

A carrier type DDS is an important contribution to the development of effective combination therapy. Nanoparticle formulations, such as micelles, liposomes, lipid nanoparticles, polyplexes, and emulsions, are frequently used for assisting immunotherapy [33, 37]. The leading candidates for immunotherapeutic materials are tumor antigens (proteins or peptides) and adjuvants that can contribute to the increase in immunogenicity in TIME. However, they are easily degraded and have a low affinity for immune cells, resulting in a failure to induce effective immune responses. These cargos can be protected from degradation and delivered to target tissues, cells or organelles when they are loaded in nanoparticles. Nanoparticles have largely contributed to the promotion of antigen presentation in DCs, the activation of DCs, targeting DCs, targeting lymph tissues, etc. A carrier type DDS therefore, represent a fundamental technology in cancer immunotherapy.

Controlling Cancer Immune Responses by a MEND

Antigen Delivery by an R8-MEND

To induce effective anti-cancer immune responses, delivering tumor-associated antigens to DCs is a major and well-established strategy. Generally, exogenous antigens that are taken up via endocytosis are degraded in lysosomes and the peptides derived from the antigens are presented on MHC class II antigens (MHC-II), resulting in antigen presentation to $CD4^+$ T cells [95]. On the other hand, endogenous antigens in the cytosol are degraded by proteasomes and the peptides are presented to MHC-I, resulting in the antigen being presented to $CD8^+$ T cells. The $CD8^+$ T cells received the antigen presentation differentiate to antigen-specific CTLs. Thus, antigen presentation on MHC-I is crucial for $CD8^+$ T cell priming. To achieve MHC-I antigen presentation, exogenous antigens must be delivered to the cytosol, because tumor antigens are almost all exogenous in nature. Fusiogenic liposomes composed of DOPE were found to enhance antigen presentation on MHC-I, namely by cross-presentation [20]. We also found that the use of a combination of R8 peptide with a fusiogenic MEND (R8-MEND) resulted in a drastic enhancement in cellular uptake and MHC-I presentation [65]. In in vitro experiments, the R8-MEND was taken up by most mouse DCs via macropinocytosis, delivering the encapsulated substance to the cytosol. Interestingly, treatment with an ovalbumin (OVA) loaded R8-MEND (R8-MEND/OVA) resulted in the induction of specific MHC-I antigen presentation, while MHC-II presentation was slight. Moreover, mice that were immunized with the R8-MEND/OVA showed a strong preventative effect against the growth of an E.G7-OVA tumor. Further study revealed that the specific MHC-I antigen presentation by the R8-MEND was due to an enhancement in C-terminal trimming of the antigen peptide via proteasomes [69].

Adjuvant Delivery by a MEND

Another important factor for the effective induction of anti-cancer immunity is DC maturation by adjuvants. Ligands for pattern-recognition receptors (PRRs) associated with innate immunity are currently used as adjuvants. These ligands induce the maturation of DCs and cytokine production. In particular, ligands for nucleic acid-sensing pathways would be expected to function as a cancer adjuvant [10]. Polyinosine-polycytidylic acid (polyI:C, a double stranded RNA, toll-like receptor (TLR) 3 and the melanoma differentiation-associated gene 5 ligand: Mda5), CpG oligodeoxynucleotides (CpG-ODNs, DNA containing unmethylated CpG motifs, the TLR9 ligand), and DNA or CDNs (cytosolic DNA sensor ligands) are frequently used for this purpose. We discuss the importance of carrier design with regard to the topology of adjuvants below [66]. We prepared two types of carriers; a polyI:C loaded R8-MEND/OVA (R8-MEND/OVA/PIC); a complex of STR-R8, OVA and polyI:C (R8/OVA/PIC-Com). Both nanoparticles had similar sizes and both were positively charged. Interestingly, when mice were immunized with the R8-MEND/OVA/PIC, a much stronger OVA-specific CTL induction was observed and antitumor effects against the E.G7-OVA tumor were more potent than that for the R8/OVA/PIC-Com. This difference can be attributed to the following reasons. In the case of the R8-MEND/OVA/PIC, polyI:C in the aqueous phase of the liposome would be expected to be in the free form, while the polyI:C in the R8/OVA/PIC-Com may not, due to electrostatic binding with STR-R8. Thus, this suggests that the polyI:C in R8-MEND/OVA/PIC would easily access the receptors compared with that in the R8/OVA/PIC-Com. Another reason is that it is possible that the polyI:C in R8-MEND/OVA/PIC might be recognized by TLR3 in endosomes and Mda5 in the cytosol, because the R8-Lip can deliver encapsulated substances to the cytosol [65].

Agonists of the cyclic-GMP-AMP synthase (cGAS)-stimulator of interferon genes (STING) pathway (STING agonist) have attracted considerable interest. In the Cancer-Immunity Cycle, DC maturation is induced by recognizing DNA derived from dead cancer cells via the cGAS-STING pathway [13, 88]. Moreover, recent reports have clearly shown that the cGAS-STING pathway, but not other innate sensors, such as TLR, is essential for innate immune sensing against immunogenic tumors [100]. Therefore, STING agonists represent promising cancer adjuvants. However, applying STING agonists as a cancer adjuvant represents a more formidable hurdle, because STING agonists must be delivered to the cytosol. We succeeded in efficiently delivering cyclic di-GMP (cdGMP), a STING agonist, by encapsulating it in a YSK05-MEND (Fig. 5) [59]. Treatment with the cdGMP

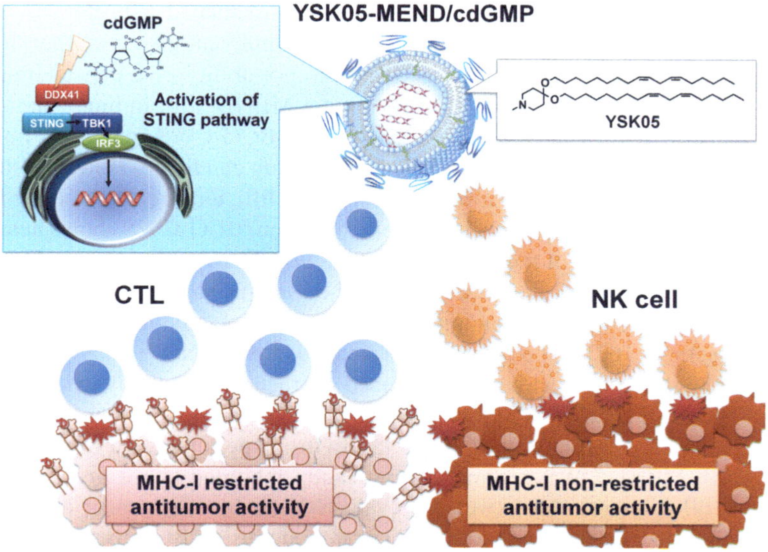

Fig. 5 Concept of cancer immunotherapy by a YSK05-MEND/cdGMP
A STING agonist, cyclic di-GMP (cdGMP), loaded in YSK05-MEND (YSK05-MEND/cdGMP) efficiently activates the STING pathway, resulting in the induction of both MHC class I (MHC-I) restricted antitumor immunity via CTL activation and MHC-I non-restricted antitumor immunity via NK cells

loaded YSK05-MEND (YSK05-MEND/cdGMP) promoted the efficient production of type I interferon (IFN) from RAW264.7 cells compared with that of a conventional transfection reagent loaded with cdGMP, suggesting that the enhancement is due to the efficient cytosolic delivery by YSK05-MEND. Treatment of mice with a mixture of OVA and YSK05-MEND/cdGMP resulted in the induction of an OVA-specific CTL and an antitumor effect against EG.7-OVA tumors, indicating that the YSK05-MEND/cdGMP has the potential to function as a potent inducer of antigen-specific tumor immunity. Meanwhile, cancer cells avoid being killed by CTL by changing the level of MHC-I expression [1]. In this case, natural killer (NK) cells can function as major reliable effector cells. Thus, we investigated the effect of YSK05-MEND/cdGMP on NK cells [70]. The IV administration of YSK05-MEND/cdGMP to mice resulted in the activation of NK cells in the spleen and prevented the growth of B16-F10 melanomas, that lack MHC-I expression, in a lung metastasis mouse model. This antitumor effect was dependent on NK cells, and NK cells infiltrated the lung with B16-F10 melanomas. The findings suggest that YSK05-MEND/cdGMP induced MHC-I non-restricted antitumor immunity via NK cells.

The intravesical instillation of a live Bacillus Calmette-Guerin (BCG) is the gold standard for therapy against superficial bladder cancer and carcinomas in situ [75]. BCG therapy can induce a very effective response and is the most successful cancer immunotherapy available to date. Thus, BCG therapy would be expected to be applicable to other types of cancer. However, applying it to other cancers is quite impossible, because of the serious adverse effects associated with the use of live mycobacteria [53]. Such adverse effects pose a significant concern when locally instilled intravesically. Thus, we considered the possibility of a cancer immunotherapy based on an artificial BCG by loading a MEND with immune-simulative components of BCG. The BCG cell wall skeleton (BCG-CWS), the main immune active component of BCG [8], would be a potent drug candidate. However, the use of BCG-CWS in a drug formulation has been quite difficult, because of its unfavorable physical characteristics. In a hydrophilic solution, BCG-CWS shows heterogeneity with particle sizes of tens of micrometers [96]. To overcome the problem of a drug formulation, we developed a novel method for packaging BCG-CWS into a nanoparticle, namely by producing a liposome evaporated via emulsified lipid (LEEL) method [67]. The BCG-CWS loaded R8-MEND (R8-MEND/CWS) achieved a high uniform and water-dispersed character without any evident aggregation. The R8-MEND/CWS was internalized into mouse bladder cancer cells (MBT-2 cells) at an efficiency of 100% in in vitro experiments. In in vivo experiments, a R8-MEND/CWS treatment significantly inhibited tumor growth in mice bearing MBT-2 tumors. Moreover, we evaluated anti-bladder cancer activity in a carcinogen-induced urinary bladder cancer rat model using BCG therapy at a similar therapeutic setting in a clinic. As a result, the intravesical administration of R8-MEND/CWS resulted in the shrinking of bladder tumors, suggesting that R8-MEND/CWS is fully capable of delivering BCG-CWS to bladder cancer cells and immune cells in the intravesical environment. We also found that the anti-bladder tumor mechanism of R8-MEND/CWS was similar to that of live BCG [68]. Collectively, the R8-MEND/CWS would be expected to be a potent candidate for an anti-bladder cancer immunotherapeutic drug. In addition, in contrast to BCG, the R8-MEND/CWS can be applied to other types of cancers, because of its non-infectiveness characteristics.

Control of Gene Expression in Immune Cells by a MEND

In terms of the fight for immune suppression in TIME, immune cells in addition to DCs, such as T cells, macrophages, NK cells, B cells etc., represent an important target. Controlling the functions of immune cells at the gene level could be a promising strategy for overcoming the immune suppression in TIME, because the current strategies such as antibodies and small molecules are difficult to deal with in cases of immune-suppressive factors. RNA interference by siRNA is a promising tool for controlling gene expression. To reliably control gene expression by siRNA in immune cells, we recently developed a MEND composed of the YSK12-C4 lipid (YSK12-MEND) (Fig. 6) [98]. The YSK12-C4 lipid has cationic properties and a high fusiogenic activity, leading to efficient cellular uptake and endosomal escape. In mouse DCs, the siRNA-loaded YSK12-MEND showed an effective gene silencing in excess of 90%, with a median effective concentration (EC_{50}) of 1.5 nM, while the gene silencing efficiency of Lipofectamine RNAiMAX (RNAiMAX), a powerful siRNA transfection reagent, was less than 60% and the EC_{50} was 25 nM [98].

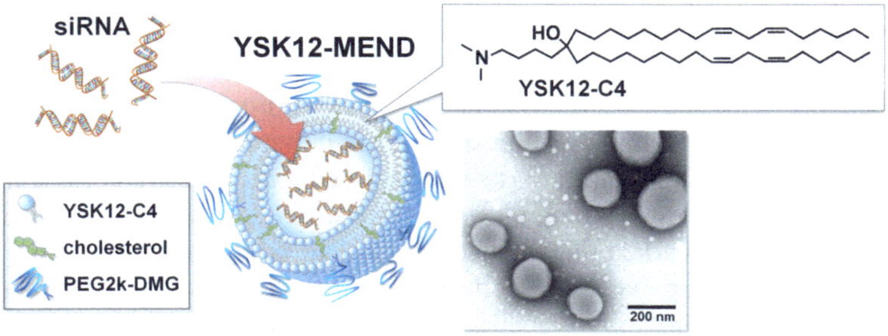

Fig. 6 Concept of siRNA-loaded YSK12-MEND
The YSK12-MEND is composes of YSK12-C4, cholesterol and 1,2-Dimirystoyl-sn-glycerol methoxyethyleneglycol 2000 ether (PEG2k-DMG). In TEM observations, sub-200 nm sized NPs were observed, and the population appeared to be uniform. The inside of the particles appeared to be filled with lipid

This suggests that the efficient gene silencing by the YSK12-MEND was due to its high activity for endosomal escape. Furthermore, a suppressor of cytokine signaling 1 (SOCS1) silenced mouse DCs by the YSK12-MEND significantly facilitated cytokine production and the efficacy of DC-based therapy against E.G7-OVA tumors. These results indicate that the YSK12-MEND is a highly potent carrier for delivering siRNA to DCs and is also a powerful technology for controlling DC functions in DC-based therapy.

We subsequently applied the YSK12-MEND to the delivery of siRNA to human immune cell lines [71]. Jurkat (human T cells), THP-1 (human monocytes), KG-1 (human macrophages) and NK-92 (human NK cells) were used and the gene silencing efficiency by YSK12-MEND was evaluated. The maximum gene silencing efficiencies of the YSK12-MEND at a siRNA dose range of 1–30 nM were 96% in Jurkat, 96% in THP-1, 91% in KG-1 and 75% in the case of NK-92. In contrast, the corresponding values for RNAiMAX were 37%, 56%, 43% and 19%, respectively. Significant cytotoxicity, however, was observed at siRNA doses required for gene silencing in NK-92 cells. To reduce the cytotoxicity in NK-92 cells, we decreased the amount of lipid in the YSK12-MEND by introducing a core formed by electrostatic interactions of siRNA with a polycation [72]. The use of the YSK12-MEND containing a siRNA core significantly reduced the cytotoxicity in NK-92 cells without an appreciable loss of gene silencing activity. The facts clearly indicate that the YSK12-MEND is also a highly promising siRNA delivery system in human immune cell lines.

Perspective

Our originally designed MEND system for efficient drug and gene delivery has been subjected to several optimization steps, since it first appeared in 2004. Various versions of the MEND system have been developed with improved efficiency and reduced toxicity (Fig. 1 and Table 1). During the journey associated with developing this system, we have learned that controlling biodistribution in the body and intracellular trafficking inside cells are both important for increasing the efficiency of such systems. Improving biodistribution relies mainly on developing systems that are stable in the circulation and actively target a certain tissue in the body. The efficient delivery to target cells is not sufficient for eliciting a therapeutic effect. Intracellular trafficking, especially strategies for enhancing endosomal escape are urgently needed. Two components of the MEND system had a remarkable effect on improving overall activity. The first, the R8 peptide as one of the CPPs, can be used either alone in non-specific systems or combined with other targeting ligands to form targeted dual ligand systems. Second, the pH-sensitive YSK lipid series, which significantly improved endosomal escape and allowed the preparation of non-positively charged systems with improved biodistribution. The MEND system discussed in this chapter has considerable promise as a platform for efficient and safe gene therapy for the treatment of currently incurable diseases in the future. Our efforts to rationally improve the MEND system continues towards moving from preclinical studies to clinical testing, which is the current challenge to address the dream of achieving a complete cure of currently incurable diseases including cancer.

Acknowledgments This work was supported by the Special Education and Research Expenses from the Ministry of Education, Culture, Sports, Science and Technology (MEXT) - Japan. We wish to thank Dr. Milton Feather for his helpful advice in editing the English in this manuscript.

References

1. Ahmad M, Rees RC, Ali SA (2004) Escape from immunotherapy: possible mechanisms that influence tumor regression/progression. Cancer Immunol Immunother 53(10):844–854. https://doi.org/10.1007/s00262-004-0540-x
2. Aili D, Mager M, Roche D, Stevens MM (2011) Hybrid nanoparticle-liposome detection of phospholipase activity. Nano Lett 11(4):1401–1405. https://doi.org/10.1021/nl1024062
3. Akita H, Kudo A, Minoura A, Yamaguti M, Khalil IA, Moriguchi R et al (2009) Multi-layered nanoparticles for penetrating the endosome and nuclear membrane via a step-wise membrane fusion process. Biomaterials 30(15):2940–2949. https://doi.org/10.1016/j.biomaterials.2009.02.009
4. Akita H, Ishiba R, Hatakeyama H, Tanaka H, Sato Y, Tange K et al (2013) A neutral envelope-type nanoparticle containing pH-responsive and SS-cleavable lipid-like material as a carrier for plasmid DNA. Adv Healthc Mater 2(8):1120–1125. https://doi.org/10.1002/adhm.201200431

5. Akita H, Ishiba R, Togashi R, Tange K, Nakai Y, Hatakeyama H et al (2015) A neutral lipid envelope-type nanoparticle composed of a pH-activated and vitamin E-scaffold lipid-like material as a platform for a gene carrier targeting renal cell carcinoma. J Control Release 200:97–105. https://doi.org/10.1016/j.jconrel.2014.12.029
6. Ara MN, Hyodo M, Ohga N, Hida K, Harashima H (2012) Development of a novel DNA aptamer ligand targeting to primary cultured tumor endothelial cells by a cell-based SELEX method. PLoS One 7(12):e50174. https://doi.org/10.1371/journal.pone.0050174
7. Ara MN, Matsuda T, Hyodo M, Sakurai Y, Hatakeyama H, Ohga N et al (2014) An aptamer ligand based liposomal nanocarrier system that targets tumor endothelial cells. Biomaterials 35(25):7110–7120. https://doi.org/10.1016/j.biomaterials.2014.04.087
8. Azuma I, Seya T (2001) Development of immunoadjuvants for immunotherapy of cancer. Int Immunopharmacol 1(7):1249–1259
9. Binnewies M, Roberts EW, Kersten K, Chan V, Fearon DF, Merad M et al (2018) Understanding the tumor immune microenvironment (TIME) for effective therapy. Nat Med 24(5):541–550. https://doi.org/10.1038/s41591-018-0014-x
10. Brubaker SW, Bonham KS, Zanoni I, Kagan JC (2015) Innate immune pattern recognition: a cell biological perspective. Annu Rev Immunol 33:257–290. https://doi.org/10.1146/annurev-immunol-032414-112240
11. Chen DS, Mellman I (2013) Oncology meets immunology: the cancer-immunity cycle. Immunity 39(1):1–10. https://doi.org/10.1016/j.immuni.2013.07.012
12. Chen DS, Mellman I (2017) Elements of cancer immunity and the cancer-immune set point. Nature 541(7637):321–330. https://doi.org/10.1038/nature21349
13. Corrales L, Matson V, Flood B, Spranger S, Gajewski TF (2017) Innate immune signaling and regulation in cancer immunotherapy. Cell Res 27(1):96–108. https://doi.org/10.1038/cr.2016.149
14. Couzin-Frankel J (2013) Breakthrough of the year 2013. Cancer immunotherapy. Science 342(6165):1432–1433. https://doi.org/10.1126/science.342.6165.1432
15. Davis ME, Chen ZG, Shin DM (2008) Nanoparticle therapeutics: an emerging treatment modality for cancer. Nat Rev Drug Discov 7(9):771–782. https://doi.org/10.1038/nrd2614
16. El-Sayed A, Khalil IA, Kogure K, Futaki S, Harashima H (2008) Octaarginine- and octalysine-modified nanoparticles have different modes of endosomal escape. J Biol Chem 283(34):23450–23461. https://doi.org/10.1074/jbc.M709387200
17. El-Sayed A, Masuda T, Khalil I, Akita H, Harashima H (2009) Enhanced gene expression by a novel stearylated INF7 peptide derivative through fusion independent endosomal escape. J Control Release 138(2):160–167. https://doi.org/10.1016/j.jconrel.2009.05.018
18. Ewert KK, Zidovska A, Ahmad A, Bouxsein NF, Evans HM, McAllister CS et al (2010) Cationic liposome-nucleic acid complexes for gene delivery and silencing: pathways and mechanisms for plasmid DNA and siRNA. Top Curr Chem 296:191–226
19. Gentile F, Curcio A, Indolfi C, Ferrari M, Decuzzi P (2008) The margination propensity of spherical particles for vascular targeting in the microcirculation. J Nanobiotechnol 6:9. https://doi.org/10.1186/1477-3155-6-9
20. Harding CV, Collins DS, Kanagawa O, Unanue ER (1991) Liposome-encapsulated antigens engender lysosomal processing for class II MHC presentation and cytosolic processing for class I presentation. J Immunol 147(9):2860–2863
21. Harris TJ, von Maltzahn G, Derfus AM, Ruoslahti E, Bhatia SN (2006) Proteolytic actuation of nanoparticle self-assembly. Angew Chem Int Ed Engl 45(19):3161–3165. https://doi.org/10.1002/anie.200600259
22. Hashiba K, Sato Y, Harashima H (2017) pH-labile PEGylation of siRNA-loaded lipid nanoparticle improves active targeting and gene silencing activity in hepatocytes. J Control Release 262:239–246. https://doi.org/10.1016/j.jconrel.2017.07.046
23. Hatakeyama H, Akita H, Kogure K, Oishi M, Nagasaki Y, Kihira Y et al (2007) Development of a novel systemic gene delivery system for cancer therapy with a tumor-specific cleavable PEG-lipid. Gene Ther 14(1):68–77. https://doi.org/10.1038/sj.gt.3302843

24. Hatakeyama H, Ito E, Akita H, Oishi M, Nagasaki Y, Futaki S et al (2009) A pH-sensitive fusogenic peptide facilitates endosomal escape and greatly enhances the gene silencing of siRNA-containing nanoparticles in vitro and in vivo. J Control Release 139(2):127–132. https://doi.org/10.1016/j.jconrel.2009.06.008
25. Hatakeyama H, Akita H, Harashima H (2011a) A multifunctional envelope type nano device (MEND) for gene delivery to tumours based on the EPR effect: a strategy for overcoming the PEG dilemma. Adv Drug Deliv Rev 63(3):152–160. https://doi.org/10.1016/j.addr.2010.09.001
26. Hatakeyama H, Akita H, Ito E, Hayashi Y, Oishi M, Nagasaki Y et al (2011b) Systemic delivery of siRNA to tumors using a lipid nanoparticle containing a tumor-specific cleavable PEG-lipid. Biomaterials 32(18):4306–4316. https://doi.org/10.1016/j.biomaterials.2011.02.045
27. Hatakeyama H, Akita H, Harashima H (2013) The polyethyleneglycol dilemma: advantage and disadvantage of PEGylation of liposomes for systemic genes and nucleic acids delivery to tumors. Biol Pharm Bull 36(6):892–899
28. Hayashi Y, Yamauchi J, Khalil IA, Kajimoto K, Akita H, Harashima H (2011) Cell penetrating peptide-mediated systemic siRNA delivery to the liver. Int J Pharm 419(1–2):308–313. https://doi.org/10.1016/j.ijpharm.2011.07.038
29. Hayashi Y, Mizuno R, Ikramy KA, Akita H, Harashima H (2012) Pretreatment of hepatocyte growth factor gene transfer mediated by octaarginine peptide-modified nanoparticles ameliorates LPS/D-galactosamine-induced hepatitis. Nucleic Acid Ther 22(5):360–363. https://doi.org/10.1089/nat.2012.0352
30. Higashi T, Khalil IA, Maiti KK, Lee WS, Akita H, Harashima H et al (2009) Novel lipidated sorbitol-based molecular transporters for non-viral gene delivery. J Control Release 136(2):140–147. https://doi.org/10.1016/j.jconrel.2009.01.024
31. Hossen MN, Kajimoto K, Akita H, Hyodo M, Ishitsuka T, Harashima H (2013) Therapeutic assessment of cytochrome C for the prevention of obesity through endothelial cell-targeted nanoparticulate system. Mol Ther 21(3):533–541. https://doi.org/10.1038/mt.2012.256
32. Immordino ML, Dosio F, Cattel L (2006) Stealth liposomes: review of the basic science, rationale, and clinical applications, existing and potential. Int J Nanomedicine 1(3):297–315
33. Irvine DJ, Hanson MC, Rakhra K, Tokatlian T (2015) Synthetic nanoparticles for vaccines and immunotherapy. Chem Rev 115(19):11109–11146. https://doi.org/10.1021/acs.chemrev.5b00109
34. Ishida T, Kirchmeier MJ, Moase EH, Zalipsky S, Allen TM (2001) Targeted delivery and triggered release of liposomal doxorubicin enhances cytotoxicity against human B lymphoma cells. Biochim Biophys Acta 1515(2):144–158
35. Ishitsuka T, Akita H, Harashima H (2011) Functional improvement of an IRQ-PEG-MEND for delivering genes to the lung. J Control Release 154(1):77–83. https://doi.org/10.1016/j.jconrel.2011.05.012
36. Kakudo T, Chaki S, Futaki S, Nakase I, Akaji K, Kawakami T et al (2004) Transferrin-modified liposomes equipped with a pH-sensitive fusogenic peptide: an artificial viral-like delivery system. Biochemistry 43(19):5618–5628. https://doi.org/10.1021/bi035802w
37. Kapadia CH, Perry JL, Tian S, Luft JC, DeSimone JM (2015) Nanoparticulate immunotherapy for cancer. J Control Release 219:167–180. https://doi.org/10.1016/j.jconrel.2015.09.062
38. Khalil IA, Harashima H (2018) An efficient PEGylated gene delivery system with improved targeting: synergism between octaarginine and a fusogenic peptide. Int J Pharm 538(1–2):179–187. https://doi.org/10.1016/j.ijpharm.2018.01.007
39. Khalil IA, Kogure K, Akita H, Harashima H (2006a) Uptake pathways and subsequent intracellular trafficking in nonviral gene delivery. Pharmacol Rev 58(1):32–45. https://doi.org/10.1124/pr.58.1.8
40. Khalil IA, Kogure K, Futaki S, Harashima H (2006b) High density of octaarginine stimulates macropinocytosis leading to efficient intracellular trafficking for gene expression. J Biol Chem 281(6):3544–3551. https://doi.org/10.1074/jbc.M503202200

41. Khalil IA, Kogure K, Futaki S, Hama S, Akita H, Ueno M et al (2007) Octaarginine-modified multifunctional envelope-type nanoparticles for gene delivery. Gene Ther 14(8):682–689. https://doi.org/10.1038/sj.gt.3302910
42. Khalil IA, Kogure K, Futaki S, Harashima H (2008) Octaarginine-modified liposomes: enhanced cellular uptake and controlled intracellular trafficking. Int J Pharm 354(1–2):39–48. https://doi.org/10.1016/j.ijpharm.2007.12.003
43. Khalil IA, Hayashi Y, Mizuno R, Harashima H (2011) Octaarginine- and pH sensitive fusogenic peptide-modified nanoparticles for liver gene delivery. J Control Release 156(3):374–380. https://doi.org/10.1016/j.jconrel.2011.08.012
44. Khalil IA, Kimura S, Sato Y, Harashima H (2018) Synergism between a cell penetrating peptide and a pH-sensitive cationic lipid in efficient gene delivery based on double-coated nanoparticles. J Control Release 275:107–116. https://doi.org/10.1016/j.jconrel.2018.02.016
45. Kibria G, Hatakeyama H, Ohga N, Hida K, Harashima H (2011) Dual-ligand modification of PEGylated liposomes shows better cell selectivity and efficient gene delivery. J Control Release 153(2):141–148. https://doi.org/10.1016/j.jconrel.2011.03.012
46. Kibria G, Hatakeyama H, Ohga N, Hida K, Harashima H (2013) The effect of liposomal size on the targeted delivery of doxorubicin to integrin αvβ3-expressing tumor endothelial cells. Biomaterials 34(22):5617–5627. https://doi.org/10.1016/j.biomaterials.2013.03.094
47. Kogure K, Moriguchi R, Sasaki K, Ueno M, Futaki S, Harashima H (2004) Development of a non-viral multifunctional envelope-type nano device by a novel lipid film hydration method. J Control Release 98(2):317–323. https://doi.org/10.1016/j.jconrel.2004.04.024
48. Kogure K, Akita H, Harashima H (2007) Multifunctional envelope-type nano device for non-viral gene delivery: concept and application of programmed packaging. J Control Release 122(3):246–251. https://doi.org/10.1016/j.jconrel.2007.06.018
49. Kogure K, Akita H, Yamada Y, Harashima H (2008) Multifunctional envelope-type nano device (MEND) as a non-viral gene delivery system. Adv Drug Deliv Rev 60(4–5):559–571. https://doi.org/10.1016/j.addr.2007.10.007
50. Kulkarni PS, Haldar MK, Nahire RR, Katti P, Ambre AH, Muhonen WW et al (2014) Mmp-9 responsive PEG cleavable nanovesicles for efficient delivery of chemotherapeutics to pancreatic cancer. Mol Pharm 11(7):2390–2399. https://doi.org/10.1021/mp500108p
51. Kusumoto K, Akita H, Ishitsuka T, Matsumoto Y, Nomoto T, Furukawa R et al (2013) Lipid envelope-type nanoparticle incorporating a multifunctional peptide for systemic siRNA delivery to the pulmonary endothelium. ACS Nano 7(9):7534–7541. https://doi.org/10.1021/nn401317t
52. Kusumoto K, Akita H, Santiwarangkool S, Harashima H (2014) Advantages of ethanol dilution method for preparing GALA-modified liposomal siRNA carriers on the in vivo gene knockdown efficiency in pulmonary endothelium. Int J Pharm 473(1–2):144–147. https://doi.org/10.1016/j.ijpharm.2014.07.007
53. Lamm DL, van der Meijden PM, Morales A, Brosman SA, Catalona WJ, Herr HW et al (1992) Incidence and treatment of complications of bacillus Calmette-Guerin intravesical therapy in superficial bladder cancer. J Urol 147(3):596–600
54. Li W, Nicol F, Szoka FC (2004) GALA: a designed synthetic pH-responsive amphipathic peptide with applications in drug and gene delivery. Adv Drug Deliv Rev 56(7):967–985. https://doi.org/10.1016/j.addr.2003.10.041
55. Li W, Huang Z, MacKay JA, Grube S, Szoka FC (2005) Low-pH-sensitive poly(ethylene glycol) (PEG)-stabilized plasmid nanolipoparticles: effects of PEG chain length, lipid composition and assembly conditions on gene delivery. J Gene Med 7(1):67–79. https://doi.org/10.1002/jgm.634
56. Maeda H, Sawa T, Konno T (2001) Mechanism of tumor-targeted delivery of macromolecular drugs, including the EPR effect in solid tumor and clinical overview of the prototype polymeric drug SMANCS. J Control Release 74(1–3):47–61

57. Matsumura Y, Maeda H (1986) A new concept for macromolecular therapeutics in cancer chemotherapy: mechanism of tumoritropic accumulation of proteins and the antitumor agent smancs. Cancer Res 46(12 Pt 1):6387–6392
58. Mishra S, Webster P, Davis ME (2004) PEGylation significantly affects cellular uptake and intracellular trafficking of non-viral gene delivery particles. Eur J Cell Biol 83(3):97–111. https://doi.org/10.1078/0171-9335-00363
59. Miyabe H, Hyodo M, Nakamura T, Sato Y, Hayakawa Y, Harashima H (2014) A new adjuvant delivery system 'cyclic di-GMP/YSK05 liposome' for cancer immunotherapy. J Control Release 184:20–27. https://doi.org/10.1016/j.jconrel.2014.04.004
60. Moriguchi R, Kogure K, Akita H, Futaki S, Miyagishi M, Taira K et al (2005) A multifunctional envelope-type nano device for novel gene delivery of siRNA plasmids. Int J Pharm 301(1–2):277–285. https://doi.org/10.1016/j.ijpharm.2005.05.021
61. Mudhakir D, Akita H, Khalil IA, Futaki S, Harashima H (2005) Pharmacokinetic analysis of the tissue distribution of octaarginine modified liposomes in mice. Drug Metab Pharmacokinet 20(4):275–281
62. Mudhakir D, Akita H, Tan E, Harashima H (2008) A novel IRQ ligand-modified nano-carrier targeted to a unique pathway of caveolar endocytic pathway. J Control Release 125(2):164–173. https://doi.org/10.1016/j.jconrel.2007.10.020
63. Nakamura Y, Kogure K, Yamada Y, Futaki S, Harashima H (2006) Significant and prolonged antisense effect of a multifunctional envelope-type nano device encapsulating antisense oligodeoxynucleotide. J Pharm Pharmacol 58(4):431–437. https://doi.org/10.1211/jpp.58.4.0002
64. Nakamura Y, Kogure K, Futaki S, Harashima H (2007) Octaarginine-modified multifunctional envelope-type nano device for siRNA. J Control Release 119(3):360–367. https://doi.org/10.1016/j.jconrel.2007.03.010
65. Nakamura T, Moriguchi R, Kogure K, Shastri N, Harashima H (2008) Efficient MHC class I presentation by controlled intracellular trafficking of antigens in octaarginine-modified liposomes. Mol Ther 16(8):1507–1514. https://doi.org/10.1038/mt.2008.122
66. Nakamura T, Moriguchi R, Kogure K, Harashima H (2013) Incorporation of polyinosine-polycytidylic acid enhances cytotoxic T cell activity and antitumor effects by octaarginine-modified liposomes encapsulating antigen, but not by octaarginine-modified antigen complex. Int J Pharm 441(1–2):476–481. https://doi.org/10.1016/j.ijpharm.2012.11.006
67. Nakamura T, Fukiage M, Higuchi M, Nakaya A, Yano I, Miyazaki J et al (2014a) Nanoparticulation of BCG-CWS for application to bladder cancer therapy. J Control Release 176:44–53. https://doi.org/10.1016/j.jconrel.2013.12.027
68. Nakamura T, Fukiage M, Suzuki Y, Yano I, Miyazaki J, Nishiyama H et al (2014b) Mechanism responsible for the antitumor effect of BCG-CWS using the LEEL method in a mouse bladder cancer model. J Control Release 196:161–167. https://doi.org/10.1016/j.jconrel.2014.10.007
69. Nakamura T, Ono K, Suzuki Y, Moriguchi R, Kogure K, Harashima H (2014c) Octaarginine-modified liposomes enhance cross-presentation by promoting the C-terminal trimming of antigen peptide. Mol Pharm 11(8):2787–2795. https://doi.org/10.1021/mp500147y
70. Nakamura T, Miyabe H, Hyodo M, Sato Y, Hayakawa Y, Harashima H (2015) Liposomes loaded with a STING pathway ligand, cyclic di-GMP, enhance cancer immunotherapy against metastatic melanoma. J Control Release 216:149–157. https://doi.org/10.1016/j.jconrel.2015.08.026
71. Nakamura T, Kuroi M, Fujiwara Y, Warashina S, Sato Y, Harashima H (2016) Small-sized, stable lipid nanoparticle for the efficient delivery of siRNA to human immune cell lines. Sci Rep 6:37849. https://doi.org/10.1038/srep37849
72. Nakamura T, Yamada K, Fujiwara Y, Sato Y, Harashima H (2018) Reducing the cytotoxicity of lipid nanoparticles associated with a Fusogenic cationic lipid in a natural killer cell line by introducing a Polycation-based siRNA Core. Mol Pharm 15(6):2142–2150. https://doi.org/10.1021/acs.molpharmaceut.7b01166

73. Nakase I, Kogure K, Harashima H, Futaki S (2011) Application of a fusiogenic peptide GALA for intracellular delivery. Methods Mol Biol 683:525–533. https://doi.org/10.1007/978-1-60761-919-2_37
74. Napoli A, Boerakker MJ, Tirelli N, Nolte RJ, Sommerdijk NA, Hubbell JA (2004) Glucose-oxidase based self-destructing polymeric vesicles. Langmuir 20(9):3487–3491
75. Redelman-Sidi G, Glickman MS, Bochner BH (2014) The mechanism of action of BCG therapy for bladder cancer--a current perspective. Nat Rev Urol 11(3):153–162. https://doi.org/10.1038/nrurol.2014.15
76. Remaut K, Lucas B, Braeckmans K, Demeester J, De Smedt SC (2007) Pegylation of liposomes favours the endosomal degradation of the delivered phosphodiester oligonucleotides. J Control Release 117(2):256–266. https://doi.org/10.1016/j.jconrel.2006.10.029
77. Sakurai Y, Hatakeyama H, Sato Y, Hyodo M, Akita H, Ohga N et al (2014) RNAi-mediated gene knockdown and anti-angiogenic therapy of RCCs using a cyclic RGD-modified liposomal-siRNA system. J Control Release 173:110–118. https://doi.org/10.1016/j.jconrel.2013.10.003
78. Sakurai Y, Matsuda T, Hada T, Harashima H (2015) Efficient packaging of plasmid DNA using a pH sensitive cationic lipid for delivery to hepatocytes. Biol Pharm Bull 38(8):1185–1191. https://doi.org/10.1248/bpb.b15-00138
79. Santiwarangkool S, Akita H, Nakatani T, Kusumoto K, Kimura H, Suzuki M et al (2017) PEGylation of the GALA peptide enhances the lung-targeting activity of Nanocarriers that contain encapsulated siRNA. J Pharm Sci 106(9):2420–2427. https://doi.org/10.1016/j.xphs.2017.04.075
80. Sasaki K, Kogure K, Chaki S, Kihira Y, Ueno M, Harashima H (2005) Construction of a multifunctional envelope-type nano device by a SUV∗-fusion method. Int J Pharm 296(1–2):142–150. https://doi.org/10.1016/j.ijpharm.2005.02.020
81. Sato Y, Hatakeyama H, Sakurai Y, Hyodo M, Akita H, Harashima H (2012) A pH-sensitive cationic lipid facilitates the delivery of liposomal siRNA and gene silencing activity in vitro and in vivo. J Control Release 163(3):267–276. https://doi.org/10.1016/j.jconrel.2012.09.009
82. Sato Y, Hatakeyama H, Hyodo M, Harashima H (2016a) Relationship between the physicochemical properties of lipid nanoparticles and the quality of siRNA delivery to liver cells. Mol Ther 24(4):788–795. https://doi.org/10.1038/mt.2015.222
83. Sato Y, Note Y, Maeki M, Kaji N, Baba Y, Tokeshi M et al (2016b) Elucidation of the physicochemical properties and potency of siRNA-loaded small-sized lipid nanoparticles for siRNA delivery. J Control Release 229:48–57. https://doi.org/10.1016/j.jconrel.2016.03.019
84. Sato Y, Matsui H, Yamamoto N, Sato R, Munakata T, Kohara M et al (2017) Highly specific delivery of siRNA to hepatocytes circumvents endothelial cell-mediated lipid nanoparticle-associated toxicity leading to the safe and efficacious decrease in the hepatitis B virus. J Control Release 266:216–225. https://doi.org/10.1016/j.jconrel.2017.09.044
85. Sato Y, Matsui H, Sato R, Harashima H (2018) Neutralization of negative charges of siRNA results in improved safety and efficient gene silencing activity of lipid nanoparticles loaded with high levels of siRNA. J Control Release 284:179–187. https://doi.org/10.1016/j.jconrel.2018.06.017
86. Shaheen SM, Akita H, Nakamura T, Takayama S, Futaki S, Yamashita A et al (2011) KALA-modified multi-layered nanoparticles as gene carriers for MHC class-I mediated antigen presentation for a DNA vaccine. Biomaterials 32(26):6342–6350. https://doi.org/10.1016/j.biomaterials.2011.05.014
87. Shin J, Shum P, Thompson DH (2003) Acid-triggered release via dePEGylation of DOPE liposomes containing acid-labile vinyl ether PEG-lipids. J Control Release 91(1–2):187–200
88. Spranger S (2016) Mechanisms of tumor escape in the context of the T-cell-inflamed and the non-T-cell-inflamed tumor microenvironment. Int Immunol 28(8):383–391. https://doi.org/10.1093/intimm/dxw014

89. Takahashi Y, Nishikawa M, Takakura Y (2009) Nonviral vector-mediated RNA interference: its gene silencing characteristics and important factors to achieve RNAi-based gene therapy. Adv Drug Deliv Rev 61(9):760–766. https://doi.org/10.1016/j.addr.2009.04.006
90. Takara K, Hatakeyama H, Ohga N, Hida K, Harashima H (2010) Design of a dual-ligand system using a specific ligand and cell penetrating peptide, resulting in a synergistic effect on selectivity and cellular uptake. Int J Pharm 396(1–2):143–148. https://doi.org/10.1016/j.ijpharm.2010.05.002
91. Takara K, Hatakeyama H, Kibria G, Ohga N, Hida K, Harashima H (2012) Size-controlled, dual-ligand modified liposomes that target the tumor vasculature show promise for use in drug-resistant cancer therapy. J Control Release 162(1):225–232. https://doi.org/10.1016/j.jconrel.2012.06.019
92. Tam YY, Chen S, Cullis PR (2013) Advances in lipid nanoparticles for siRNA delivery. Pharmaceutics 5(3):498–507. https://doi.org/10.3390/pharmaceutics5030498
93. Tanaka H, Akita H, Ishiba R, Tange K, Arai M, Kubo K et al (2014) Neutral biodegradable lipid-envelope-type nanoparticle using vitamin A-scaffold for nuclear targeting of plasmid DNA. Biomaterials 35(5):1755–1761. https://doi.org/10.1016/j.biomaterials.2013.11.016
94. Terada T, Iwai M, Kawakami S, Yamashita F, Hashida M (2006) Novel PEG-matrix metalloproteinase-2 cleavable peptide-lipid containing galactosylated liposomes for hepatocellular carcinoma-selective targeting. J Control Release 111(3):333–342. https://doi.org/10.1016/j.jconrel.2005.12.023
95. Trombetta ES, Mellman I (2005) Cell biology of antigen processing in vitro and in vivo. Annu Rev Immunol 23:975–1028. https://doi.org/10.1146/annurev.immunol.22.012703.104538
96. Uenishi Y, Kawabe K, Nomura T, Nakai M, Sunagawa M (2009) Morphological study on Mycobacterium bovis BCG Tokyo 172 cell wall skeleton (SMP-105). J Microbiol Methods 77(2):139–144. https://doi.org/10.1016/j.mimet.2009.01.019
97. Ukawa M, Akita H, Hayashi Y, Ishiba R, Tange K, Arai M et al (2014) Neutralized nanoparticle composed of SS-cleavable and pH-activated lipid-like material as a long-lasting and liver-specific gene delivery system. Adv Healthc Mater 3(8):1222–1229. https://doi.org/10.1002/adhm.201300629
98. Warashina S, Nakamura T, Sato Y, Fujiwara Y, Hyodo M, Hatakeyama H et al (2016) A lipid nanoparticle for the efficient delivery of siRNA to dendritic cells. J Control Release 225:183–191. https://doi.org/10.1016/j.jconrel.2016.01.042
99. Watanabe T, Hatakeyama H, Matsuda-Yasui C, Sato Y, Sudoh M, Takagi A et al (2014) In vivo therapeutic potential of dicer-hunting siRNAs targeting infectious hepatitis C virus. Sci Rep 4:4750. https://doi.org/10.1038/srep04750
100. Woo SR, Fuertes MB, Corrales L, Spranger S, Furdyna MJ, Leung MY et al (2014) STING-dependent cytosolic DNA sensing mediates innate immune recognition of immunogenic tumors. Immunity 41(5):830–842. https://doi.org/10.1016/j.immuni.2014.10.017
101. Xu D, Jin G, Chai D, Zhou X, Gu W, Chong Y et al (2018) The development of CAR design for tumor CAR-T cell therapy. *Oncotarget* 9(17):13991–14004. https://doi.org/10.18632/oncotarget.24179
102. Yamada Y, Kogure K, Nakamura Y, Inoue K, Akita H, Nagatsugi F et al (2005) Development of efficient packaging method of oligodeoxynucleotides by a condensed nano particle in lipid envelope structure. Biol Pharm Bull 28(10):1939–1942
103. Yamamoto N, Sato Y, Munakata T, Kakuni M, Tateno C, Sanada T et al (2016) Novel pH-sensitive multifunctional envelope-type nanodevice for siRNA-based treatments for chronic HBV infection. J Hepatol 64(3):547–555. https://doi.org/10.1016/j.jhep.2015.10.014

104. Yin H, Kanasty RL, Eltoukhy AA, Vegas AJ, Dorkin JR, Anderson DG (2014) Non-viral vectors for gene-based therapy. Nat Rev Genet 15(8):541–555. https://doi.org/10.1038/nrg3763
105. Zalipsky S, Qazen M, Walker JA, Mullah N, Quinn YP, Huang SK (1999) New detachable poly(ethylene glycol) conjugates: cysteine-cleavable lipopolymers regenerating natural phospholipid, diacyl phosphatidylethanolamine. Bioconjug Chem 10(5):703–707
106. Zheng J, Wan Y, Elhissi A, Zhang Z, Sun X (2014) Targeted paclitaxel delivery to tumors using cleavable PEG-conjugated solid lipid nanoparticles. Pharm Res 31(8):2220–2233. https://doi.org/10.1007/s11095-014-1320-8

Part IV
Cancer Stromal Targeting (CAST) Therapy and Diagnosis

Principle of CAST Strategy

Yasuhiro Matsumura

Abstract This chapter describes the conceptual basis of a novel form of cancer therapy that we have developed, which uses structural components of the stroma of a tumor as an anchor and launchpad for disseminating chemotherapeutic agents within a tumor. It also utilizes a phenomenon called the Enhanced Permeability and Retention (EPR) effect, described in this chapter and elsewhere in this book, to localize and concentrate the effect of the therapy on the lesion and minimize damage to healthy organs. We refer to this combined strategy as cancer (CA) stroma (S) targeting (T), CAST therapy. The concept of CAST therapy is that stroma-targeting immunconjugates extravasate selectively from tumor vessels (but not from normal counterparts), due to the EPR effect, bind to the stroma to create a scaffold, from which sustained release of cytotoxic agents occurs and subsequently diffuse throughout the tumor tissue to damage both tumor cells and tumor vessels. Various anticancer agents, gene-delivery systems, and monoclonal antibody (mAb)-based therapies have been produced, based on the enhanced permeability and retention (EPR) effect, but the current combination of EPR with cancer stroma targeting for delivery of a drug is original and new. As one example to illustrate how we utilize the CAST concept, we present a version using an antibody to insoluble fibrin. We have reported that hemorrhage caused by cancer erosion can induce fibrin clot formation that creates a mass of insoluble fibrin (IF) in the tumor stroma. Along with that, Influx of fibroblasts and inflammatory cells into the tumor tissue also occurs. Consequently, most cancers possess a fibrin-rich stroma that hinders the distribution of macromolecules, including antibody-drug conjugates (ADCs). Therefore, the EPR effect is not clinically and efficiently functioning in such situations. We have developed an anti-IF mAb that does not react with fibrinogen or soluble fibrin and have successfully created and optimized anti-IF ADCs. An ADC is an anti-IF antibody conjugated with monomethyl auristatin E (MMAE), a potent microtubule inhibitor, via a Val-Leu-Lys linker that is severed specifically by plasmin (PLM). Free MMAE is only released when the IF- ADC is bound to the epitopes in the IF

Y. Matsumura (✉)
Division of Developmental Therapeutics, Exploratory Oncology Research & Clinical Trial Center, National Cancer Center, Kashiwa, Japan
e-mail: yhmatsum@east.ncc.go.jp

Y. Matsumura, D. Tarin (eds.), *Cancer Drug Delivery Systems Based on the Tumor Microenvironment*, https://doi.org/10.1007/978-4-431-56880-3_11

because plasmin is only active on IF and is neutralized by the innate α2-plasmin inhibitor circulating in the blood. Free MMAE may therefore easily reach cancer cells by diffusing through the stromal barrier. MMAE released from IF-ADC may also attack the tumor vascular endothelial cells resulting in tumor necrosis.

Keywords EPR effect · CAST therapy · Drug delivery system · Tumor stromal barrier · Blood coagulation · Insoluble fibrin · Plasmin · Antibody drug conjugate (ADC)

Introduction

We cannot always reliably distinguish normal human cells from cancerous counterparts, despite the fact that pleomorphism and anaplasia are signature histological characteristics of malignant tumors. This is due to marked heterogeneity within tumor cell populations which complicates the task of destroying malignant growths without damaging healthy organs and this difficulty has greatly handicapped effective anti-cancer therapy. For this reason, there has been considerable incentive to develop a drug delivery system (DDS) that can selectively deliver and concentrate toxic anticancer drugs within cancer containing tissue and organs. It is now widely recognised that this can be achieved by exploiting the Enhanced Permeability and Retention (EPR) effect first reported in 1986 [1]. This phenomenon results from the leakiness of morphologically ill-formed blood vessels within tumors, which allow large molecules to seep through their walls into the extracellular space, together with the absence of draining lymphatics in such growths. Histopathological studies of normal and neoplastic biopsy samples demonstrate the presence of lymphatic vessels in normal tissue and the relative absence of such channels in tumors (Fig. 1). Recognition and confirmation of this difference occurred relatively late in the study of tumor pathology because of the unavailability of suitable reagents to unequivocally distinguish lymphatic vessels from blood vascular capillaries.

The EPR effect can briefly be explained as follows: low molecular weight substances easily leak out even from normal blood vessels and this fact is directly linked to adverse drug events in the case of low molecular weight anticancer drugs. On the other hand, macromolecules are too large to pass through the normal vessel walls and circulate in the blood for a long time, unless they are trapped by the reticuloendothelial system in various organs. Solid tumors generally possess several pathophysiological characteristics which influence the accumulation of macromolecular drugs, namely, hypervasculature, secretion of vascular permeability factors stimulating extravasation of macromolecules within the cancer, and absence of effective lymphatic drainage from tumors that impedes the efficient clearance of macromolecules accumulated in solid tumor tissues. As a result, polymeric substances with high biocompatibility selectively accumulate in solid tumors and stay longer. These characteristics of solid tumors are the basis of the EPR effect. The first paper on EPR has now been cited more than 5500 times (as of August 2018 from Google Scholar).

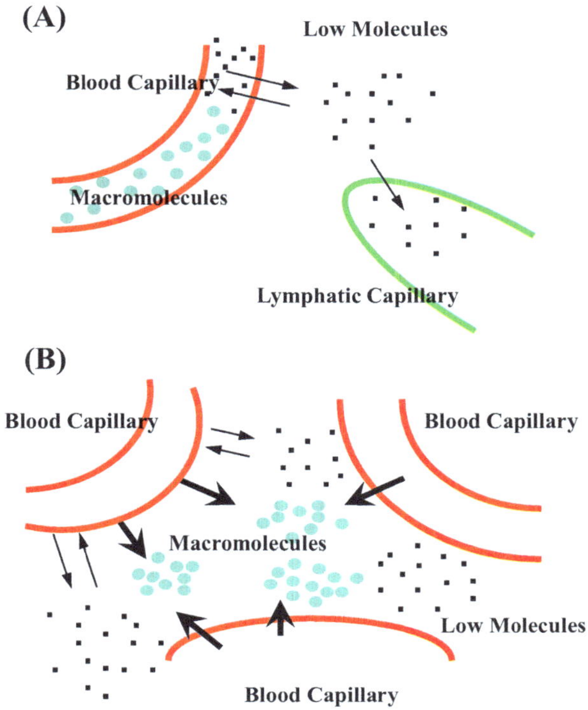

Fig. 1 Diagram of EPR effect [1] (**a**) Low molecular weight substances easily leak out even from normal blood vessels. In the case of small molecule anticancer drugs, it is directly linked to adverse events. (**b**) High molecular weight substances do not leak out of normal vessels but easily leak from tumor vessels with enhanced vascular permeability

The EPR effect has been therefore verified worldwide in non-clinical experiments; however, it has not been widely used in clinics. Many DDS including various ACA formulations and gene delivery systems based on this effect have been produced and used in clinics [2–4]. However, as yet, therapy using DDS is not a mainstream technique in the field of oncology. This is so despite the fact that MCC-465 is a DDS produced by Mitsubishi Pharma Corporation more than 20 years ago (Fig. 2). The formulation consists of doxorubicin-incorporated liposomes conjugated with polyethylene glycol (PEG) and anti-GAH antibodies that specifically bind to stomach cancer cells. MCC-465 was expected to be highly effective in the area of cancer treatment because the medication utilize both the EPR effect and an active targeting mechanism, resulting from the anti-GAH antibodies [5]. This expectation was reinforced by the fact that animal experiments revealed a remarkable antitumor effect in 2 different stomach cancer xenograft models [5] and in others [6]. Unexpectedly, however, no antitumor effect could be confirmed in the phase 1 clinical trial specializing in patients with gastric cancer [7]. The deviation between non-clinical experimental results of MCC 465 and those of its clinical trial served as a trigger to seriously rethink the optimum method of implementation of a new DDS.

These unexpected and disappointing results and subsequent intensive studies led to the discovery of cancer stromal targeting (CAST) therapy described below [8].

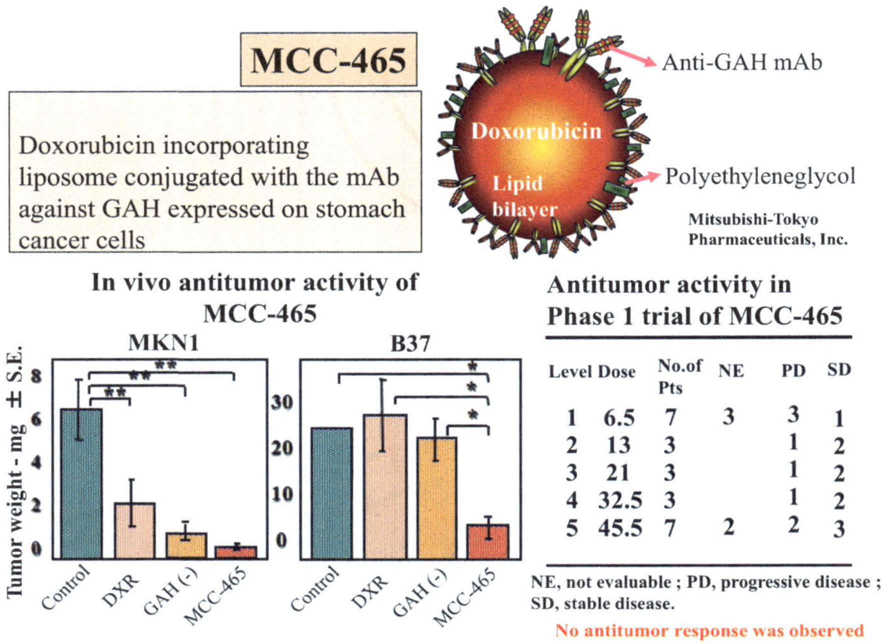

Level	Dose	No.of Pts	NE	PD	SD
1	6.5	7	3	3	1
2	13	3		1	2
3	21	3		1	2
4	32.5	3		1	2
5	45.5	7	2	2	3

Fig. 2 Divergence between mouse experiment and clinical study in gastric cancer [5–7]. MCC-465 is a doxorubicin-encapsulated liposome to which PEG and GAH antibodies (that specifically bind to stomach cancer cells) are added. MCC-465 can use both EPR effect and active targeting. MCC-465 showed a remarkable antitumor effect in two types of gastric cancer xenografts model. In a clinical trial specialized for gastric cancer, no antitumor effect could be confirmed at all

Tumor Stroma as a Barrier of DDS

With regard to the pitfall of DDS, We examined pathophysiologically in detail focusing on human pancreatic cancer. We compared the clinical pancreatic cancer tissues with human pancreatic xenografts in nude mice and found a clear difference between the two. The most important point is that clinical pancreatic cancer possesses abundant stroma, whereas xenograft does not have it (Fig. 3). By the way, tumor vessels that supply oxygen and nutrients to tumor cells exist in the tumor stroma. Naturally, drugs for damaging tumor cells are also delivered via their tumor vessels. It is, therefore, reasonable to hypothesize that abundant stroma may be a barrier hindering the distribution of macromolecules, including various DDS formulations, within cancer tissues. On the contrary, there are few tumor stroma in human pancreatic cancer xenograft which is an in vivo experimental cancer, only blood vessels and tumor cells. These facts suggest that DDS has distinctive effects by distributing throughout the tumor in experimental pancreatic cancer. On the other hand, DDS is ineffective in clinical human pancreatic cancer because the tumor stromal barrier prevents DDS from reaching the cancer cells within the tumor

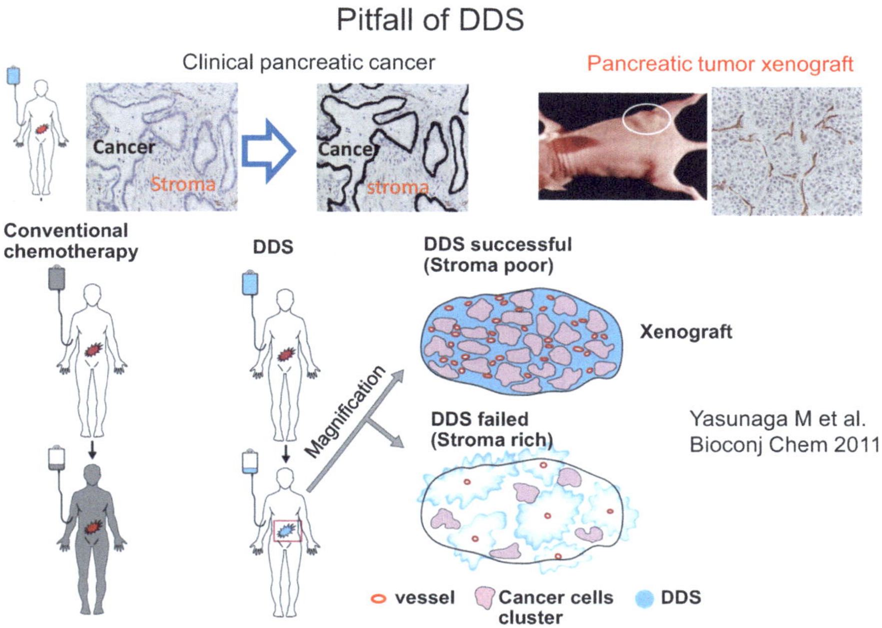

Fig. 3 Pitfalls of DDS [8]. Strong vascular invasive cancers such as pancreatic and stomach cancers have abundant stroma with consequences of hypercoagulability. Abundant stroma prevents the penetration of macromolecules including DDS formulations within the tumor. In contrast, pancreatic tumor xenografts contained only tumor cells and no stroma. DDS has a high antitumor effect in xenografts model because of the lack of tumor stroma, which may allow DDS to be distributed throughout the tumor. On the other hand, DDS is ineffective in clinical human pancreatic cancer because tumor stroma prevents DDS from reaching tumor cells

tissue [8]. So-called intractable cancer such as stomach cancer and pancreatic cancer is known to have abundant interstitium. It is also known that normal solid cancers maintain tumor interstitia more or less as compared with blood-based tumors such as lymphoma, and the more interstitial, the more malignant and drug-resistant. Therefore, the problem of interstitial barrier is a problem to be considered also in other solid cancers such as breast cancer and colon cancer..

Tumor Stroma Generated by Coagulation of Extravasated Blood

To understand how the stroma of many tumors becomes dense and impenetrable to large molecules, it is appropriate to consider why and how the stroma is formed and expands in the field of a growing tumor. This is pertinent to tumor biology as well as to the improvement of cancer therapy.

In addition to recruitment of stromal cells from neighboring tissue and repair of local tissue damage by collagen deposition, delicate new blood vessels entering the zone are prone to rupture causing microhemorrhages. Also, as cancerous invasion involves active destruction of interstitial tissue and of adjacent normal or tumor vessels, further hemorrhage occurs, coagulation zymogens encounter Tissue Factor and fibrin clots immediately form in situ to stop the bleeding. These clots are subsequently replaced by collagen, as occurs in the normal wound healing process. Although there are some similarities between wound healing and the mechanisms underlying cancer-induced stroma, a fundamental difference between the two is that the fibrin clot formation in cancer lasts for as long as cancer cells survive in the body. Therefore, the "malignant cycle of blood coagulation" is a versatile repetitive process of cancer stroma generation consisting of cancer invasion into vessels, hemorrhage, insoluble fibrin (IF) formation, and replacement with collagenous tissue [8]. Consequently, cancer induced blood coagulation generates IF rich tumor stroma which contributes to enhancing the barrier preventing DDS from reaching cancer cells [8].

However, the EPR effect on fluid dynamics in tumor tissues can be exploited to enable macromolecular targeted therapeutic agents to reach malignant cells through dense tumor stroma. In the EPR effect, the increased tumor vascular permeability is the most important factor responsible for the leakage of macromolecules into the interstitial space. At the time that we described the EPR effect, we also succeeded in purifying two types of kinin (bradykinin and hydroxyprolyl^3-bradykinin) from the ascitic fluid of a patient with gastric cancer and pleural effusion of a patient with lung cancer [9, 10].

We also clarified that the kinin production system in cancer is due to the activation of Hageman factor, an intrinsic coagulation factor XII by cancer cells and the kinins caused the ascites. [11].

At the same time, Dvorak et al. discovered that vascular permeability factor (VPF) was involved in tumor vascular permeability [12]. Later, it was found that VPF was identical to vascular endothelial growth factor (VEGF) [13]. Recently, an extrinsic coagulation factor, namely a tissue factor (TF), appeared to activate VEGF production [14]. In conclusion, both intrinsic and extrinsic coagulation factors are involved in tumor vascular permeability as well as tumor-induced blood coagulation. When blood coagulation and vascular hyperpermeability occur in a cancer tissue, they can cause inflow of macromolecular proteins and edema in the tumor tissue followed by the accumulation of cancer-associated fibroblasts and other cells [15–18].

For DDS, it was mentioned above that tumor stroma based on blood coagulation is a barrier. As a matter of fact, it is known that some bacterial colonies encircled with fibrin filament are resistant to antibacterial agents because of the fibrin barrier [19]. Thus, various diseases can produce fibrin clots and we have to keep in mind that this insoluble protein can generate profits for the body but can cause several inconvenient outcomes simultaneously.

In the nineteenth century, French physician Armand Trousseau described thrombophlebitis in patients with stomach cancer [20]. Today, it is clinically proven that hypercoagulability is increased and thrombosis occurs in many cancers, especially in invasive cancers such as GBM, pancreatic cancers, and stomach cancers [21]. However, this Trousseau syndrome is a thrombosis that occurs as a result of a systemic hypercoagulable state induced by cancer, and we will stress that fibrin clot formation in cancer tissue is only coagulation enhancement within the tumor tissue.

Alteration or Disruption of the Tumor Stroma to Facilitate DDS

Kano et al. proposed that low dose TGFβ receptor inhibitor treatment could alter a cancer-associated neovasculature to exhibit more leakiness for macromolecules, with less pericyte-coverage and greater endothelial area resulting in enhancement of DDS therapy [22].

Olive et al. reported that Hedgehog inhibitor (HHI) could enhance the antitumor activity of gemcitabin (GM) by destroying tumor stroma and preparing the way for GM to attack cancer cells in pancreatic ductal adenocarcinoma (PDA). They proposed that HHI could deplete tumor stromal tissue of PDA and the combination therapy of GM with HHI improved the delivery and efficacy of therapeutics in patients with PDA [23]. The idea was novel and reasonable as a new combination therapy of GM with HHI against pancreatic cancer, which is a stroma-rich cancer.

A similar strategy involving extra cellular matrix- degrading co-treatments, like the reduction of hyaluronan, which could help to promote nanomedicine accumulation and efficacy [24] has been proposed.

Utilization of the Tumor Stroma for DDS

Instead of regarding the tumor stroma as simply an obstacle to therapy, we can exploit it as an aid to improving drug delivery by adopting a novel strategy described below:

Antibody Drug Conjugate (ADC)

Instead of disrupting the tumor stroma to allow payloads to penetrate through the stroma smoothly and reach cancer cells in the tumor tissue, there is a way to use the cancer stroma as a scaffold from which the payloads can be launched continuously. In hypervascular tumors, drug formulations categorized as a DDS can accumulate in the tumors effectively, encounter tumor cells readily and exert their antitumor effect directly. However, in hypovascular tumors, there several impediments to direct encounters between cytotoxic agents and cancer cells. For example, polyethylene glycol (PEG)-liposomes can be delivered efficiently to the tumor but free drug is not transferred sufficiently to cancer cells, particularly those that are distant from the tumor vessels, because the formulations are too large to move through the tumor interstitium and are too stable to allow the free drug within them to be released easily [25]. In fact, Doxil, a doxorubicin (DXR) incorporated PEG-liposome is known to be clinically effective against hypervascular cancers such as ovarian cancer and breast cancer, whereas it is not effective against stomach and pancreatic cancers, both of which are characterized by a low density of tumor microvessels and dense stroma. On the other hand, a micelle carrier system, composed of DXR incorporated in PEG-poly (aspartic acid) block copolymer, is small (10–80 nm) compared with liposomes, and can be distributed more uniformly in tumor tissue. Furthermore, compared with liposomes that are extremely stable in the body, a micelle carrier system for a drug has the potential to allow the effective sustained release of the drug inside a tumor following the accumulation of micelles into tumor tissue. In our drug distribution experiments using an in vitro solid tumor model, when spheroids formed from several cancer cell lines were exposed to DXR incorporated in micelles, DXR was distributed throughout the spheroids similar to the case when free DXR was exposed to spheroids. On the other hand, when the spheroids were exposed to doxil, DXR was distributed only to the surface of the spheroids (Fig. 4). It has been suggested that Doxil can deliver DXR to a solid tumor more efficiently than DXR

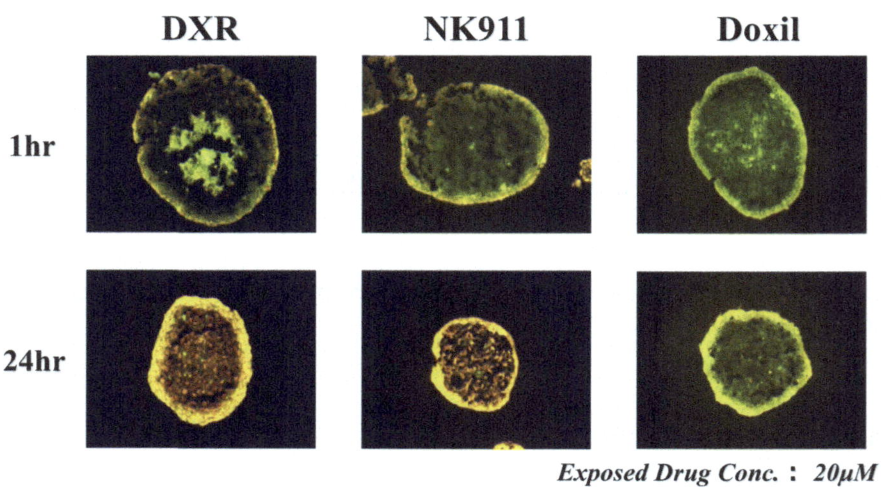

Fig. 4 Distribution of DXR in spheroids composed of HT-29 cells after exposure to free DXR, NK911, and doxil [26]. Spheroids were exposed to free DXR, NK911 (DXR incorporated micells), or doxil (DXR incorporated pegylated liposome) at a DXR dose-equivalent of 20 μM for 1hr and 24 h at 37 °C. The frozen sections were examined using under a fluorescence microscope at the excitation wavelength of 470 nm and emission wavelength of 560nm. In the liposome formulation, the distribution of DXR in spheroids can hardly be confirmed

incorporated micelles via the EPR effect because Doxil may be more stable in plasma than the micelles. On the other hand, DXR packed in the micelles may be distributed by diffusion to cancer cells distant from the tumor vessel because DXR incorporated micelles can leak out of the tumor vessel and may be able to release free DXR more efficiently than Doxil. It is, therefore, suggested that drug carrier systems such as liposomes and micelles should be selected appropriately bearing in mind the pathological characteristics of the tumor interstitium and the tumor vasculature [26].

Some groups including us reported that some enzymes present in tumor stroma are responsible for the extracellular hydrolysis of some linkers connecting payload with antibody. ADCs do not always need to reach to tumor cells and to be internalized into tumor cells to achieve their toxic effects [27–30].

Based on the considerations described above, cancer stromal targeting (CAST) therapy was invented (Fig. 5) as a novel way to insert macromolecular ADCs deep into dense fibrous tumors and facilitate the malignant cells within [8].

To date, we have devised and constructed two types of ADC categorized as CAST therapy. These use antibodies against structural components of the tumor stroma. One consists of an anti-collagen 4 antibody conjugated with SN-38 via an ester bond [26]. For the construction of this ADC, an ester bond was used for grad-

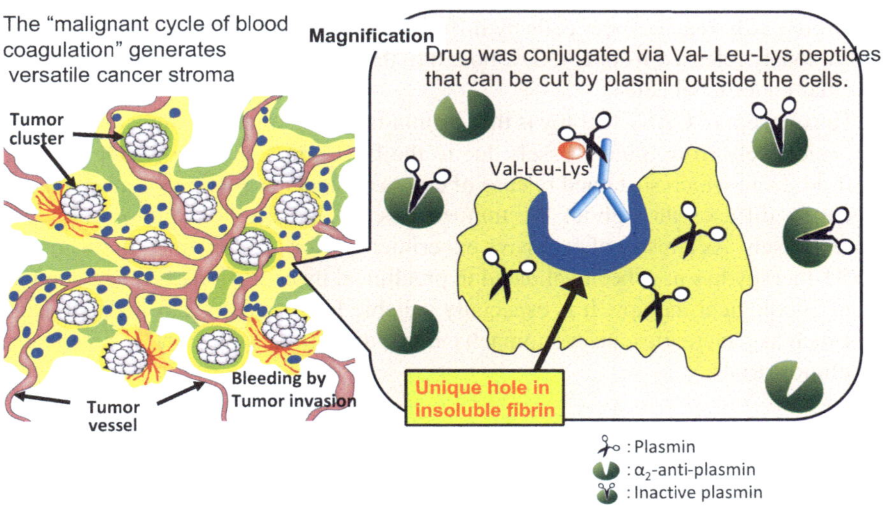

Fig. 5 Diagram of Cancer Stromal Targeting (CAST) Therapy [8, 32]. The anti-IF mAb conjugated with anticancer drugs (IF-ADC) ADC selectively leaks into the tumor interstitium by the tumor vascular hyperpermeability mechanism, bind to the specific dent structure in the fibrin clot, and create a scaffold from which effective sustained release of the free drug may occur. Free drug is only released when the IF- ADC is bound to the epitopes in the insoluble fibrin because plasmin is only active on IF and is neutralized by the innate α2-plasmin inhibitor circulating in the blood. The free anticancer drug released from ADC easily passes through the tumor stroma and reaches tumor cells to attack cancer cells. It also damages interstitial tumor blood vessels

ual cleavage in the extra-cellular space of the tumor. The anti-collagen 4 ADC are too large to pass through the normal vessel wall but they easily extravasate from leaky tumor vessels. Thus, the anti-collagen 4 ADC can accumulate selectively in the tumor tissue and then exert sustained-release of SN-38. Our basic strategy for overcoming the stromal barrier as a protective shield for cancer cells is to make use of the stroma as a scaffold on which to assemble ADC, followed by the release of payload, SN-38 to the cancer cells. This free SN-38 can easily reach the cancer cells by diffusion, although the carrier mAb can hardly penetrate the stroma.

The second consists of an antibody which specifically distinguishes insoluble fibrin (IF) conjugated with MMAE. As described above, clinical solid cancers are rich in cancer stroma, and IF is precipitated in the interstitial substance. Moreover, it was also confirmed that the higher the malignancy of the cancer, the more precipitation of IF. In view of these pathophysiological data, it was conceived to produce an antibody that specifically recognizes IF. Consequently, we succeeded in obtaining a mAb which distinguished IF from fibrinogen, soluble fibrin (precursor of fibrin clot), and D-dimer (degradation product of fibrin clot), which are all soluble proteins [28, 31]. We then succeeded in developing an anti-IF ADC conjugated with MMAE via a Val-Leu-Lys linker that is severed specifically by plasmin (PLM) [32]. PLM is known to be produced exclusively on fibrin clots from its zymogen precursor plasminogen (PLG) by plasminogen activator (PA) and to be neutralized by α2-AP in the blood stream. Free MMAE released from the conjugate by PLM may therefore easily reach cancer cells by diffusing through the stromal barrier. MMAE continuously released from IF-ADC bound to the IF scaffold also attacks the tumor vascular endothelial cells.

The concept of CAST therapy is that stroma-targeting immunconjugates extravasate selectively from tumor vessels due to the EPR effect and bind to the stromal scaffold, from which sustained release of cytotoxic agent occurs over time and subsequently diffuses throughout the tumor tissue to damage both tumor cells and tumor vessels. As a result of extensive experimental work documented in this book, CAST therapy has now been validated in preclinical investigations as a new modality of oncological therapy. It is especially suitable for refractory, stromal-rich cancers such as pancreatic cancer, stomach cancyr, and glioblastoma and is now ready for clinical trials..

Radioimmune Therapy (RIT)

To date, RIT has been developed mainly for some hematological malignancies. ^{90}yttrium-Ibritumomab [33] and ^{131}I-Tositumomab [34] have been approved for follicular Non-Hodgkin lymphoma. These RIT agents target the CD20 epitope on B cells that are destroyed via production of β particles. Several RIT agents that emit α particles are now under clinical evaluation for various cancers including glioma, ovarian cancer, and prostate cancer [35].

Although α-rays are powerful, the flight distance can be as short as one cell. In order to reliably deliver a sufficient dose of α-rays to cancer cells it is necessary couple the RIT agent with a cancer specific antibody. Therefore, we are planning to develop an RIT that adds β rays which have a relatively long flight distance to the anti-IF mAb described in an earlier paragraph, for use as an additional cancer therapeutic modality.

Combination of CAST Therapy with Immuno-Checkpoint Inhibitors

The advent of antibody therapeutics of immune-checkpoint inhibitors (ICI) has drastically changed the field of oncology and the ICIs have now been a mainstream in cancer therapy [36]. There are, however, many patients who are not sensitive to a single regimen of ICIs. In order to improve the effect of ICIs, many groups have proposed the combination therapy of some chemotherapeutic drugs with ICIs [37]. Nevertheless, there is a concern that conventional low molecular anticancer drugs could induce systemic hematological toxicity which may attenuate ICIs' effect. Because CAST therapy is intended to selectively target tumor cells and regulatory T cells within tumor tissues throughout the body, it will serve as an ideal therapy in combination with ICIs..

References

1. Matsumura Y, Maeda H (1986) A new concept for macromolecular therapeutics in cancer chemotherapy: mechanism of tumoritropic accumulation of proteins and the antitumor agent smancs. Cancer Res 46:6387–6392
2. Duncan R (2003) The dawning era of polymer therapeutics. Natl Rev 2:347–360
3. Peer D, Karp JM, Hong S, Farokhzad OC, Margalit R, Langer R (2007) Nanocarriers as an emerging platform for cancer therapy. Nat Nanotechnol 2:751–760
4. Matsumura Y, Kataoka K (2009) Preclinical and clinical studies of anticancer agent-incorporating polymer micelles. Cancer Sci 100:572–579
5. Hosokawa S, Tagawa T, Niki H, Hirakawa Y, Nohga K, Nagaike K (2003) Efficacy of immunoliposomes on cancer models in a cell-surface-antigen-density-dependent manner. Brit J Cancer 89:1545–1551
6. Hamaguchi T, Matsumura Y, Nakanishi Y, Muro K, Yamada Y, Shimada Y, Shirao K, Niki H, Hosokawa S, Tagawa T, Kakizoe T (2004) Antitumor effect of MCC-465, pegylated liposomal doxorubicin tagged with newly developed monoclonal antibody GAH, in colorectal cancer xenografts. Cancer Sci 95:608–613
7. Matsumura Y, Gotoh M, Muro K, Yamada Y, Shirao K, Shimada Y, Okuwa M, Matsumoto S, Miyata Y, Ohkura H, Chin K, Baba S, Yamao T, Kannami A, Takamatsu Y, Ito K, Takahashi K (2004) Phase I and pharmacokinetic study of MCC-465, a doxorubicin (DXR) encapsulated in PEG immunoliposome, in patients with metastatic stomach cancer. Ann Oncol 15:517–525
8. Matsumura Y (2012) Cancer stromal targeting (CAST) therapy. Adv Drug Deliv Rev 64:710–719

9. Matsumura Y, Kimura M, Kato H, Yamamoto T, Maeda H (1989) In: Kinins V, Abe K, Moriya H, Fujii S (eds) Quantification, isolation and structural determination of bradykinin and hydroxyprolyl-bradykinin in tumor ascites. Plenum Publishing, New York, pp 587–592
10. Maeda H, Matsumura Y, Kato H (1988) Purification and identification of (Hydroxyprolyl3) bradykinin in ascitic fluid from a patient with gastric cancer. J Biol Chem 263:16051–16054
11. Matsumura Y, Kimura M, Yamamoto T, Maeda H (1988) Involvement of the kinin-generating cascade and enhanced vascular permeability in tumor tissue. Jpn J Cancer Res 79:1327–1334
12. Senger DR, Galli SJ, Dvorak AM, Peruzzi CA, Harvey VS, Dvorak HF (1983) Tumor cells secret a vascular permeability factor that promotes accumulation of ascites fluid. Science 21:983–985
13. Ferrara N, Hillan KJ, Gerber HP, Novotny W (2004) Discovery and development of bevacizumab, an anti-VEGF antibody for treating cancer. Nat Rev Drug Discov 3:391–400
14. Dvorak HF, Rickles FR (2006) Malignancy and hemostasis. In: Colman RW, Marder VJ, Clowes AW, George JN, Goldharber SZ (eds) Hemostasis and thrombosis: basic principles and clinical practice, 5th edn. LippinCott Williams $ Wilkins, Philadelphia, pp 851–873
15. Bussard KM, Mutkus L, Stumpf K, Gomez-Manzano C, Marini FC (2016) Tumor-associated stromal cells as key contributors to the tumor microenvironment. Breast Cancer Res 18:84
16. Nielsen MF, Mortensen MB, Detlefsen S (2016) Key players in pancreatic cancer-stroma interaction: cancer-associated fibroblasts, endothelial and inflammatory cells. World J Gastroenterol 22:2678–2700
17. Brown LF, Dvorak AM, Dvorak HF (1989) Leaky vessels, fibrin deposition, and fibrosis: a sequence of events common to solid tumors and to many other types of disease. Am Rev Respir Dis 140:1104–1107
18. Busuttil SJ, Ploplis VA, Castellino FJ, Tang L, Eaton JW, Plow EF (2004) A central role for plasminogen in the inflammatory response to biomaterials. J Thromb Haemost 2:1798–1805
19. Sawai T, Tomono K, Yanagihara K, Yamamoto Y, Kaku M, Hirakata Y, Koga H, Tashiro T, Kohno S (1997) Role of coagulase in a murine model of hematogenous pulmonary infection induced by intravenous injection of Staphylococcus aureus enmeshed in agar beads. Infect Immun 65(2):466–471
20. Trousseau A (1865) Pegmasia alba dolens, vol 3. Balliere JB et Fils, Paris
21. Stein PD, Beemath A, Meyers FA, Skaf E, Sanchez J, Olson RE (2006) Incidence of venous thromboembolism in patients hospitalized with cancer. Am J Med 119:60–68
22. Kano MR, Bae Y, Iwata C, Morishita Y, Yashiro M, Oka M, Fujii T, Komuro A, Kiyono K, Kaminishi M, Hirakawa K, Ouchi Y, Nishiyama N, Kataoka K, Miyazono K (2007) Improvement of cancer-targeting therapy, using nanocarriers for intractable solid tumors by inhibition of TGF-beta signaling. Proc Natl Acad Sci 104:3460–3465
23. Olive KP, Jacobetz MA, Davidson CJ, Gopinathan A, McIntyre D, Honess D, Madhu B, Goldgraben MA, Caldwell ME, Allard D, Frese KK, Denicola G, Feig C, Combs C, Winter SP, Ireland-Zecchini H, Reichelt S, Howat WJ, Chang A, Dhara M, Wang L, Rückert F, Grützmann R, Pilarsky C, Izeradjene K, Hingorani SR, Huang P, Davies SE, Plunkett W, Egorin M, Hruban RH, Whitebread N, McGovern K, Adams J, Iacobuzio-Donahue C, Griffiths J, Tuveson DA (2009) Inhibition of hedgehog signalling enhances delivery of chemotherapy in a mouse model of pancreatic cancer. Science 324:1457–1461
24. Thompson CB, Shepard HM, O'Connor PM, Kadhim S, Jiang P, Osgood RJ, Bookbinder LH, Li X, Sugarman BJ, Connor RJ, Nadjsombati S, Frost GI (2010) Enzymatic depletion of tumor hyaluronan induces antitumor responses in preclinical animal models. Mol Cancer Ther 9:3052–3064
25. Unezaki S, Maruyama K, Hosoda J, Magae I, Koyanagi Y, Nakata M, Ishida O, Iwatsuru M, Tsuchiya S (1996) Direct measurement of the extravasation of polyethyleneglycol-coated liposomes into solid tumor tissue by in vivo fluorescence microscopy. Int J Pharm 144:11–17
26. Tsukioka Y, Matsumura Y, Hamaguchi T, Koike H, Moriyasu F, Kakizoe T (2002) Pharmaceutical and biomedical differences between micellar doxorubicin (NK911) and liposomal doxorubicin (Doxil). Jpn J Cancer Res 93:1145–1153

27. Yasunaga M, Manabe S, Tarin D, Matsumura Y (2011) Cancer-stroma targeting therapy by cytotoxic immunoconjugate bound to the collagen 4 network in the tumor tissue. Bioconjug Chem 22:1776–1783
28. Yasunaga M, Manabe S, Matsumura Y (2011) New concept of cytotoxic immunoconjugate therapy targeting cancer-induced fibrin clots. Cancer Sci 102:1396–1402
29. Bala V, Rao S, Boyd BJ, Prestidge CA (2013) Prodrug and nanomedicine approaches for the delivery of the camptothecin analogue SN38. J Control Release 172:48–61
30. Gebleux R, Stringhini M, Casanova R, Soltermann A, Neri D (2018) Non-internalizing antibody-drug conjugates display potent anti-cancer activity upon proteolytic release of monomethyl auristatin E in the subendothelial extracellular matrix. Int J Cancer 140:1670–1679
31. Hisada Y, Yasunaga M, Hanaoka S, Saijou S, Sugino T, Tsuji A, Saga T, Tsumoto K, Manabe S, Kuroda J, Kuratsu J, Matsumura Y (2013) Discovery of an uncovered region in fibrin clots and its clinical significance. Matsumura Y Sci Rep 3:2604–2610
32. Fuchigami H, Manabe S, Yasunaga M, Matsumura Y (2018) Chemotherpy payload of anti-insoluble fibrin antibody-drug conjugate is released specifically upon binding to fibrin. Sci Rep 8:14211
33. Lerch K, Leng C, Pinto A, Linkesch W, Sill H, Linden O, Andreas Viardot A, Keller U, Hess G, Lastoria S, Frigeri F, Neik N, Pott C, Scholz CW, Pezzutto A (2016) ^{90}Yttrium-Ibritumomab Tiuxetan as first line treatment for follicular non-Hodgkin lymphoma. 5 year results from an international multicenter phase II clinical trial. Blood 128:1806
34. Kaminski MS, Tuck M, Estes J, Kolstad A, Ross CW, Zasadny K, Regan D, Kison P, Fisher S, Kroll S, Wahl RL (2005) 131I-tositumomab therapy as initial treatment for follicular lymphoma. N Engl J Med 352(5):441–449
35. Larson JA, Carrasquillo JA, Cheung NV, Press O (2015) Radioimmunotherapy of human tumours. Nat Rev Cancer 15:347–360
36. Pardoll DM (2012) Immunology beats cancer: a blueprint for successful translation. Nat Immunol 13(12):1129–1132
37. Cook AM, Lesterhuis WJ, Nowak AK, Lake RA (2016) Chemotherapy and immunotherapy: mapping the road ahead. Curr Opin Immunol 39:23–29

CAST Therapy

Masahiro Yasunaga, Shino Manabe, and Yasuhiro Matsumura

Abstract Cancer consists not only of tumor cells but also of tumor stroma. The latter has various components such as fibroblasts, mesenchymal cells, blood vessels or extracellular matrix proteins (e.g. collagen, fibronectin or fibrin). (See chapter by Tarin describing the pathogenesis of cancer and stromal changes during the process). Moreover, it also promotes tumor cell growth, survival, invasion and metastasis. It is thus increasingly important to understand the role of tumor stroma in tumorigenesis and tumor progression. In this chapter, stroma-mediated drug-resistance and methods to overcome it are described and evaluated.

Most recently, antibody-drug conjugates (ADCs), consisting of an antibody and a cytotoxic drug connected via a specialized linker have become available as a next generation of antibody therapeutics, intended to enhance focused delivery of anti-cancer agents into the internal tumor environment.

However, none of them has been proved to be effective in treating refractory cancers such as pancreatic cancer or scirrhous gastric cancer. These cancers are known to possess abundant stroma, which hinders the distribution of therapeutic antibodies and ADCs within the cancerous growth. This is the so-called tumor stromal barrier.

To overcome this drawback, we have developed a unique type of ADC, namely, cancer stromal targeting (CAST) therapy, in which stroma targeting mAbs (anti-collagen 4 or anti-insoluble fibrin (IF) mAbs) were conjugated with anticancer agents. Our results confirmed that ADCs bound to collagen 4 or IF in the stroma, from which effective sustained release of the drugs occurred. The released drug subsequently diffused through the tumor tissue, causing marked arrest of tumor growth associated with damage to tumor blood vessels and death of cancer cells.

M. Yasunaga · Y. Matsumura (✉)
Division of Developmental Therapeutics, Exploratory Oncology Research & Clinical Trial Center, National Cancer Center, Kashiwa, Japan
e-mail: yhmatsum@east.ncc.go.jp

S. Manabe
Cluster for Pioneering Research, RIKEN, Wako, Japan

Research Center for Pharmaceutical Development, Tohoku University, Sendai, Japan

Y. Matsumura, D. Tarin (eds.), *Cancer Drug Delivery Systems Based on the Tumor Microenvironment*, https://doi.org/10.1007/978-4-431-56880-3_12

Keywords Tumor stromal barrier · Collagen 4 · Insoluble fibrin · Antibody drug conjugate · Cancer stromal targeting therapy

Introduction

In recent years, the clinical application of antibody-based therapy in oncology has increased substantially. Many therapeutic antibodies designed for controlling the growth of a variety of cancers are already in clinical use. Currently, antibody-drug conjugates (ADCs), the next generation of therapeutic antibodies, are considered a promising strategy to enhance the cytotoxic effect of anti-neoplastic drugs by delivering them into and around the cancer cells more efficiently [1–4]. A most important consideration in medical oncology is the effectiveness of ADCs in treating relapsed malignant diseases. Thus, as typical examples, SGN-35 is effective for patients with CD30-positive relapsed or refractory malignant lymphoma [5–7]. T-DM1 is also effective for the patients with HER2-positive advanced or remnant breast cancer previously treated with standard drugs [6, 8].

However, the use of high molecular weight agents including ADCs presents a dilemma for cancer therapy, since the very properties that favor their high accumulation in a lesion and specificity in clinical action, also result in the low diffusion of these macromolecules within a tumor. Indeed, most human solid tumors possess abundant stroma that forms a barrier preventing macromolecular drugs from reaching and directly attacking cancer cells. For example, Matsumura Y observed that a liposomal drug categorized as a drug delivery system (DDS) had a potent antitumor effect in mice model but not in humans. He hypothesized that this was due to the greater density and fibrous composition of the extracellular matrix in most human tumors relative to most murine neoplasms. Subsequently, he investigated how the tumor stroma was formed and its molecular composition as a prelude to isolating antibodies reacting with antigens in the tumor stroma [9, 10]. As conventional ADCs depend on enzymatic cleavage following internalization into the cytoplasm or lysosome, the process of cell uptake is disturbed by the stromal barrier. Therefore, this barrier limits the effectiveness of ADCs, regardless of the internalization ability. The heterogeneity of tumor cells also prevents the development of ADC therapy based on a cell-specific antigen. To overcome these drawbacks, we created a unique type of ADC, namely, cancer stroma targeting (CAST) therapy, in which stroma targeting mAbs (anti-collagen 4 or anti-IF mAbs) were conjugated to cytotoxic anticancer agents (ACAs) [9, 11–13].

Our ADC bound to collagen 4 or IF in the stroma, from which sustained release of ACAs and their distribution throughout the tumor occurred; this had a strong antitumor effect against stroma-rich tumors compared with that of conventional ADCs.

Here, we initially outline tumor stromal and conventional ADCs, and review the emerging field of stromal targeting therapy including our CAST therapy.

Tumor Stroma

Organ or tissue consists of not only parenchymal cells but also stroma. Stroma is composed of cellular components such as fibroblasts, macrophages, blood cells or protein components such as collagen, fibrin, fibronectin and extracellular matrix. Blood vessels, lymph vessels or nerves also exist in the stroma (Fig. 1a). Stroma also plays an important role in embryogenic organogenesis and tissue remodeling, for example, pancreatic stroma can produce morphogens like activin for creating the initial pancreatic bud and provide a scaffold to support pancreatic duct morphogenesis or for the development of other branched glands such as the breast (Fig. 1b). Vasculogenesis by endothelial progenitor cells is promoted by the presence and availability of collagen 4 and VEGF, but vascular tube formation is incompetent, whereas well-organized vascular tube formation occurs in the presence of stromal cells (Fig. 1c). The tumor stroma, contains various types of cells and intercellular

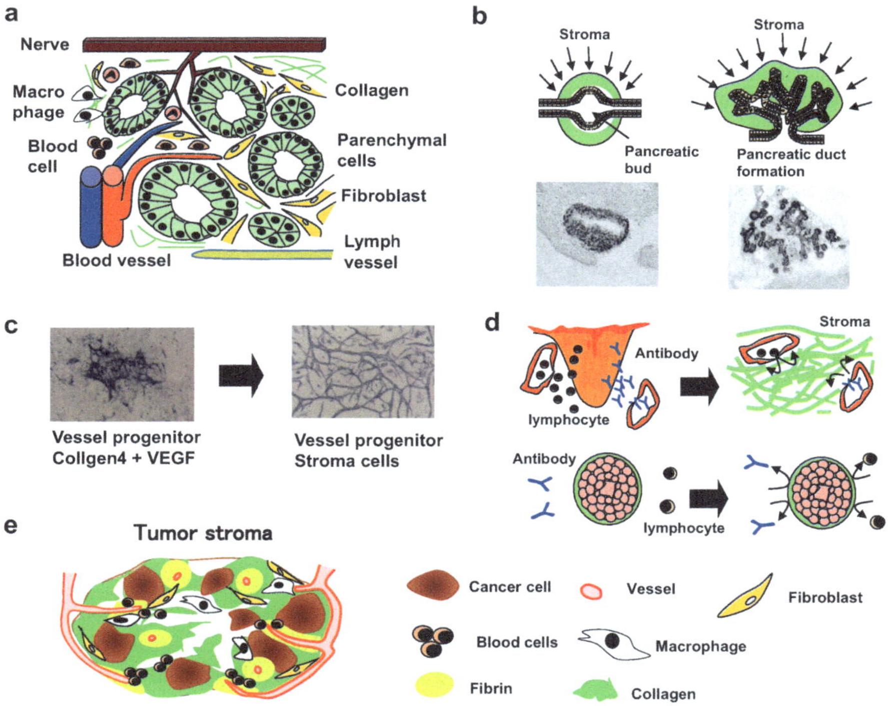

Fig. 1 Tumor stroma
Tissue has not only parenchymal cells but also various stromal cells and its components such as fibroblasts, macrophages, blood cells, nerves, vessels or collagen (**a**). Stroma regulates organogenesis (**b**) and vasculogenesis (**c**). The tumor microenvironment is also composed of various types of the stromal cells (**d**)

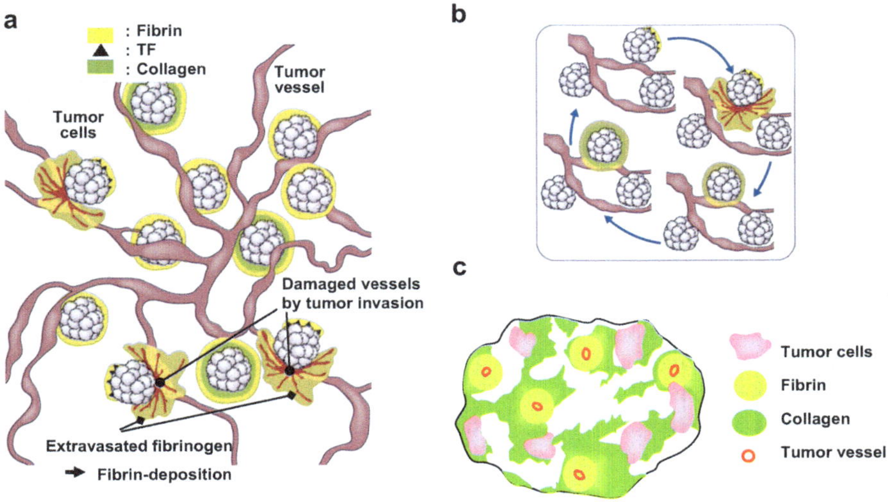

Fig. 2 Malignant coagulation cycle
In the tumor, tumor tissue factor or blood vessel damage causes fibrin formation from the fibrinogen. After the fibrin deposition, collagen formation is subsequently induced (**a**). Malignant coagulation cycle has 4 steps of (1) tumor cell invasion (2) vascular damage (3) IF deposition (4) collagen formation, and it repeats as long as cancer cells survive (**b**). Consequently, tumor stroma has dense stroma composed of IF and collagen (**c**)

materials described in the Chapter on tumor pathology. In this microenvironment, stroma promotes the growth, survival, invasion or metastasis of tumor cells (Fig. 1d).

Most recently, it has been shown that a specialized stromal component, namely the niche, has an important role in the maintenance of cancer stem-like cells (CSCs) properties [14, 15]. CSCs are a unique minor population in the tumor, but they have the ability to maintain the capacity of regeneration, proliferate extensively and form new tumors [16–20]. Recent studies indicated that cancer treatment failure is caused by the existence of CSCs having chemoresistant property [17]. The concept of CSCs is strongly associated with advances in the understanding of stem cell biology. It has been demonstrated that niche control stem cells exist in many organs and tissues such as bone marrow, liver, intestine or brain [21].

In addition to cellular events, blood coagulation cascades are activated in the tumor stroma and various substances such as IF and collagens are produced in the lesion following blood coagulation [9]. As a result of these cellular and extra cellular events, the tumor microenvironment becomes increasingly disorganized (Fig. 2).

Thus it has become increasingly important to understand the role of the stroma in tumorigenesis. There are two distinct types of tumor tissue: those with little stroma, for example, malignant lymphoma (ML), and those with dense stroma, for example, pancreatic cancer, and the latter is strongly associated with the treatment-resistant phenotype.

Antibody Drug Conjugates (ADCs)

Antibody drug conjugates (ADC) have three parts: the antibody, the linker, and the drug, and their size ranges from 10 to 20 nm. Importantly, ACAs are of very small size, below 1 nm; consequently, they can distribute equally into both tumor and normal tissues, often causing adverse effects. By contrast, an ADC is a high molecular weight (HMW) agent but this can extravasate easily from leaky tumor vessels due to their immature structure with intercellular openings. Also the permeability of tumor vessels can be increased by the influence of some "vascular permeability factors" such as VEGF [22] or Kinins, but these do not exert such effects upon normal vessels [23, 24]. In addition, because of the lack of lymphatic vessels acting as a drainage system for HMW agents, they can stay in the tumor for a long time [25]. This mechanism is specific for cancer and is called the enhanced permeability and retention (EPR) effect [25]. ADCs are also capable of active targeting of tumor constituents, depending on specific recognition and binding to the target antigen [26, 27]. A linker of ADC is stable in the bloodstream but should efficiently release the drug in the tumor cells or within their microenvironment [1, 28, 29].

The total number of molecules of the same drug conjugated with a single antibody molecule is about two or four, at maximum eight, otherwise the affinity of the mAb is diminished if too many molecules of ACA are attached to the mAb. Accordingly, unrealistic amounts of ADCs would need to be administered if the mAbs are conjugated with an ordinary ACA [10]. Therefore, highly toxic agents are strongly required [1, 4, 28, 29]. In the approved ADCs, although maytansinoid, auristatin and calicheamicin have been used as a payload, they are too toxic to be administered as single agents. ADC technologies have been proven to be useful for the targeted delivery of ACAs, while minimizing their adverse effects [30].

Pharmacologically, ADCs exert their effects via 4 sequential steps:

- distribution via the systemic circulation,
- the EPR effect including passive targeting,
- penetration within the tumor tissue, and
- action on cells, which includes active targeting and controlled release.

This series of actions is similar to the stages of action of drug delivery system (DDS) drugs, such as liposomes or micelles. The linker- severing strategy is a typical controlled release technology in DDS. It is clear therefore that ADCs should belong to the DDS drug category of onco-therapeutic agents [30].

Tumor Stromal Barrier and ADC

Before approval of T-DM1 for the treatment of breast cancer, the clinical application of ADCs was limited to malignant hematologic tumors [1, 4]. In vitro, ADCs work well for both malignant cells derived from hematologic malignancy and epithelial solid tumors [13]. In particular, when intracellular internalization of an antibody or release efficiency of a drug is efficient, cytotoxic effect of ADCs is usually higher than that of free drugs. However, in actual human solid tumors, efficacy has not yet been confirmed in clinical trials [1, 31].

We have shown that the following unique mechanism exists in cancer stroma formation [9]. That is, in the cancer locus, (1) cancer cells infiltrate into the vessel wall, (2) bleeding follows due to vascular injury, (3) fibrin monomer polymerises to form fibrin coagulum in the injured area, (4) replacement of fibrin with collagenous tissue. This is a series of cancer-host reactions similar to repair of inflammatory damage by fibrotic replacement. However, there is substantial difference between reparative fibrosis in benign non-malignant diseases such as infarction, inflammation and injury and reactive fibrosis in cancer [9, 11, 12]. The former is a transient process and is finally becomes quiescent, but the latter exhibits no cessation as long as cancer cells are present [32]. This phenomenon is the so called "malignant cycle of blood coagulation", in which focal fibrin polymerization and replacement by collagen fiber formation continue to be sporadically repeated at diffusely scattered foci in the tumor [9, 11, 33]. Platelet aggregation also occurs at this site of cancer-induced injury. Cancer-induced blood coagulation involves both intrinsic and extrinsic blood coagulation cascades [9, 22, 23]. The intrinsic blood coagulation process in the tumor tissues produces several bradykinins, and a vascular endothelial growth factor (VEGF) is known to be produced at the site of extrinsic blood coagulation. The extravasated polymerised fibrin then forms the framework for the deposition of collagen and is gradually resorbed resulting in a dense tumor stroma composed of abundant collagen and decreasing fibrin [9] which becomes infiltrated by fibroblasts and other mesenchymal cells. Consequently, surrounding cancer cells in the solid tumor become embedded in abundant dense stroma (Figs. 1d and 2). Troublingly, this dense stroma forms the so-called tumor stromal barrier [9, 11, 12] which inhibits the penetration of drugs especially macromolecules such as antibodies, liposomes or micelles within tumor tissue,. The effects of this living barrier composed of fibroblasts and collagen are to (1) impede the passage of macromolecules through the interstitial space (2) by suppressing the convection and diffusion of pharmacological agents (3) and by create greater distances between blood vessels and cancer cells, which is a disadvantage for the drug delivery. It is also possible that there might be some additional as yet unknown factors such as electrostatic charges or disturbed electro-magnetic fields contributing to disturbance of and binding the flow of ions and molecules through the stromal barrier. Therefore, further studies are needed to clarify these mechanisms.

Antibody Delivery

We first conducted in vivo imaging to confirm the behavior of the antibody as a drug delivery vehicle. Nonspecific antibody IgG can accumulate selectively in a tumor over 3 days after intra-venous administration simply as a result of the EPR effect and this is called passive targeting. On the other hand, cancer-specific IgG can accumulate and stay in the tumor for much longer, more than 7 days, because of active targeting augmenting the EPR effect or passive targeting. By contrast, cancer-specific Fab, a small fragment of antibody, can accumulate in the tumor faster but disappear faster than the complete IgG because it lost longer retention in the tumor tissue and showed rapid renal clearance like ACAs as a low molecular weight drug. More surprisingly, the accumulation of cancer-specific IgM in the tumor was very weak. Although it was cancer-specific, it was defeated by nonspecific control IgG (Fig. 3). We also evaluated the difference between extravasation and penetration activity between IgG and IgM. These properties of IgM were remarkably weaker than those of IgG. It suggests the importance of in vivo study using imaging technology. Next, we evaluated whether a stromal barrier interferes with antibody diffusion within tumor tissue. Anti-CD20 antibody and anti-EpCAM antibody, both of which were cancer-specific to lymphoma and solid tumor respectively, were administered mice with stroma-poor lymphoma or pancreatic cancer with abundant stroma.

Whole body imaging demonstrated that both antibodies accumulated well into both types of tumors. However, although the anti-CD20 antibody was distributed throughout the whole tumor area, the distribution of the anti-EpCAM antibody was restricted to zones around tumor vessels in the peripheral area, and was not seen in

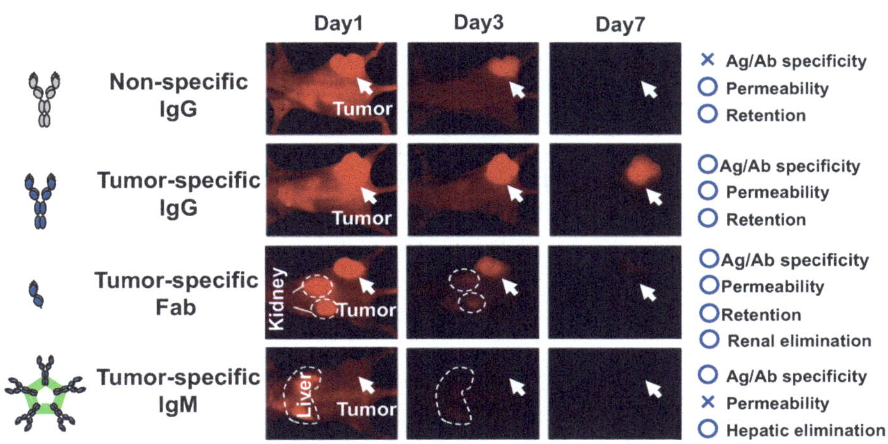

Fig. 3 Antibody delivery
Fluorescent non-specific IgG, tumor-specific IgG, tumor-specific Fab or tumor-specific IgM was administered into mice bearing human cancer xenografts. The in vivo imaging was conducted on day 1, 3 or 7 after the injection. Arrows indicate the tumor position

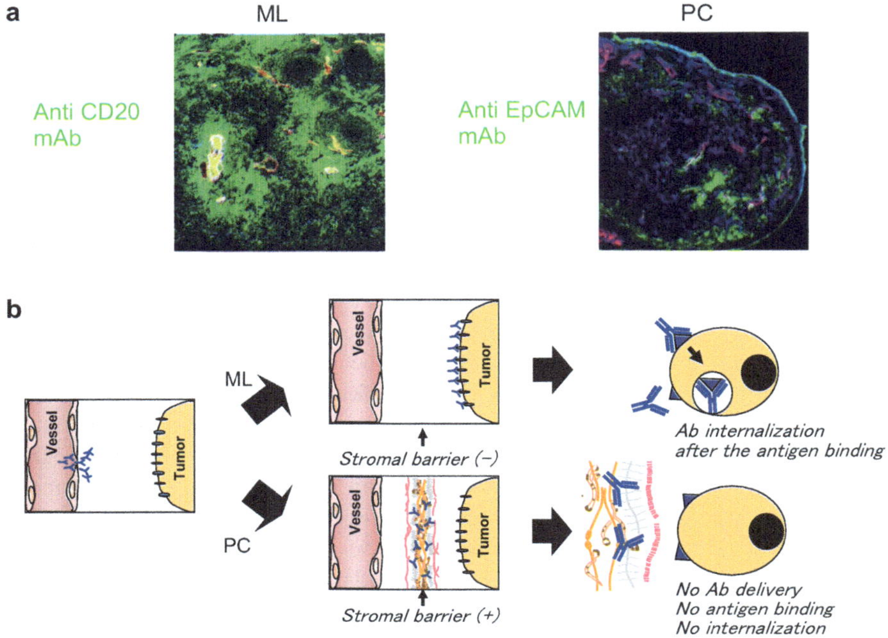

Fig. 4 Stromal barrier
Fluorescent anti-CD20 antibody and fluorescent anti-EpCAM antibody were administered into mice with malignant lymphoma (ML)-tumor and pancreatic cancer (PC)-tumor xenograft respectively. Antibody distribution within each tumor tissue is shown (**a**). Antibodies can be delivered into the tumor cells easily in a stroma-less tumor, e.g. ML. (**b**) Dense stroma suppresses antibody delivery in a stroma-rich tumor, e.g. PC

the central area (Fig. 4a). This demonstrated clearly that dense stroma inhibits the distribution of the antibody within the tumor tissue.

In addition, ADCs, need to be internalized within tumor cells for efficient cytotoxicity of the payload drug to be achieved and failure to accomplish this can be one mechanism contributing to therapeutic failure (Fig. 4b). Also, a naked antibody may not be able to exert neutralizing activity. That could be reason why antibody therapeutics including ADC does not show efficacy against refractory cancer so far.

CAST Therapy

In order to overcome the stromal barrier, we developed stroma-targeting ADCs, in order to create a novel cancer stromal targeting (CAST) therapy (Fig. 5). These agents are composed of antibodies to stromal components (collagen 4 or IF coupled with cytotoxic drugs as payloads via a specialized linker [9, 11–13]. In contrast

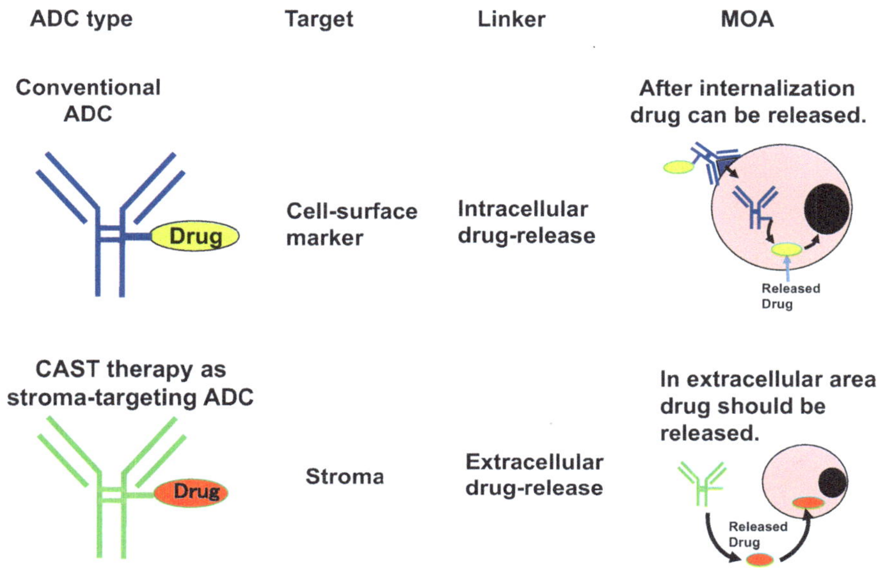

Fig. 5 Drug design of CAST therapy
The distinctive feature of CAST therapy with stromal targeting ADC compared to that of conventional ADC is shown

conventional ADCs, use a cell surface marker as a ligand for the antibody and the linker is designed to be released by intracellular cleavage. After uptake of the ADCs (internalization) into tumor cells, the bond of the linker is cleaved by intracellular enzymes, typically lysosomal enzymes, and then the drug can be released. In contrast, CAST-ADCs use antibodies fabricated to target stromal components and the chosen linker is one which can be cleaved extracellularly.

CAST-ADCs are designed to accumulate on the abundant tumor stroma scaffold, from which the drugs are released by natural physiologic conditions in the extracellular space. Subsequently, they diffuse throughout the whole tumor area, and kill both tumor cells and tumor vessels (Fig. 6). Besides the radical change from anti-cell antibody to anti-stromal antibody, characterising this new strategy, a specially designed linker technology was tailored to accomplish sustained drug release in the tumor stroma.

We developed a PEGylated ester-bond linker for this purpose [9, 11–13], because ester-bonds can be cleaved by hydrolysis outside of cells. Although an ester bond is unstable in moisture or water and is easily hydrolyzed, the neighboring PEG moiety in our construct protects the bond from rapid cleavage and this results in sustained release of the toxic drug within the tumor stroma. Another construct that we have designed uses a plasmin-sensitive peptide linker. This can be cleaved in specifically in IF-positive tumor stroma, where plasmin is locally active during fibrin degradation (fibrinolysis) associated with replacement by collagen [34].

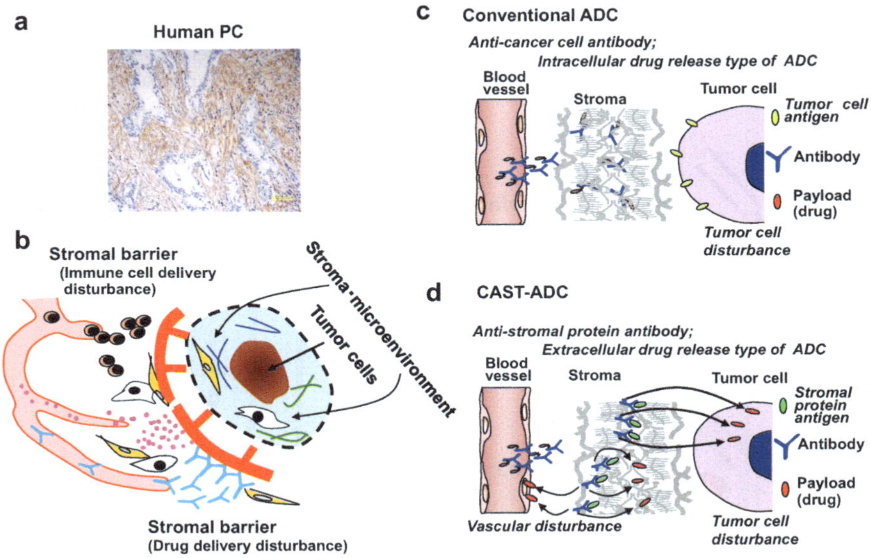

Fig. 6 Strategy of CAST therapy
Pathological features of human pancreatic cancer. Pancreatic cancer cells (blue) are surrounded by dense stroma (brown) (**a**). Schema of the stromal barrier, which disturbs both drug and immune cell delivery, is shown (**b**). Conventional ADC is composed of anti-cancer cell antibodies with a payload which can be released within the tumor cell; intracellular drug release type of ADC (**c**). CAST ADC is composed of anti-stromal protein antibodies with a payload which can be released outside the tumor cells; extracellular drug release type of ADC (**d**)

CAST Therapy Using Anti-collagen 4 Antibody

Although more than 20 types of collagen exist in the tissues of the human body, collagen types 1–5 represent about 99% of their collagen content. Among these, we focused on collagen 4, which exists within tumor tissues as two types of constituent elements. One is a component of tumor vascular basement ensleeving the endothelial tubes, the other is distributed in disorderly structureless manner in the intercellular space.

The former type also exists abundantly in normal lungs and kidneys, but its localization is restricted to the outside of the vascular endothelial cells (on the opposite side of the bloodstream) and as a basal lamina providing a substratum for epithelial cell attachment. Macromolecules such as antibodies can extravasate from leaky, incompetent tumor vessels as a consequence of the EPR effect but not from normal vessels [9]. Therefore, circulating anti-collagen 4 antibody can be seep into the tumor stroma selectively but does not exist in normal tissues and organs. Accordingly, we developed an anti-collagen 4 antibody (called AC4) which we conjugated with the cytotoxic agent SN-38 via a PEGylated ester-bond linker (Fig. 7a) [12, 13] for use in CAST therapy. In the stroma-rich pancreatic cancer xenograft model, we

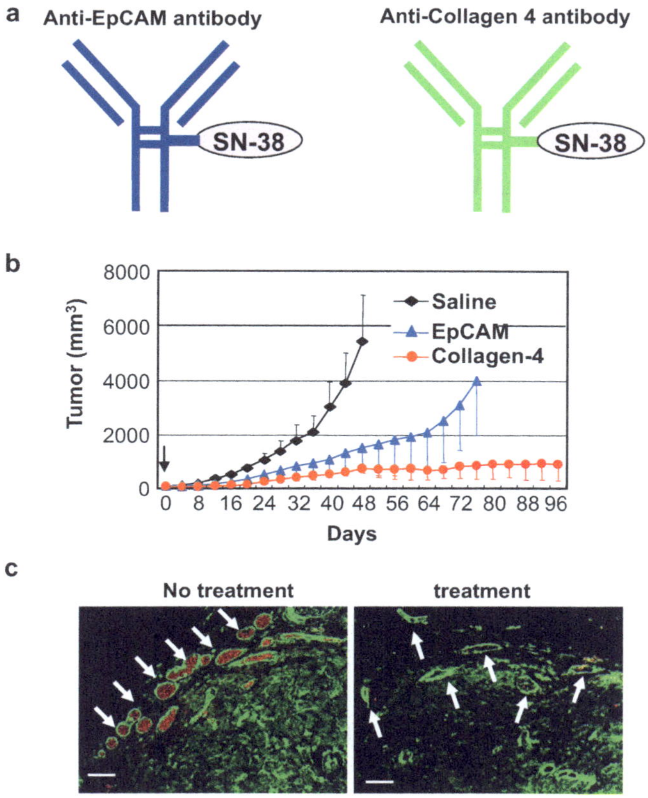

Fig. 7 Anti-collagen 4 ADC
A cell targeting anti EpCAM-ADC and a stromal-targeting anti-collagen-4-ADC are shown. SN-38 is used as a payload (**a**). Anti-tumor activity was examined in an animal model with SUIT2 pancreatic cancer xenograft. Saline, anti-EpCAM-ADC (EpCAM) or anti-collagen-4-ADC (Collagen-4) were administered on day 0 (**b**). Immunostaining of CD31 (red) and collagen-4 (green) of tumor vessels after ~~the~~ anti-collagen-4 ADC treatment was shown. No treatment sample was used as a negative control (**c**)

found that AC4-ADC CAST showed a strong anti-tumor effect, significantly different from the control group treated with anti-EpCAM-ADC, as a cell targeting control (Fig. 7b). Histopathological studies showed that, in addition to considerable tumor cell death, prominent destruction of tumor blood vessels occurred. CD31-positive vascular endothelial cells disappeared completely in tumor tissue leaving their collagen 4 envelopes intact. When cut in transversely in histological sections, these were seen as ring structures (Fig. 7c). This pathological feature therefore corresponds to the empty sleeves of tumor blood vessels, composed of their basement membranes, and is direct evidence of a good response to anti- tumor therapy [35]. On the other hand, blood and biochemical tests confirmed that there was no evidence of myelotoxicity, hepatotoxicity or renal toxicity. Efficient antitumor action and safety of Col.4-CAST therapy were, therefore, demonstrated in the preclinical setting by these experiments [12, 13].

CAST Therapy Using Anti-fibrin Antibody

Fibrin is formed by cleavage of fibrinopeptides A and B from fibrinogen by the action of activated thrombin at the disease site, and, their molecular structures change accordingly. As a result large numbers of soluble fibrin monomer molecules become insoluble polymers (Fig. 8a). In addition, we recently found that a fibrin-specific pocket-like structure which does not exist in fibrinogen forms during this process (Fig. 8b) [33].

Polymerised fibrin also exists in sites of blood coagulation caused by a variety of inflammatory and other pathological processes in vivo. Examples include pathological conditions as cerebral infarction, myocardial infarction, pancreatitis and rheumatoid arthritis and in such events it often disappears within approximately 1 week [33].

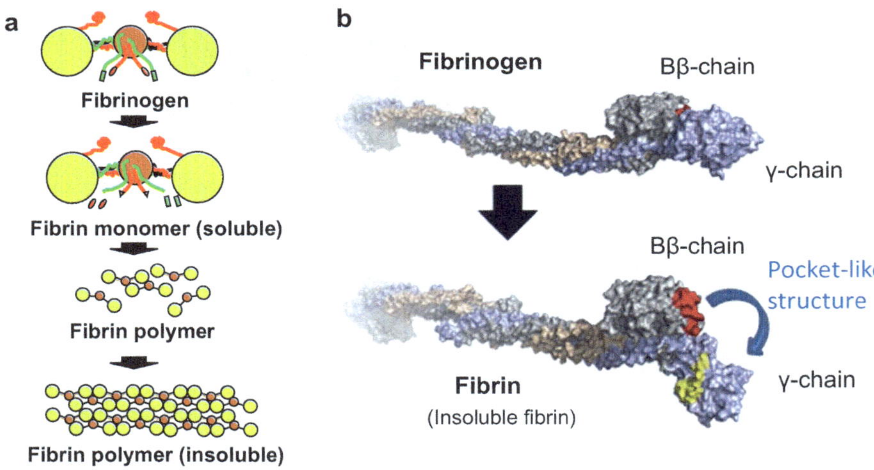

Fig. 8 Molecular and structural change from fibrinogen to fibrin
Fibrinogen exists in a normal healthy state. In a pathophysiological state such as injury, inflammation, thrombosis or cancer, activated thrombin catalyzes the conversion of fibrinogen to fibrin. During polymerization, soluble fibrin monomer converts into insoluble fibrin polymer (**a**). In the structural change in fibrin clots transformed from fibrinogen, the unique pocket-like region composed of the B-beta chain and residues 206–220 of the gamma chain is formed (**b**)

On the other hand, IF can be detected in neoplasms as long as tumor cells survive in the body. Therefore, there is a substantial difference in duration between coagulation or inflammatory-fibrin and tumor-fibrin [11].

Moreover, the patients with these diseases have symptoms such as redness, pain, or fever. In this situation, anti-cancer treatment can never be done clinically. Taken together, these findings clearly suggest long-existing IF is pathophysiologically tumor-specific [9, 33].

In order to exploit this feature for cytotoxic tumor therapy, we developed anti-IF antibodies which can bind the IF-specific pocket-like structure (Fig. 8b) [33]. In both ELISA and immunostaining, these bound IF but not fibrinogen. Moreover, we found that they were able to recognize both human and mouse IF, because the epitope of the antibody is evolutionarily conserved from fish to humans [11, 33]. Therefore, we could examine the specificity of our antibodies for binding polymerised fibrin in mouse tumor when cut in transverse section in histological sections instead of in human subject testing. For this purpose, we used a chemically-induced mouse cutaneous cancer model (Fig. 9a). Among the several anti-IF antibodies, we selected one clone and evaluated the specificity of it by in vivo imaging and immunohistochemistry (Fig. 9b and c). The fluorescent anti-IF antibody accumulated in the IF-positive areas in the tumor-stroma, but not in the cytokeratin-positive tumor cells (Fig. 9c) [11]. Next, we evaluated the specificity of the anti-IF antibody using positron emission tomography (PET) [33]. Zirconium, with a half-life of about 3 days, was used as a tracer. The anti-IF antibody selectively accumulated in the tumor. The signals obtained with the anti-IF antibody were stronger than those with the control antibody. Signal intensity was strongest at day 5 after administration. We thus confirmed that the anti-IF antibody was actually tumor-stroma specific. In the same mouse model, anti-IF CAST-ADC therapy resulted in a strong antitumor effect (Fig. 10a). Damaged tumor vessels were also observed using imaging studies in vivo. (Fig. 10b). We thus succeeded in demonstrating the efficacy of anti-IF tumor therapy using CAST-ADC conjugated with SN-38 via a PEGylated ester-bond linker [11].

For further studies, we used a variety of different tumor types, one of which was a genetically engineered mouse model (GEMM) of pancreatic cancer (PC). This model has distinctive pathophysiological findings including dense stroma, very similar to human PC [36, 37]. In addition, we incorporated 2 other mouse tumor models in this work and these were a pancreatic cancer (PC) subcutaneous xenograft and a PC- orthotropic (pancreas)-xenograft. Although the GEMM-PC model has dense stroma, the other tumor types do not. Also, GEMM-PC tumors show strong IF localization in the tumor stroma, but the other types show no, or far weaker, deposition of this material (Fig. 11a). In our initial studies, we evaluated anti-IF antibody delivery by fluorescent imaging. The results, confirmed that the anti-IF antibody accumulated in the fibrin-positive tumor-stroma selectively (Fig. 11b).

Next, we prepared an anti-IF-ADC agent with the cytotoxic agent MMAE attached to the antibody via a plasmin cleavable linker (PCL) composed of Val-Leu-Lys (now known as ant-IF-CAST-ADC with MMAE-PCL, Fig. 12a) [34]. The

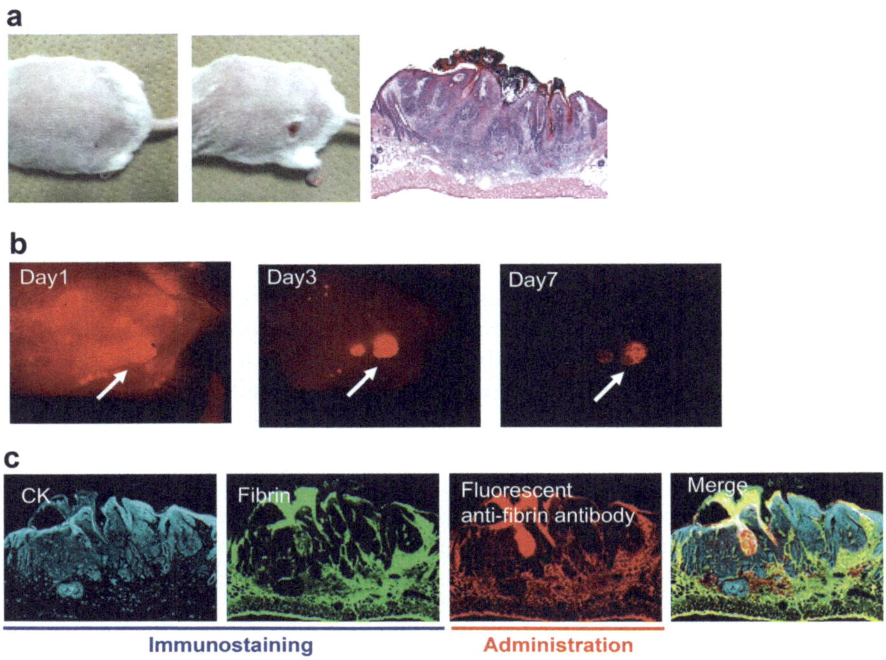

Fig. 9 Stromal targeting by anti-IF antibody in a chemically-induced mouse skin cancer
Mice bearing chemically induced skin tumors are shown in the 2 panels on the left and a histological section of the tumor stained with hematoxylin and eosin (H.E) is shown in the panel on the right (**a**). In vivo systemic imaging analysis of fluorescent anti-IF antibody tumor staining on Days 1, 3 and 7 after administration. Arrows indicate the position of the tumor in each panel (**b**). Histological examination of the distribution of injected fluorescent anti-IF antibody (red) within tumor tissue stained with cytokeratin antibody (right-blue) or IF antibody (green). Overlap of the images obtained with injected anti-IF antibody and deposited IF (**c**) shows co-localization of antigen and antibody as yellow colored areas

rationale for this molecular design, was that, in general, plasmin is activated exclusively on IF but is inactive in the blood and healthy organs, because this enzyme is completely neutralized by its innate inhibitor, α2-antiplasmin. Additionally, as described earlier, IF is produced from fibrinogen in the vicinity of invading cancer cells and it is subsequently degraded by plasmin-dependent fibrinolysis. Thus, the plasmin cleavable linker in the ADC would be expected to be lysed by activated plasmin in the vicinity of polymerized fibrin, freeing the cytotoxic drug to diffuse and poison neighboring cancer cells. From such work it has become apparent that PCL is superior to an ester-bond linker in CAST-ADCs designed for clinical use. Regarding choice of payload, we used MMAE instead of SN-38 to enhance the efficacy, because the cytotoxic effect of MMAE is much stronger than that of SN-38. In addition, it is already successfully utilized as a payload in another clinically-used ADC, namely Brentuximab vedotin.

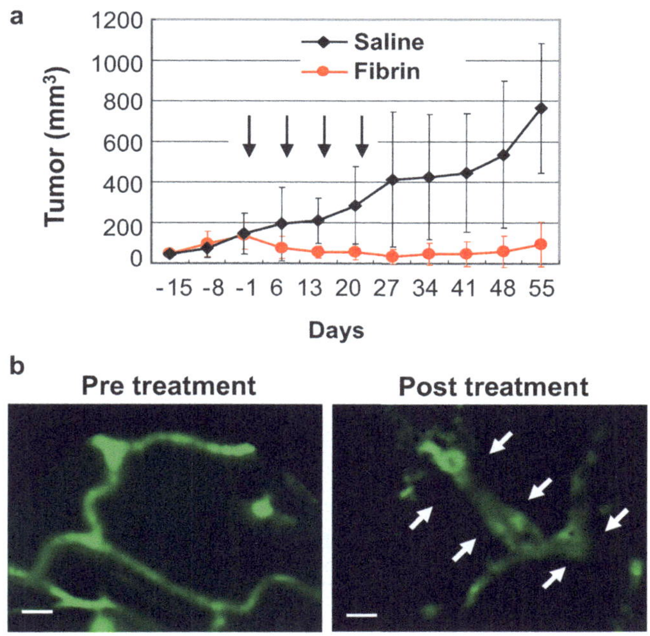

Fig. 10 Anti-tumor effect of anti-fibrin CAST-ADC Anti-tumor activity was examined in mice bearing chemically-induced skin tumors. Anti-fibrin ADC with SN-38 via ester-bond or saline was administrated on Day 0, 7, 14,1nd 21 (**a**). The changes in tumor vessels were visualized with FITC-dextran before and after the injection of the anti-fibrin ADC. Arrows indicate the damaged tumor vessels after treatment (**b**)

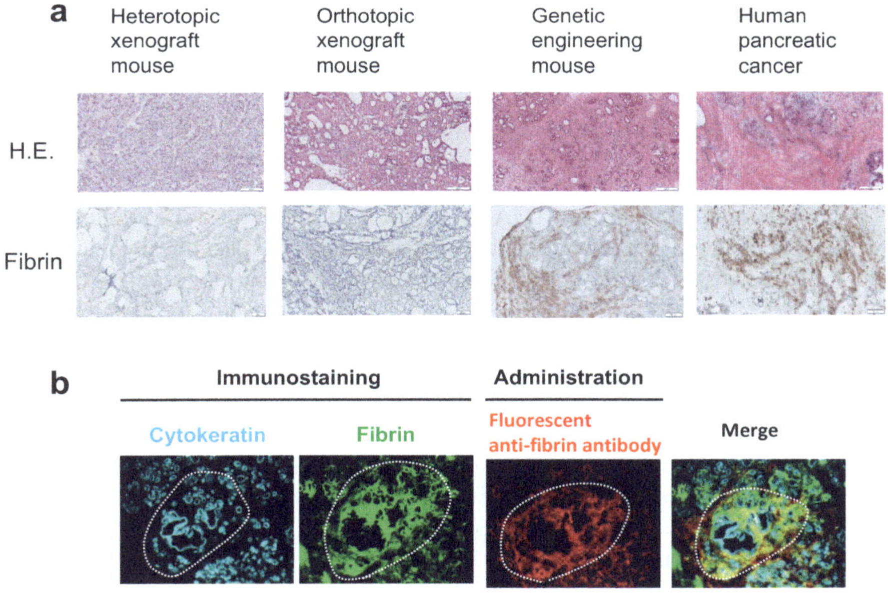

Fig. 11 The genetic engineering mouse model (GEMM) of pancreatic cancer
H.E. staining (upper) and IF staining (lower) of heterotopic (skin) xenograft, orthotopic (pancreas) xenograft, GEMM or human sample of pancreatic cancer (**a**)
Histological examination of the distribution of injected fluorescent anti-IF antibody (red) within GEMM tumor tissue stained with cytokeratin (blue) or IF (green). Yellow indicates the areas of colocalized anti-IF antibody and deposited IF in overlapped images (**b**)

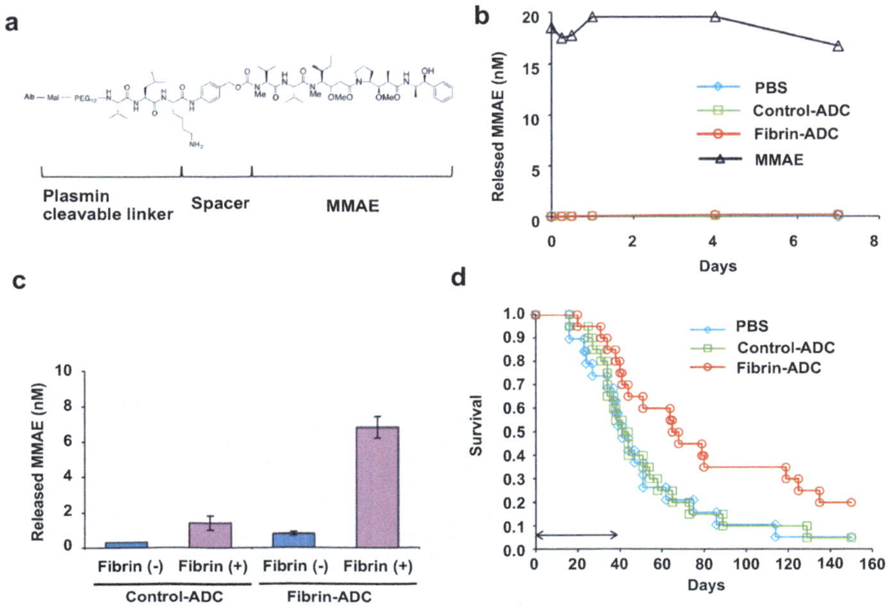

Fig. 12 Anti-fibrin-CAST-ADC
Drug design of anti-fibrin-ADC with MMAE via plasmin cleavable linker (fibrin-ADC) is shown. (**a**) Stability of fibrin-ADC in human plasma. Fibrin ADC or control-ADC; non-specific IgG with MMAE via plasmin cleavable linker was incubated in plasma for 0–7 days, and free MMAE in plasma was measured. PBS and MMAE were used as negative and positive control respectively. (**b**) Survival of GEMM in the fibrin-ADC treatment. ADCs (20 mg/kg) were administered twice in a week, total 12 times. Saline or control-ADC was used as control (**c**)

We measured the release of MMAE from the ADC on fibrin-coated plates in the presence of plasminogen activator, plasminogen and α2-antiplasmin, an innate plasmin inhibitor. These experiments thus mimicked in vivo conditions, to test whether the ant-IF-ADC-MMAE-PL molecules were cleaved by plasmin specifically in the presence of IF. The results, demonstrated that, even in the presence of α2-AP, anti-IF-ADC was cleaved to release significantly higher amounts of MMAE on IF-coated plates compared to control-ADC (Fig. 12b). Collectively, these experiments indicated that the combination of anti-IF mAb and PCL could be severed exclusively on IF deposits in a tumor for effective sustained release of free MMAE. To confirm this we conducted experiments to evaluate the therapeutic efficacy of our new anti-IF-CAST-ADC/MMAE-PCL agent against murine GEMM-PC cancers in a preclinical setting [34]. The results, displayed in the Kaplan-Meier curves (Fig. 12c) showed that the group treated with anti-IF-CAST-ADC with MMAE-PCL exhibited significantly extended survival of the mice as compared to control groups treated with PBS or control-ADC. Figure 13 shows our concept of anti-IF-CAST-ADC therapy. CAST-ADC extravasates from leaky tumor vessels as a result of the EPR effect and binds selectively to the IF network around the tumor vessels near invading cancer cells to create the scaffold. Subsequently, the linker is cleaved and the toxic payload

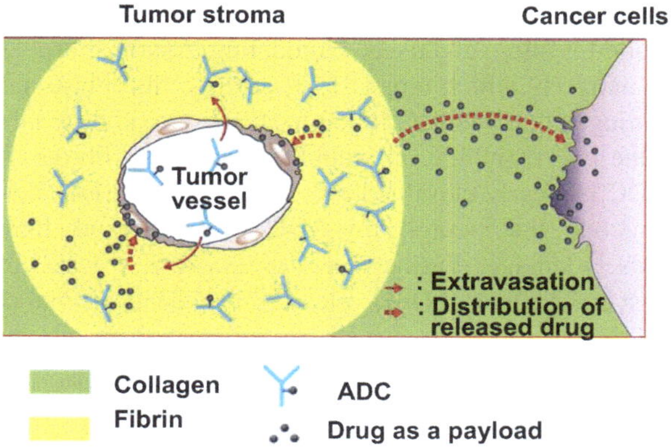

Fig. 13 The schema of anti-fibrin CAST-ADC. CAST-ADC extravasates from leaky tumor vessels (EPR effect), binds selectively to the fibrin network around the tumor vessels to create the scaffold. Consequently, the drug as a payload can be released from ADC and distributed throughout the entire tumor area. It can damage both cancer cells and tumor vessels

is released from the bound ADC to diffuse throughout the entire tumor area. Unlike an antibody, which is a large molecule, the released drug is a small molecule which can spread through the stromal barrier according to the physico-chemical principles of passive diffusion. Thus it can move from a high concentration area to a low concentration area even in a static state where there is no actual flow of fluid. The beneficial outcome is that the toxic agent damages both the cancer cells and tumor vessels.

Other Stroma Targeting Therapy

In addition to our new concept of CAST therapy described above, several other anti-stroma targeting therapies have been proposed. In all of them the proposed major MOA (mode of action) is to improve drug delivery or weaken the stromal barrier.

There were high expectations that some based on inhibition of stromal cell signaling molecules such as hedgehog and JAK would damage stromal cells and thereby reduce tumor stromal density but they have so far failed to produce satisfactory results in clinical trials [38, 39]. Meanwhile, an alternative approach proposed that hyaluronidase can reduce the amount of the stroma by degradation of hyaluronic acid [40, 41] and a clinical trial of this agent is ongoing.

In different study, therapy with a naked anti-PDGFαR antibody was investigated [42]. This agent neutralizes and kills many PDGFαR-high positive stromal cells that exist in tumor stroma. It showed significant anti-tumor effect even in a xenograft model with PDGFα receptor positive tumor cells. This result clearly demonstrated the effectiveness of stroma targeting therapy against stroma-rich tumor such as pancreatic cancer. However, in some cases, it showed no effect. It might be effective on highly stroma-dependent tumor cells. Anti-CSF1 (Colony-stimulating Factor-1) antibody, which targets monocytes, macrophages or myeloid cells in the tumor microenvironment has also been developed for clinical use [43].

FAP-α is a serine protease and is known to be highly expressed in organogenesis, wound healing, and tissue remodeling. An anti-fibroblast activation protein-α (FAP-α) antibody which targets FAP-positive fibroblasts in tumor stroma was therefore conjugated via SMCC linker to the cytotoxic drug DM1 [44], in an effort to determine if FAP-inhibition would reduce tumor fibrosis Unlike our CAST-ADC, this ADC is clearly classified as a cell targeting ADC; nevertheless its MOA is stroma targeting. Interestingly, this ADC showed not only fibroblast damage but also tumor vessel damage. Furthermore, mitotic arrest was observed in cancer cells as a bystander effect of DM1 released from dead stromal cells.

Conclusions

We described the biological significance of therapeutic targeting of tumor stroma. In addition, the development of ADC and our recent research works using DDS and molecular imaging were reviewed. Finally, to overcome the stromal barrier, we succeeded in developing CAST-ADC based on our unique DDS technology. The primary mode of action; damage of both cancer cells and tumor vessels caused by the released drug from CAST-ADC on the tumor stroma, was also indicated.

References

1. Diamantis N, Banerji U (2016) Antibody-drug conjugates–an emerging class of cancer treatment. Br J Cancer 114:362–367. https://doi.org/10.1038/bjc.2015.435
2. Mack F, Ritchie M, Sapra P (2014) The next generation of antibody drug conjugates. Semin Oncol 41:637–652. https://doi.org/10.1053/j.seminoncol.2014.08.001
3. Senter PD, Sievers EL (2012) The discovery and development of brentuximab vedotin for use in relapsed Hodgkin lymphoma and systemic anaplastic large cell lymphoma. Nat Biotechnol 30:631–637. https://doi.org/10.1038/nbt.2289
4. Thomas A, Teicher BA, Hassan R (2016) Antibody-drug conjugates for cancer therapy. Lancet Oncol 17:e254–e262. https://doi.org/10.1016/s1470-2045(16)30030-4
5. Pro B et al (2012) Brentuximab vedotin (SGN-35) in patients with relapsed or refractory systemic anaplastic large-cell lymphoma: results of a phase II study. J Clin Oncol Off J Am Soc Clin Oncol 30:2190–2196. https://doi.org/10.1200/jco.2011.38.0402
6. Sau S, Alsaab HO, Kashaw SK, Tatiparti K, Iyer AK (2017) Advances in antibody-drug conjugates: a new era of targeted cancer therapy. Drug Discov Today 22:1547–1556. https://doi.org/10.1016/j.drudis.2017.05.011
7. Younes A et al (2012) Results of a pivotal phase II study of brentuximab vedotin for patients with relapsed or refractory Hodgkin's lymphoma. J Clin Oncol Off J Am Soc Clin Oncol 30:2183–2189. https://doi.org/10.1200/jco.2011.38.0410
8. Verma S et al (2012) Trastuzumab emtansine for HER2-positive advanced breast cancer. N Engl J Med 367:1783–1791. https://doi.org/10.1056/NEJMoa1209124
9. Matsumura Y (2012) Cancer stromal targeting (CAST) therapy. Adv Drug Deliv Rev 64:710–719. https://doi.org/10.1016/j.addr.2011.12.010

10. Matsumura Y et al (2004) Phase I and pharmacokinetic study of MCC-465, a doxorubicin (DXR) encapsulated in PEG immunoliposome, in patients with metastatic stomach cancer. Ann Oncol Off J Eur Soc Med Oncol 15:517–525
11. Yasunaga M, Manabe S, Matsumura Y (2011a) New concept of cytotoxic immunoconjugate therapy targeting cancer-induced fibrin clots. Cancer Sci 102:1396–1402. https://doi.org/10.1111/j.1349-7006.2011.01954.x
12. Yasunaga M, Manabe S, Tarin D, Matsumura Y (2011b) Cancer-stroma targeting therapy by cytotoxic immunoconjugate bound to the collagen 4 network in the tumor tissue. Bioconjug Chem 22:1776–1783. https://doi.org/10.1021/bc200158j
13. Yasunaga M, Manabe S, Tarin D, Matsumura Y (2013) Tailored immunoconjugate therapy depending on a quantity of tumor stroma. Cancer Sci 104:231–237. https://doi.org/10.1111/cas.12062
14. Shimokawa M et al (2017) Visualization and targeting of LGR5(+) human colon cancer stem cells. Nature 545:187–192. https://doi.org/10.1038/nature22081
15. Visvader JE, Lindeman GJ (2008) Cancer stem cells in solid tumours: accumulating evidence and unresolved questions. Nat Rev Cancer 8:755–768. https://doi.org/10.1038/nrc2499
16. Al-Hajj M, Wicha MS, Benito-Hernandez A, Morrison SJ, Clarke MF (2003) Prospective identification of tumorigenic breast cancer cells. Proc Natl Acad Sci U S A 100:3983–3988. https://doi.org/10.1073/pnas.0530291100
17. Hanahan D, Weinberg RA (2011) Hallmarks of cancer: the next generation. Cell 144:646–674. https://doi.org/10.1016/j.cell.2011.02.013
18. Lapidot T et al (1994) A cell initiating human acute myeloid leukaemia after transplantation into SCID mice. Nature 367:645–648. https://doi.org/10.1038/367645a0
19. O'Brien CA, Pollett A, Gallinger S, Dick JE (2007) A human colon cancer cell capable of initiating tumour growth in immunodeficient mice. Nature 445:106–110. https://doi.org/10.1038/nature05372
20. Ricci-Vitiani L, Lombardi DG, Pilozzi E, Biffoni M, Todaro M, Peschle C, De Maria R (2007) Identification and expansion of human colon-cancer-initiating cells. Nature 445:111–115. https://doi.org/10.1038/nature05384
21. Reya T, Morrison SJ, Clarke MF, Weissman IL (2001) Stem cells, cancer, and cancer stem cells. Nature 414:105–111. https://doi.org/10.1038/35102167
22. Senger DR, Galli SJ, Dvorak AM, Perruzzi CA, Harvey VS, Dvorak HF (1983) Tumor cells secrete a vascular permeability factor that promotes accumulation of ascites fluid. Science (New York, NY) 219:983–985
23. Matsumura Y, Kimura M, Yamamoto T, Maeda H (1988) Involvement of the kinin-generating cascade in enhanced vascular permeability in tumor tissue. Jpn J Cancer Res 79:1327–1334
24. Nagy JA, Benjamin L, Zeng H, Dvorak AM, Dvorak HF (2008) Vascular permeability, vascular hyperpermeability and angiogenesis. Angiogenesis 11:109–119. https://doi.org/10.1007/s10456-008-9099-z
25. Matsumura Y, Maeda H (1986) A new concept for macromolecular therapeutics in cancer chemotherapy: mechanism of tumoritropic accumulation of proteins and the antitumor agent smancs. Cancer Res 46:6387–6392
26. Adams GP, Weiner LM (2005) Monoclonal antibody therapy of cancer. Nat Biotechnol 23:1147–1157. https://doi.org/10.1038/nbt1137
27. Damelin M, Zhong W, Myers J, Sapra P (2015) Evolving strategies for target selection for antibody-drug conjugates. Pharm Res 32:3494–3507. https://doi.org/10.1007/s11095-015-1624-3
28. Doronina SO et al (2003) Development of potent monoclonal antibody auristatin conjugates for cancer therapy. Nat Biotechnol 21:778–784. https://doi.org/10.1038/nbt832
29. Lyon RP et al (2015) Reducing hydrophobicity of homogeneous antibody-drug conjugates improves pharmacokinetics and therapeutic index. Nat Biotechnol 33:733–735. https://doi.org/10.1038/nbt.3212

30. Yasunaga M et al (2017) Development of antibody-drug conjugates using DDS and molecular imaging. Bioengineering (Basel, Switzerland) 4(3):78. https://doi.org/10.3390/bioengineering4030078
31. Dan N, Setua S, Kashyap VK, Khan S, Jaggi M, Yallapu MM, Chauhan SC (2018) Antibody-drug conjugates for cancer therapy: chemistry to clinical implications. Pharmaceuticals (Basel, Switzerland) 11(2):32. https://doi.org/10.3390/ph11020032
32. Dvorak HF (1986) Tumors: wounds that do not heal. Similarities between tumor stroma generation and wound healing. N Engl J Med 315:1650–1659. https://doi.org/10.1056/nejm198612253152606
33. Hisada Y et al (2013) Discovery of an uncovered region in fibrin clots and its clinical significance. Sci Rep 3:2604. https://doi.org/10.1038/srep02604
34. Fuchigami H, Manabe S, Yasunaga M, Matsumura Y (2018) Chemotherapy payload of anti-insoluble fibrin antibody-drug conjugate is released specifically upon binding to fibrin. Sci Rep 8:14211. https://doi.org/10.1038/s41598-018-32601-0
35. Baluk P, Morikawa S, Haskell A, Mancuso M, McDonald DM (2003) Abnormalities of basement membrane on blood vessels and endothelial sprouts in tumors. Am J Pathol 163:1801–1815. https://doi.org/10.1016/s0002-9440(10)63540-7
36. Hingorani SR et al (2005) Trp53R172H and KrasG12D cooperate to promote chromosomal instability and widely metastatic pancreatic ductal adenocarcinoma in mice. Cancer Cell 7:469–483. https://doi.org/10.1016/j.ccr.2005.04.023
37. Obonai T, Fuchigami H, Furuya F, Kozuka N, Yasunaga M, Matsumura Y (2016) Tumour imaging by the detection of fibrin clots in tumour stroma using an anti-fibrin Fab fragment. Sci Rep 6:23613. https://doi.org/10.1038/srep23613
38. Catenacci DV et al (2015) Randomized phase Ib/II study of gemcitabine plus placebo or vismodegib, a hedgehog pathway inhibitor, in patients with metastatic pancreatic cancer. J Clin Oncol Off J Am Soc Clin Oncol 33:4284–4292. https://doi.org/10.1200/jco.2015.62.8719
39. Hurwitz H et al (2018) Ruxolitinib + capecitabine in advanced/metastatic pancreatic cancer after disease progression/intolerance to first-line therapy: JANUS 1 and 2 randomized phase III studies. Investig New Drugs 36:683–695. https://doi.org/10.1007/s10637-018-0580-2
40. Hingorani SR et al (2016) Phase Ib study of PEGylated recombinant human hyaluronidase and gemcitabine in patients with advanced pancreatic cancer. Clin Cancer Res 22:2848–2854. https://doi.org/10.1158/1078-0432.ccr-15-2010
41. Olive KP et al (2009) Inhibition of Hedgehog signaling enhances delivery of chemotherapy in a mouse model of pancreatic cancer. Science (New York, NY) 324:1457–1461. https://doi.org/10.1126/science.1171362
42. Laing N et al (2013) Inhibition of platelet-derived growth factor receptor alpha by MEDI-575 reduces tumor growth and stromal fibroblast content in a model of non-small cell lung cancer. Mol Pharmacol 83:1247–1256. https://doi.org/10.1124/mol.112.084079
43. Cannarile MA, Weisser M, Jacob W, Jegg AM, Ries CH, Ruttinger D (2017) Colony-stimulating factor 1 receptor (CSF1R) inhibitors in cancer therapy. J Immunother Cancer 5:53. https://doi.org/10.1186/s40425-017-0257-y
44. Ostermann E et al (2008) Effective immunoconjugate therapy in cancer models targeting a serine protease of tumor fibroblasts. Clin Cancer Res 14:4584–4592. https://doi.org/10.1158/1078-0432.ccr-07-5211

CAST Diagnostic Imaging

Atsushi B. Tsuji and Tsuneo Saga

Abstract Cancer stromal targeting (CAST) therapy has the potential to overcome drug resistance and improve therapeutic efficacy, even in cancer with abundant stroma that inhibits the penetration of anticancer agents. Many types of cancer form varying degrees of stroma, which generally increases with disease progression. Noninvasive imaging methods for assessing the degree of stroma formation in individual cancers are useful for diagnosing malignancy potential and selecting patients who could potentially benefit from CAST therapy. Antibodies against several types of stromal targets, including insoluble fibrin and tissue factor, have been developed and could be good candidate platforms – not only for CAST therapy, but also for imaging to assess the degree of cancer stroma formation. This section focuses primarily on antibody-based nuclear medicine imaging targeting cancer stroma because it has the highest sensitivity and quantitative ability among the noninvasive imaging techniques applicable to clinical practice, and CAST therapy has been based on antibody-drug conjugates.

Keywords Nuclear medicine imaging · Immuno-PET · Immuno-SPECT · Molecular imaging · Noninvasive imaging · Antibody · Radionuclide · Positron-emitter · Single gamma-emitter

A. B. Tsuji (✉)
Department of Molecular Imaging and Theranostics, National Institute of Radiological Sciences, National Institutes for Quantum and Radiological Science and Technology (QST-NIRS), Chiba, Japan
e-mail: tsuji.atsushi@qst.go.jp

T. Saga
Department of Diagnostic Radiology, Kyoto University Hospital, Kyoto, Japan
e-mail: saga@kuhp.kyoto-u.ac.jp

Y. Matsumura, D. Tarin (eds.), *Cancer Drug Delivery Systems Based on the Tumor Microenvironment*, https://doi.org/10.1007/978-4-431-56880-3_13

Abbreviations

^{18}F-FDG	2-deoxy-2-[^{18}F]fluoro-D-glucose
ADC	antibody-drug conjugate
CAST	cancer stromal targeting
CT	computed tomography
IgG	immunoglobulin G
MRI	magnetic resonance imaging
PET	positron emission tomography
SPECT	single photon emission computed tomography

Introduction

Cancer stromal targeting (CAST) therapy proposed by Matsumura Y has the potential to overcome drug resistance and improve therapeutic efficacy, even in cancers with abundant stroma, which inhibits the penetration of anticancer agents [36]. Stroma formation occurs in many types of cancer, and the degree of stroma formation is very diverse and is positively correlated with degree of malignancy [26]. Noninvasive imaging methods for assessing the degree of stroma formation in individual cancer tissues are useful for diagnosing malignancy potential and selecting patients who could potentially benefit from CAST therapy. Antibodies against several types of stromal targets, including insoluble fibrin, tissue factor, and collagen IV, have been developed and could be good candidate platforms not only for CAST therapy, but also for imaging to assess the extent of cancer stroma formation.

CAST therapy is a kind of antibody-drug conjugate (ADC) in which anticancer agents are released within the cancer stroma where more drugs are delivered to cancer cells compared with conventional ADCs [21, 36, 66, 70]. Matsumura and colleagues reported three types of CAST therapeutic agents using antibodies that recognize insoluble fibrin clots [46, 67], tissue factor [30], and collagen IV [68, 69]. Because the antibody has high target specificity, the same antibody used for CAST therapy can be used for noninvasive imaging to select appropriate patients for CAST therapy. Nuclear medicine imaging with radiolabeled antibodies, immuno-positron emitting tomography (immuno-PET), and immuno-single-photon emission computed tomography (immuno-SPECT), are most suitable for this purpose because they are highly quantitative and sensitive compared with other noninvasive imaging techniques applicable to clinical practice, such as magnetic resonance imaging (MRI) and ultrasound [35]. Furthermore, antibody radiolabeling methods are well established [52]. In this section, we first provide an overview of radiolabeled antibody imaging and then focus on CAST imaging using radiolabeled antibodies that are specific for cancer stroma.

Radiolabeled Antibody Imaging

Antibody

Advances in antibody engineering have enabled the development of various antibodies with high affinity and modifications of the forms of antibodies ranging from intact immunoglobulin G (IgG), which is the most common form used in clinical practice, to small-sized antibody fragments (Fig. 1). IgG is a high molecular-weight protein of approximately 150 kDa comprising several variable and constant domains (Fig. 1). The IgG form can deliver various agents, including anti-cancer agents and radionuclides, to target lesions. Both uptake in the target lesion and blood clearance of IgG, however, are slow (Fig. 2, left panel) and it generally takes more than 24 h for the tumor-to-blood ratio to become greater than one (Fig. 2, right panel). It is therefore necessary to wait several days (typically 2–4 days) to obtain images with sufficiently high contrast. Although IgG-based imaging is suitable for evaluating the pharmacokinetics of IgG-based therapy, including CAST therapy as a companion diagnostic, a shorter waiting time for imaging is preferable in the clinic for both patient convenience and efficient through-put of the

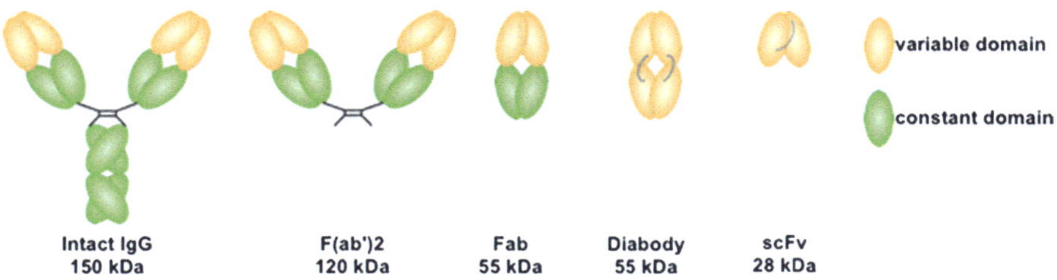

Fig. 1 Schematic structures of some antibody forms used for imaging. The corresponding molecular weight is shown under each name

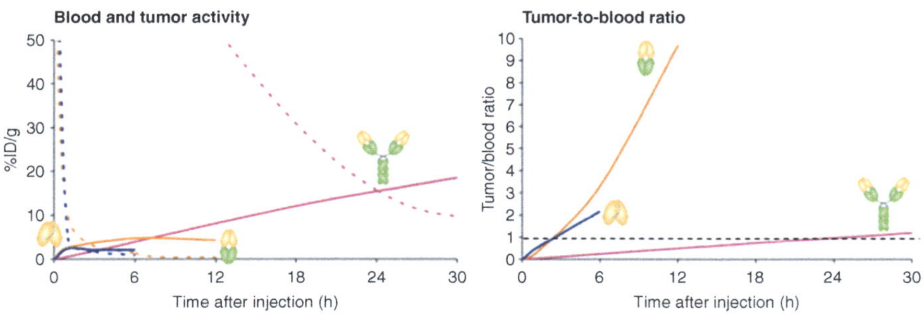

Fig. 2 Typical time-activity curves (left panel) and tumor-to-blood ratios (right panel) after intravenous injection of radiolabeled antibody in tumor-bearing mice. The solid and dashed lines in the left panel indicate radioactivity (% injected dose per gram [%ID/g]) in tumor and blood, respectively. The magenta, orange, and blue lines indicate IgG, Fab, and scFv, respectively

examination. Antibody fragments with smaller molecular weights have been employed, such as F(ab′)$_2$, Fab, diabody, and scFv, to accelerate blood clearance and tissue penetration (Fig. 1). Although tumor uptake of radiolabeled antibody fragments is less than that of IgG (Fig. 2, left panel), sufficiently high tumor-to-blood ratios can be obtained within several hours due to the rapid decrease in background radioactivity (Fig. 2, right panel). Small-sized antibody fragments, however, have physiologic uptake patterns in various normal (non-neoplastic tissues) that differ from IgG (e.g., Fab and scFv show markedly high renal uptake). Thus, the selection of an appropriate antibody form should be based on the antibody properties, target molecules, and target disease.

Radionuclide and Radiolabeling

For nuclear medicine imaging with radiolabeled antibody, antibodies (or their fragments) should be labeled with positron-emitting radionuclides for PET imaging and with gamma ray-emitting radionuclides for SPECT imaging. Although both PET and SPECT are highly sensitive imaging modalities, PET has higher sensitivity (10^{-11}–10^{-12} mole/L) than SPECT (10^{-10}–10^{-11} mole/L) [35]. Various radionuclides suitable for PET/SPECT imaging with radiolabeled antibodies are currently available and their properties are summarized in Table 1. As mentioned above, intact IgG is eliminated from the blood slowly and therefore requires radionuclides with a longer half-life, such as Zr-89 and I-124 for PET, and In-111 and Ga-67 for SPECT. Although In-111 is most commonly used in clinical practice, several clinical studies using ^{89}Zr-labeled antibodies are currently being conducted because Zr-89 emits positrons and has higher sensitivity than In-111, which emits gamma-rays. For labeling engineered antibody fragments with a smaller size, that clear the blood more rapidly than IgG, radionuclides with a shorter half-life are more suitable, so as to minimize the dose absorbed by the patients, e.g., Cu-64 for PET, and Tc-99 m and I-123 for SPECT.

Radiolabeling methods for IgG and its fragments are well established [52]. Various labeling methods with radiohalogens, such as I-123, I-124, and Br-76, are reported and can be categorized as direct or indirect. The direct labeling method is simple and widely used to radiolabel biomolecules, including antibodies. This method employs oxidizing agents such as chloramine-T and 1,3,4,6-tetrachloro-3α,6α-diphenyl glycoluril (Iodogen), which oxidize radiohalogens to react with the tyrosine residues in antibodies. Although this method is simple and can rapidly produce radiolabeled antibodies, free radiohalogen is quickly released *in vivo* and redistributes to organs such as the thyroid and stomach. To address this problem, indirect methods were developed using linkers such as *N*-succinimidyl-3-(tri-n-butylstannyl)benzoate, which is resistant to dehalogenation, and maleimidoethyl 3-(tri-n-butylstannyl)hippurate, whose metabolite is readily excreted into the urine [2, 72]. To label radiometals such as Tc-99 m, Cu-64, In-111, and Zr-89, bifunctional chelating agents that react with amino acid residues in proteins on one side

Table 1 Representative radionuclides for antibody-based imaging

Nuclide	Hafl-life (h)	Decay mode (branching)	Energy of γ-ray (yield)	Imaging
^{99m}Tc	6	IT	0.14 MeV (89%)	SPECT
^{64}Cu	13	EC (41%) β^- (40%) β^+ (19%)	1.35 MeV (1%)	PET
^{123}I	13	EC	0.16 MeV (83%)	SPECT
^{86}Y	15	EC (66%) β^+ (34%)	0.63 MeV (33%) 0.70 MeV (15%) 1.08 MeV (83%) 1.15 MeV (31%) 1.85 MeV (17%) 1.92 MeV (21%)	PET
^{76}Br	16	β^+ (57%) EC (43%)	0.56 MeV (74%) 0.66 MeV (16%) 1.85 MeV (17%)	PET
^{111}In	67	EC	0.17 MeV (88%) 0.25 MeV (94%)	SPECT
^{67}Ga	78	EC	0.09 MeV (38%) 0.19 MeV (24%) 0.30 MeV (19%)	SPECT
^{89}Zr	78	EC (78%) β^+ (22%)	0.90 MeV (100%)	PET
^{124}I	100	EC (75%) β^+ (25%)	0.60 MeV (61%)	PET
^{177}Lu	161	β^-	0.12 MeV (7%) 0.21 MeV (11%)	SPECT
^{131}I	193	β^-	0.36 MeV (81%)	SPECT

Data are based on Table of Isotopes 7th edition (edited by C. Michael Lederer and Virginia S. Shirle, John Wiley & Sons, New York, USA, 1978)
EC electron capture, *IT* isomeric transition

and form complexes with radiometals on the other side, are commonly used. The radiometal-chelate complex is needed for stability *in vivo*, and many studies have revealed the appropriate combination for antibody imaging (Table 2). Most combinations have already been used in both preclinical and clinical studies.

CAST Diagnostic Imaging

Through an intensive study of cancer induced blood coagulation mechanism, Matsumura Y focused on extracellular matrix components in tumor stroma as suitable ligands for binding and holding anti-neoplastic drugs in proximity to tumor cells and succeeded in developing antibodies against insoluble fibrin, collagen IV, and tissue factor. These antibodies have been described as CAST therapeutic agents and three specific variants hold high promise for clinical deployment, namely:

Table 2 Representative chelates used for labeling of antibody with radiometals

Chelate	Structural formula	Applicable radionuclide
CHX-A"-DTPA	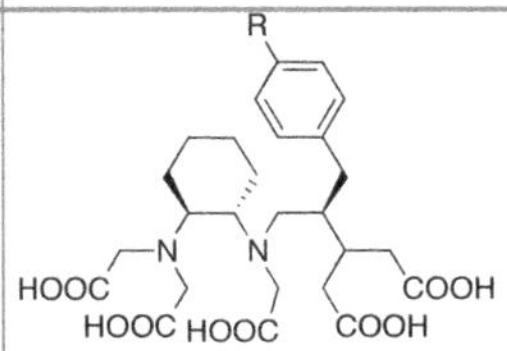	^{86}Y, ^{111}In, ^{177}Lu
Tiuxetan	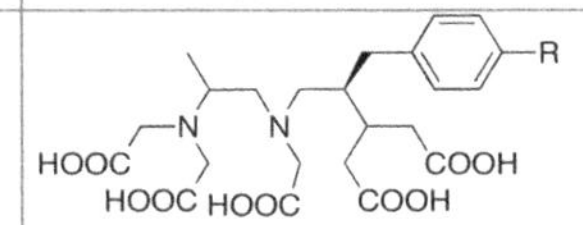	^{86}Y, ^{111}In, ^{177}Lu
DOTA	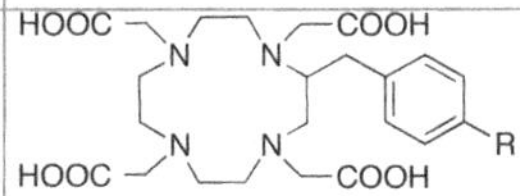	^{64}Cu, ^{111}In, ^{177}Lu
NOTA	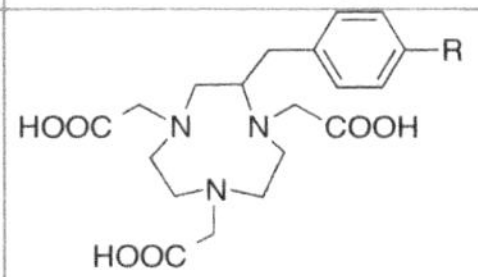	^{64}Cu, ^{67}Ga
PCTA	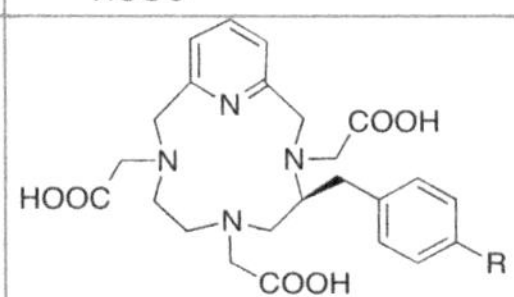	^{64}Cu, ^{67}Ga
TETA	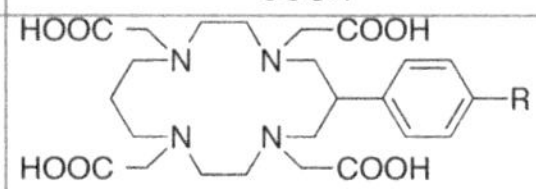	^{64}Cu
Cyclen	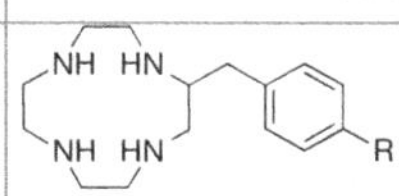	^{64}Cu
SarAr	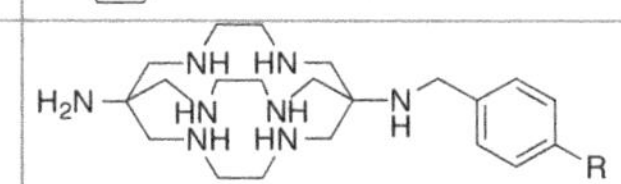	^{64}Cu
MAG3	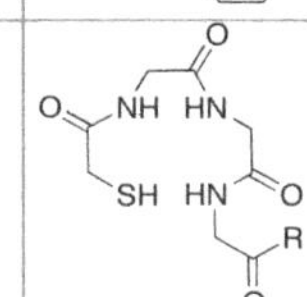	^{99m}Tc
HYNIC	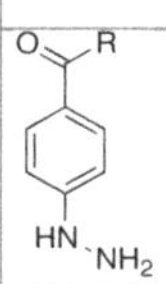	^{99m}Tc

(continued)

Table 2 (continued)

Chelate	Structural formula	Applicable radionuclide
Deferoxamine	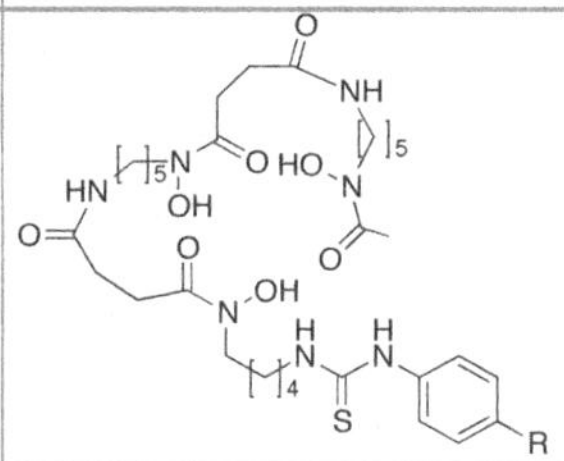	^{67}Ga, ^{89}Zr

antibody 102-10 for insoluble fibrin [67], antibody 1849 for tissue factor [30], and antibody 35-4 for collagen IV [68]. In the following subsections, we describe the use of these antibodies for CAST imaging. Because cancer stroma comprises various components other than the three targets mentioned above, we then describe other antibodies and peptides targeting those components.

Fibrin-Targeted Imaging

In cancer tissues, there is a "malignant cycle of blood coagulation", in which cancer cells invade blood vessels, resulting in hemorrhage and immediate fibrin clot formation that is subsequently replaced by collagenous stroma, and this cycle is continuously repeated in cancer tissue [36]. Based on this concept, this team of investigators succeeded in producing a monoclonal antibody that was capable of distinguishing insoluble fibrin from soluble fibrinogen, soluble fibrin monomer, and fibrin degradation products (FDP) Although fibrin clots (insoluble form of fibrin) are observed in various disease conditions in addition to cancer, such as cerebral infarction, skin wounds, and arthritis, fibrin clots in these disease conditions are transient and continuous fibrin clot formation is highly specific to cancer tissue [26, 67]. Furthermore, the insoluble form of fibrin is observed in most cancers and its degree depends on disease progression. Hisada et al. found a specific region only present in insoluble fibrin, not in fibrinogen, soluble fibrin, and D-dimer (a fibrin degradation product) [26]. Therefore, this specific region of insoluble fibrin would be a promising cancer stromal target. An antibody named 102-10 that recognizes this region was developed and evaluated as a CAST therapeutic agent. In this study, the antibody 102-10 was conjugated with an anti-tumor agent, SN-38, linked via an ester bond, which can release SN-38 in the tumor stroma and exhibits more significant anti-tumor effects compared with unbound SN-38 in a chemically-induced skin cancer mouse model with rich stroma [67]. Antibody 102-10 was then studied for its suitability for noninvasive imaging applicable to clinical practice by labeling with a positron-emitter Zr-89 for PET imaging. PET with ^{89}Zr-labeled 102-10 successfully depicted skin cancer in mice and tumor uptake of ^{89}Zr-labeled 102-10 was higher than that of ^{89}Zr-labeled cetuximab as a control antibody (Fig. 3) [26]. The autoradiographic

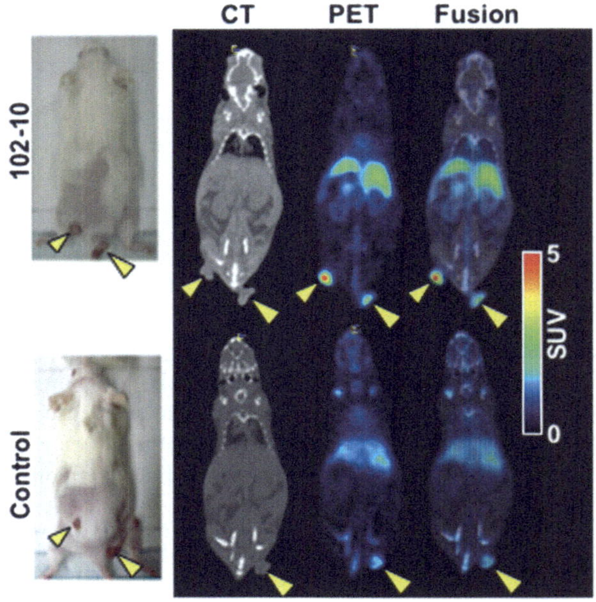

Fig. 3 PET/CT coronal images with ^{89}Zr-labeled anti-fibrin antibody 102-10 and control antibody cetuximab at day 5 after intravenous injection in spontaneous tumor models. Arrowheads show tumors [26]

analysis indicated that the radioactivity of ^{89}Zr-labeled 102-10 was localized in the cancer stroma, but not in cancer cells [26]. Therefore, radiolabeled 102-10 is a promising imaging probe to noninvasively evaluate the degree of fibrin deposition in cancer tissues. Obonai et al. reported that immunohistochemical and ex vivo imaging revealed that 102-10 Fab was distributed selectively to fibrin clots in pancreatic ductal adenocarcinoma tumors 3 h after the injection and had disappeared from the body after 24 h. This study indicated that fibrin deposition in the cancer stroma is a promising target for tumor imaging and that 102-10 Fab has great potential as an imaging agent for detecting cancer tissue in vivo [46]. The results of this study strongly support the potential for CAST in diagnosis and treatment.

There are several reports of other nuclear medicine imaging and MRI probes targeting soluble fibrin and D-dimer, and these probes target mainly thrombi, but not cancer. Spuentrup et al. published clinical trials of thrombus imaging in patients using a gadolinium-based MRI contrast agent EP-2104R comprising a fibrin-specific peptide with four gadolinium-chelate complexes [57]. Consistent with that study, Uppal et al. reported that EP-2104R enhanced the tumor periphery with high concentrations of fibrin in a breast cancer mouse model [63], and Starmans et al. reported SPECT imaging with ^{111}In-labeled peptides based on EP-2104R in a carotid artery thrombosis rat model [58] and in a breast cancer mouse model [59]. Caravan and colleagues used different peptides labeled with a positron-emitter, Cu-64, and reported PET studies of imaging thrombus formation in animal thrombosis models [7, 15]. As D-dimer-targeted imaging, an anti-D-dimer Fab' fragment DI80B3 labeled with a gamma-emitter Tc-99 m was developed, it was applied in preclinical and clinical studies to assess patients with suspected deep vein thrombosis [18, 40, 41].

Tissue Factor-Targeted Imaging

Cancer is intrinsically erosive and destructive, and continuously damages the cancer-affected tissues, resulting in a "malignant cycle of blood coagulation" [36]. Tissue factor is a transmembrane glycoprotein (47 kDa) expressed on the surfaces of many types of cells and is a key element in the initiation of the extrinsic coagulation cascade [36]. Moreover, high tissue factor expression is observed in many types of cancer, while expression is low in healthy tissues [19, 23, 29, 36, 44, 61, 65]. Sato et al. developed a new anti-tissue factor neutralizing antibody named 1849 [55], and Koga et al. applied this antibody to a tissue factor-targeted CAST therapeutic agent by conjugating antibody 1849 with monomethyl auristatin E (MMAE) [30]. This CAST therapeutic agent significantly suppressed tumor growth in a mouse pancreatic cancer model BxPC-3 [30]. Takashima et al. reported that tissue factor is profoundly expressed in glioma tissues based on immunohistochemical analysis using antibody 1849 as a primary antibody. ^{111}In-Labeled 1849 showed markedly high accumulation in orthotopic glioma xenografts, allowing the xenografts to be clearly visualized using immuno-SPECT/CT imaging (Fig. 4) [61]. In the case of tissue factor, the target is expressed not only in the stroma but also on cancer cells, such as pancreatic cancer and glioma. Therefore, the tissue factor-targeted CAST agent is expected to allow for highly effective delivery of anti-cancer agents to cancer cells, resulting in high therapeutic efficacy.

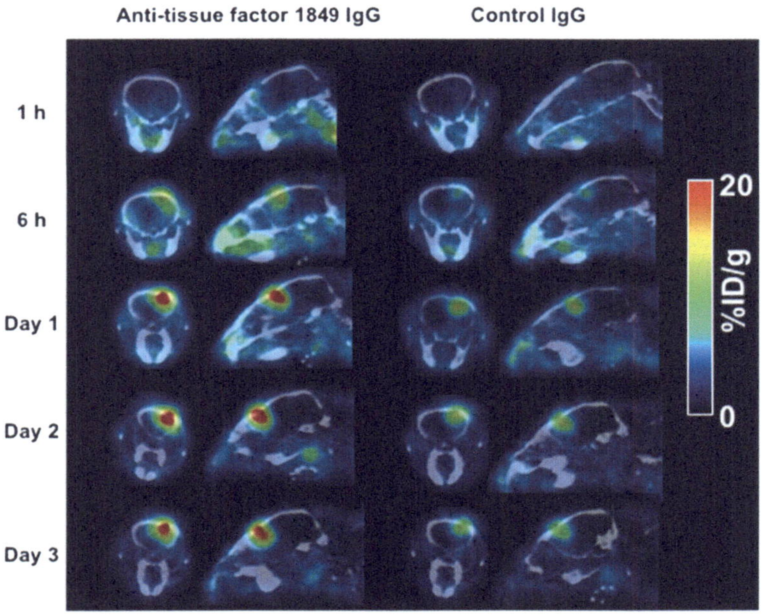

Fig. 4 Serial SPECT/CT images (transverse and sagittal) with ^{111}In-labeled anti-tissue factor antibody 1849 (left panels) and control antibody (right panels) in U87MG orthotopic glioma models [61]

An agent called TF-011-MMAE, comprising an anti-tissue factor antibody TF-011 and MMAE, was developed as a conventional ADC agent, but not a CAST agent [10]. TF-011-MMAE has antitumor activity in patient-derived xenograft models derived from seven different solid cancers [10], suggesting that tissue factor-targeted therapy could be effective for a wide range of solid tumors. To date, there are no reports of noninvasive imaging with TF-001.

Cai and colleagues reported several immuno-PET studies with anti-tissue factor antibody ALT-836 in a phase 1 clinical trial in subjects with acute lung injury or acute respiratory distress syndrome, revealing an acceptably safe dose of 0.08 mg/kg in patients [39]. Although a phase 2 trial (ClinicalTrials.gov Identifier, NCT00879606) was conducted, the results are not yet reported. The first report of an immuno-PET studies with ALT-836, employed the IgG form labeled with Cu-64 and evaluated pancreatic cancer mouse models [27]. The half-life of Cu-64 is 13 h, which is not optimal for IgG-based imaging due to the slow blood clearance of IgG (Fig. 2). Another study generated the Fab form of ALT-836, to adjust for the brief ~~the~~ half-life of Cu-64, and evaluated its properties as a PET probe in a triple negative breast cancer model, MDA-MB-231 [56]. Luo et al. generated a heterodimer of ALT-836 Fab with anti-CD105 Fab and evaluated the ^{64}Cu-labeled heterodimer in a pancreatic cancer model, BxPC-3 [34]. This heterodimer provided better images compared with the Fab PET probes [34]. Hernandez et al. employed another approach, namely, they labeled ALT-836 IgG with Zr-89 with a longer half-life of 78 h compared to Cu-64 [24]. ^{89}Zr-labeled ALT-836 IgG [24] showed higher BxPC-3 tumor uptake compared with ^{64}Cu-labeled ALT-836 Fab [56] and better image quality compared with ^{64}Cu-labeled ALT-836 IgG [27].

Nielsen et al. reported a unique approach using factor VIIa as a PET probe [42]. Factor VIIa is an initiator in the extrinsic coagulation pathway and binds to tissue factor with high affinity. Nielsen et al. radiolabeled factor VIIa with a positron-emitter F-18, and its tumor uptake correlated with tissue factor expression levels in three tumor models [42]. Compared with the antibody-based imaging mentioned above, ^{18}F-labeled factor VIIa showed lower tumor uptake but higher contrast images in a shorter period of time after the injection [42]. Although further clinical evaluation is required, it could be more suitable for assessing tissue factor expression status correlated with malignancies. Nielsen et al. also reported a PET study with ^{64}Cu-labeled factor VIIa in pancreatic cancer models [43]. Cu-64 has a longer half-life than F-18, therefore, ^{64}Cu-labeled factor VIIa enabled delayed imaging, which improved tumor uptake and image contrast [43].

Collagen-Targeted Imaging

At the final stage of the "malignant cycle of blood coagulation", fibrin clots are replaced with collagenous tissues [36]. Yasunaga et al. generated a collagen-targeted CAST therapeutic agent using an anti-collagen IV antibody 35-4 conjugated with a topoisomerase I inhibitor SN-38 [68]. This collagen-targeted CAST therapy showed a significant anti-tumor effect in pancreatic cancer mouse models PSN-1 and SUIT-2 [68]. Unfortunately, there are no reports of noninvasive imaging applicable to clinical practice using the antibody 35-4, although optical imaging with IRDye 800-labeled 35-4 is reported to show high tumor uptake and low uptake in healthy organs [68]. The antibody 35-4 is considered to be applicable to a noninvasive imaging probe for cancer.

Several collagen-targeted MRI studies with EP-3533 peptide-based probes binding to collagen I have been reported. Caravan and colleagues developed a gadolinium-based MRI agent and conducted several preclinical studies using a myocardial infarction mouse model [12], liver fibrosis rat and mouse models [48], a pulmonary fibrosis mouse model [13], and a pancreatic cancer model [49]. EP-3533 MRI visualized and quantified fibrosis in all the models. Chen et al. developed another MRI contrast agent based on high-density lipoprotein nanoparticles coupled with EP-3533 and evaluated in an atherosclerosis mouse model [14]. MR signal enhancements were found in atherosclerotic plaques in aortic vessel walls and correlated with macrophage and collagen contents [14]. Caravan and colleagues also developed a collagen-targeted PET probe, CBP7, to overcome the sensitivity limitations of MRI [17]. Desogere et al. compared five types of ^{64}Cu-labeled peptide probes, including a core motif that binds to collagen with high affinity and demonstrated that ^{64}Cu-labeled CBP7 is a promising probe for noninvasive imaging of pulmonary fibrosis because the probe showed the highest uptake in fibrotic lungs and the highest target-to-background ratio [17]. Despite positive results of the MRI imaging probe based on EP-3533, the collagen-targeted PET probe CBP7 developed as a PET imaging probe offers a shorter and more economical path to clinical translation than MRI probes [17].

Other Stroma-Targeted Imaging

Cancer stroma has many components and expresses many molecules, providing several target candidates for CAST therapy [19, 31, 36, 38, 50]. Tenascin-C is an extracellular matrix glycoprotein that is specifically and transiently expressed depending on the type of tissue injury, and has a role in tissue rebuilding [38]. High tenascin-C expression is observed in many types of cancer with progressive tissue damage [38, 53]. Many preclinical and clinical studies of tenascin-C-targeted imaging and therapy are reported. Reulen et al. employed two antibodies (BC-4 and BC-24) against the epidermal growth factor-like sequence of tenascin-C and evaluated the therapeutic efficacy of antibodies radiolabeled with therapeutic β-emitters I-131 or Y-90 in patients with glioma [53]. The median survival of grade III and IV glioma patients treated with radioimmunotherapy using those radiolabeled antibodies exceeded that of historical controls treated with conventional radiotherapy and chemotherapy [53]. Although they did not report imaging with those antibodies, substituting an imaging radionuclide as a γ-emitter such as I-123, I-124, or In-111 enables noninvasive imaging in patients. Zalutsky and colleagues developed an anti-tenascin-C antibody, 81C6, and conducted clinical studies with antibodies radiolabeled with I-131 [8, 16, 37, 51, 71] or a therapeutic α-emitter, At-211 [73], in patients with brain tumors. The ^{131}I-labeled 81C6 highly accumulated in glioma and the combination therapy with external beam radiotherapy and adjuvant chemotherapy encouraged survivals of patients with glioma compared with historical controls [51]. To enhance the efficacy, they labeled 81C6 with At-211, which were associated with a promising benefit in patients with glioma in a feasible clinical study [73]. Heuveling et al. reported a microdosing immuno-PET study with an antibody, F16SIP, that recognizes the extra domain A1 of tenascin-C radiolabeled with a positron-emitter I-124 in patients with head and neck cancer [25]. The ^{124}I-labeled F16SIP visualized tumors on PET images at 24 h after injection [25]. Radioimmunotherapy with ^{131}I-labeled F16SIP has been reported in patients with refractory Hodgkin's lymphoma [1]. The radioimmunotherapy produced objective responses in some patients [1].

Integrins are expressed in many types of cancer and interact with extracellular matrix proteins such as fibrin, collagen, and tenascin-C [20, 31]; hence, integrins are considered potential target molecules for CAST imaging and therapy. Integrin $\alpha_V\beta_3$ is the most extensively studied among the integrin family members, and cyclic peptides containing Arg-Gly-Asp (RGD) amino acid sequence are commonly used for targeted imaging [33]. Because monomeric cyclic RGD imaging probes have low tumor uptake, multimeric cyclic RGD imaging probes were developed to overcome this limitation [9, 32]. ^{64}Cu-Cyclam-RAFT-c(−RGD-)$_4$ is a tetrameric cyclic RGD-based PET imaging probe [28] with higher tumor uptake in an orthotopic pancreatic cancer mouse model, BxPC-3, having abundant stroma, compared with ^{18}F-FDG, which is widely used in clinical practice (Fig. 5) [3]. Although most integrin-targeted imaging studies use RGD-based probes, there are several reports of antibody-based probes. Cai et al. developed ^{64}Cu-labeled antibody Abegrin against integrin $\alpha_V\beta_3$ and evaluated its properties as a PET tracer using tumor mouse models U87MG, MDA-MB-435, and PC-3 [11]. PET with ^{64}Cu-labeled Abegrin showed marked high tumor uptake in U87MG tumors having very high integrin $\alpha_V\beta_3$ expression, whereas it showed lower tumor uptake in MDA-MB-435 and PC-3 tumors with lower integrin $\alpha_V\beta_3$ expression [11]. Consistent with this research, Veeravagu et al. radiolabeled Abegrin with a therapeutic radionuclide, Y-90, and evaluated its efficacy using U87MG (high integrin $\alpha_V\beta_3$ expression) and HT-29 (low integrin $\alpha_V\beta_3$ expression) tumor mouse models [64]. The ^{90}Y-labeled Abegrin significantly suppressed U87MG tumors with control groups, but not HT-29 tumors [64]. PET with ^{18}F-FDG and ^{18}F-FLT revealed a reduction of cell proliferation, metabolic activity, and DNA synthesis in U87MG treated with ^{90}Y-labeled Abegrin [64]. Saga et al.

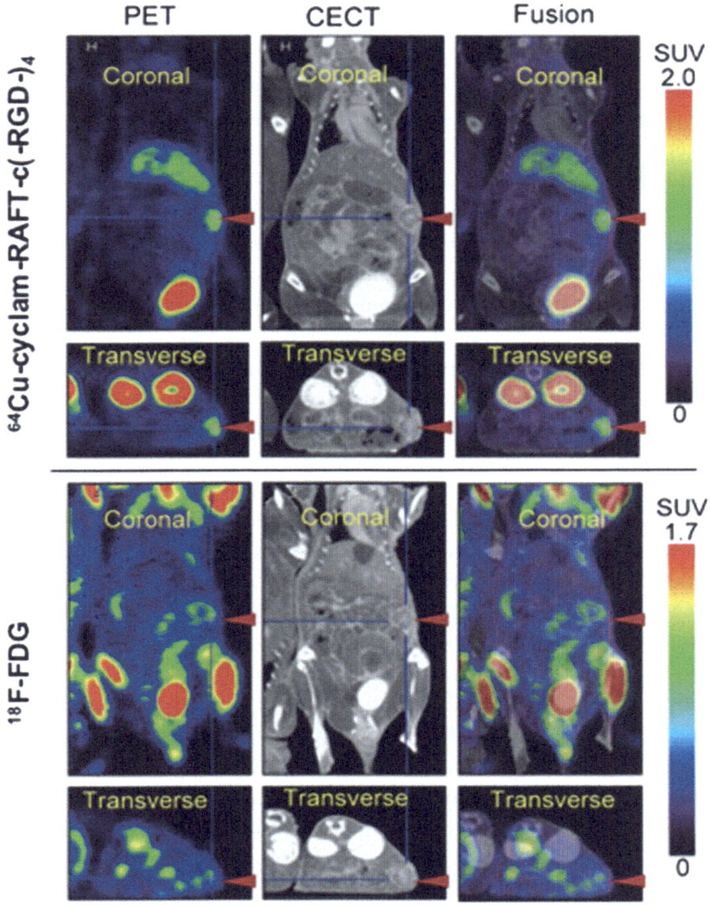

Fig. 5 PET/CT images (coronal and transverse) of BxPC-3 orthotopic pancreatic cancer mouse models. ^{64}Cu-Cyclam-RAFT-c(-RGD-)$_4$ PET/CT images are shown in the upper panels and ^{18}F-FDG PET/CT images are shown in the lower panels. Arrowheads show tumors [3]

reported the therapeutic efficacy of ^{111}In-labeled antibody GA17, which recognizes integrin α_3 in a glioblastoma mouse model [54]. The ^{111}In-labeled antibody GA17 internalized into tumor cells and retained within the cancer cells [54]. High dose of ^{111}In-labeled antibody GA17 suppressed tumor growth [54]. Aung et al. conducted immuno-SPECT studies with ^{111}In-labeled antibody against integrin $\alpha_6\beta_4$ in a pancreatic cancer mouse model, BxPC-3 [4]. BxPC-3 tumor uptake of the ^{111}In-labeled antibody increased with time, and the highest tumor uptake of 43% ID/g was found at 96 h after injection [4]. In addition, the same antibody was applied to radioimmunotherapy with ^{90}Y-labeled antibody [5, 6]. The ^{90}Y-labeled antibody suppressed BxPC-3 tumors [5], and combined therapy with a PI3K/mTOR inhibitor BEZ235 enhanced the therapeutic efficacy [6].

Perspective

Radiolabeled antibodies recognizing cancer stroma targets can provide important information for CAST therapy, such as the degree of the cancer stroma in each patient. Many promising antibodies including anti-insoluble fibrin, anti-tenascin-C, and anti-integrins have been developed, thereby producing new CAST therapeutic agents and diagnostic imaging agents. These agents are expected to improve the prognosis of refractory cancer and provide useful information for deeply understanding the role of cancer stroma in clinical oncology.

Tumor accumulation of released anti-tumor agents from CAST therapeutic agents cannot be directly evaluated by noninvasive imaging using radiolabeled antibodies of CAST diagnostic imaging. Mass-spectrometric imaging of tumor sections administered CAST agents can be used to assess the accumulation of drugs in tumor cells [21, 66, 70]. Unfortunately, this method does not apply to noninvasive imaging and cannot be used to evaluate drug accumulation in whole tumor tissues. Therefore, the development of a novel noninvasive imaging method applicable to cancer patients is needed. The same linker used for CAST therapeutic agents can conjugate not only anti-tumor agents, but also chelates to antibodies for imaging; hence, the radionuclide-chelate complex can be released in cancer stroma, like anti-tumor agents of CAST therapeutic agents. Such radiolabeled antibodies might facilitate noninvasive and direct evaluation of drug accumulation in each patient.

The size and structure of stromal cells affect *in vivo* water diffusion, and one MRI method, diffusion-weighted imaging, provides information about the stromal cell size and structure because diffusion-weighted imaging can measure some index related to them by changing diffusion-weighted imaging parameters such as magnetic gradient shape and diffusion time [45, 47], and this technique is applied to clinical diagnosis [22, 60, 62]. Further clinical studies to compare nuclear medicine imaging and diffusion-weighted MRI are required to clarify their roles in stroma-targeted diagnostic imaging. More importantly, it is pivotal to know the pathological and molecular aspects of cancer stroma in order to fulfill the CAST imaging strategies (See the chapter "The Cancer Stroma and Its Relevance to Tumor Survival and Treatment" by Tarin D).

References

1. Aloj L, D'Ambrosio L, Aurilio M, Morisco A, Frigeri F, Caraco C, Di Gennaro F, Capobianco G, Giovannoni L, Menssen HD, Neri D, Pinto A, Lastoria S (2014) Radioimmunotherapy with Tenarad, a 131I-labelled antibody fragment targeting the extra-domain A1 of tenascin-C, in patients with refractory Hodgkin's lymphoma. Eur J Nucl Med Mol Imaging 41:867–877. https://doi.org/10.1007/s00259-013-2658-6
2. Arano Y, Wakisaka K, Ohmomo Y, Uezono T, Mukai T, Motonari H, Shiono H, Sakahara H, Konishi J, Tanaka C (1994) Maleimidoethyl 3-(tri-n-butylstannyl)hippurate: a useful radioiodination reagent for protein radiopharmaceuticals to enhance target selective radioactivity localization. J Med Chem 37:2609–2618

3. Aung W, Jin Z-H, Furukawa T, Claron M, Boturyn D, Sogawa C, Tsuji AB, Wakizaka H, Fukumura T, Fujibayashi Y, Dumy P, Saga T (2013) Micro-positron emission tomography/contrast-enhanced computed tomography imaging of orthotopic pancreatic tumor-bearing mice using the $\alpha v\beta_3$ integrin tracer 64Cu-labeled cyclam-RAFT-c(-RGDfK-)$_4$. Mol Imaging 12:376–387. https://doi.org/10.2310/7290.2013.00054
4. Aung W, Tsuji AB, Sudo H, Sugyo A, Furukawa T, Ukai Y, Kurosawa Y, Saga T (2016a) Immunotargeting of integrin α6β4 for single-photon emission computed tomography and near-infrared fluorescence imaging in a pancreatic Cancer model. Mol Imaging 15:1536012115624917. https://doi.org/10.1177/1536012115624917
5. Aung W, Tsuji AB, Sudo H, Sugyo A, Ukai Y, Kouda K, Kurosawa Y, Furukawa T, Saga T (2016b) Radioimmunotherapy of pancreatic cancer xenografts in nude mice using 90Y-labeled anti-α6β4 integrin antibody. Oncotarget 7:38835–38844. https://doi.org/10.18632/oncotarget.9631
6. Aung W, Tsuji AB, Sudo H, Sugyo A, Ukai Y, Kouda K, Kurosawa Y, Furukawa T, Saga T, Higashi T (2017) Combined treatment of pancreatic cancer xenograft with 90Y-ITGA6B4-mediated radioimmunotherapy and PI3K/mTOR inhibitor. World J Gastroenterol 23:7551–7562. https://doi.org/10.3748/wjg.v23.i42.7551
7. Ay I, Blasi F, Rietz TA, Rotile NJ, Kura S, Brownell A-L, Day H, Oliveira BL, Looby RJ, Caravan P (2014) In vivo molecular imaging of thrombosis and thrombolysis using a fibrin-binding positron emission tomographic probe. Circ Cardiovasc Imaging 7:697–705. https://doi.org/10.1161/CIRCIMAGING.113.001806
8. Bigner DD, Brown MT, Friedman AH, Coleman RE, Akabani G, Friedman HS, Thorstad WL, McLendon RE, Bigner SH, Zhao XG, Pegram CN, Wikstrand CJ, Herndon JE, Vick NA, Paleologos N, Cokgor I, Provenzale JM, Zalutsky MR (1998) Iodine-131-labeled antitenascin monoclonal antibody 81C6 treatment of patients with recurrent malignant gliomas: phase I trial results. J Clin Oncol 16:2202–2212
9. Boturyn D, Coll J-L, Garanger E, Favrot M-C, Dumy P (2004) Template assembled cyclopeptides as multimeric system for integrin targeting and endocytosis. J Am Chem Soc 126:5730–5739. https://doi.org/10.1021/ja049926n
10. Breij ECW, de Goeij BECG, Verploegen S, Schuurhuis DH, Amirkhosravi A, Francis J, Miller VB, Houtkamp M, Bleeker WK, Satijn D, Parren PWHI (2014) An antibody-drug conjugate that targets tissue factor exhibits potent therapeutic activity against a broad range of solid tumors. Cancer Res 74:1214–1226. https://doi.org/10.1158/0008-5472.CAN-13-2440
11. Cai W, Wu Y, Chen K, Cao Q, Tice DA, Chen X (2006) In vitro and in vivo characterization of 64Cu-labeled Abegrin TM, a humanized monoclonal antibody against integrin v 3. Cancer Res 66:9673–9681. https://doi.org/10.1158/0008-5472.CAN-06-1480
12. Caravan P, Das B, Dumas S, Epstein FH, Helm PA, Jacques V, Koerner S, Kolodziej A, Shen L, Sun W-C, Zhang Z (2007) Collagen-targeted MRI contrast agent for molecular imaging of fibrosis. Angew Chem Int Ed 46:8171–8173. https://doi.org/10.1002/anie.200700700
13. Caravan P, Yang Y, Zachariah R, Schmitt A, Mino-Kenudson M, Chen HH, Sosnovik DE, Dai G, Fuchs BC, Lanuti M (2013) Molecular magnetic resonance imaging of pulmonary fibrosis in mice. Am J Respir Cell Mol Biol 49:1120–1126. https://doi.org/10.1165/rcmb.2013-0039OC
14. Chen W, Cormode DP, Vengrenyuk Y, Herranz B, Feig JE, Klink A, Mulder WJM, Fisher EA, Fayad ZA (2013) Collagen-specific peptide conjugated HDL nanoparticles as MRI contrast agent to evaluate compositional changes in atherosclerotic plaque regression. JACC Cardiovasc Imaging 6:373–384. https://doi.org/10.1016/j.jcmg.2012.06.016
15. Ciesienski KL, Yang Y, Ay I, Chonde DB, Loving GS, Rietz TA, Catana C, Caravan P (2013) Fibrin-targeted PET probes for the detection of thrombi. Mol Pharm 10:1100–1110. https://doi.org/10.1021/mp300610s
16. Cokgor I, Akabani G, Kuan CT, Friedman HS, Friedman AH, Coleman RE, McLendon RE, Bigner SH, Zhao XG, Garcia-Turner AM, Pegram CN, Wikstrand CJ, Shafman TD, Herndon JE, Provenzale JM, Zalutsky MR, Bigner DD (2000) Phase I trial results of iodine-131-labeled antitenascin monoclonal antibody 81C6 treatment of patients with newly diagnosed malignant gliomas. J Clin Oncol 18:3872

17. Désogère P, Tapias LF, Rietz TA, Rotile N, Blasi F, Day H, Elliott J, Fuchs BC, Lanuti M, Caravan P (2017) Optimization of a collagen-targeted PET probe for molecular imaging of pulmonary fibrosis. J Nucl Med 58:1991–1996. https://doi.org/10.2967/jnumed.117.193532
18. Douketis JD, Ginsberg JS, Haley S, Julian J, Dwyer M, Levine M, Eisenberg PR, Smart R, Tsui W, White RH, Morris TA, Kaatz S, Comp PC, Crowther MA, Kearon C, Kassis J, Bates SM, Schulman S, Desjardins L, Taillefer R, Begelman SM, Gerometta M (2012) Accuracy and safety of 99mTc-labeled anti-D-dimer (DI-80B3) Fab' fragments (ThromboView®) in the diagnosis of deep vein thrombosis: a phase II study. Thromb Res 130:381–389. https://doi.org/10.1016/j.thromres.2012.05.011
19. Drake TA, Morrissey JH, Edgington TS (1989) Selective cellular expression of tissue factor in human tissues. Implications for disorders of hemostasis and thrombosis. Am J Pathol 134:1087–1097
20. Dvorak HF (2015) Tumors: wounds that do not heal-redux. Cancer Immunol Res 3:1–11. https://doi.org/10.1158/2326-6066.CIR-14-0209
21. Fujiwara Y, Furuta M, Manabe S, Koga Y, Yasunaga M, Matsumura Y (2016) Imaging mass spectrometry for the precise design of antibody-drug conjugates. Sci Rep 6:srep24954. https://doi.org/10.1038/srep24954
22. Goudarzi B, Kishimoto R, Komatsu S, Ishikawa H, Yoshikawa K, Kandatsu S, Obata T (2010) Detection of bone metastases using diffusion weighted magnetic resonance imaging: comparison with 11C-methionine PET and bone scintigraphy. Magn Reson Imaging 28:372–379. https://doi.org/10.1016/j.mri.2009.12.008
23. Haas SL, Jesnowski R, Steiner M, Hummel F, Ringel J, Burstein C, Nizze H, Liebe S, Löhr JM (2006) Expression of tissue factor in pancreatic adenocarcinoma is associated with activation of coagulation. World J Gastroenterol. Baishideng Publishing Group Inc 12:4843–4849. https://doi.org/10.3748/wjg.v12.i30.4843
24. Hernandez R, England CG, Yang Y, Valdovinos HF, Liu B, Wong HC, Barnhart TE, Cai W (2017) ImmunoPET imaging of tissue factor expression in pancreatic cancer with 89Zr-Df-ALT-836. J Control Release. https://doi.org/10.1016/j.jconrel.2017.08.029
25. Heuveling DA, de Bree R, Vugts DJ, Huisman MC, Giovannoni L, Hoekstra OS, Leemans CR, Neri D, van Dongen GAMS (2013) Phase 0 microdosing PET study using the human mini antibody F16SIP in head and neck cancer patients. J Nucl Med 54:397–401. https://doi.org/10.2967/jnumed.112.111310
26. Hisada Y, Yasunaga M, Hanaoka S, Saijou S, Sugino T, Tsuji A, Saga T, Tsumoto K, Manabe S, Kuroda J-I, Kuratsu J-I, Matsumura Y (2013) Discovery of an uncovered region in fibrin clots and its clinical significance. Sci Rep 3:2604. https://doi.org/10.1038/srep02604
27. Hong H, Zhang Y, Nayak TR, Engle JW, Wong HC, Liu B, Barnhart TE, Cai W (2012) Immuno-PET of tissue factor in pancreatic cancer. J Nucl Med 53:1748–1754. https://doi.org/10.2967/jnumed.112.105460
28. Jin Z-H, Furukawa T, Galibert M, Boturyn D, Coll J-L, Fukumura T, Saga T, Dumy P, Fujibayashi Y (2011) Noninvasive visualization and quantification of tumor αVβ3 integrin expression using a novel positron emission tomography probe, 64Cu-cyclam-RAFT-c(-RGDfK-)4. Nucl Med Biol 38:529–540. https://doi.org/10.1016/j.nucmedbio.2010.11.008
29. Khorana AA, Ahrendt SA, Ryan CK, Francis CW, Hruban RH, Hu YC, Hostetter G, Harvey J, Taubman MB (2007) Tissue factor expression, angiogenesis, and thrombosis in pancreatic cancer. Clin Cancer Res 13: 2870–2875. https://doi.org/10.1158/1078-0432.CCR-06-2351
30. Koga Y, Manabe S, Aihara Y, Sato R, Tsumura R, Iwafuji H, Furuya F, Fuchigami H, Fujiwara Y, Hisada Y, Yamamoto Y, Yasunaga M, Matsumura Y (2015) Antitumor effect of antitissue factor antibody-MMAE conjugate in human pancreatic tumor xenografts. Int J Cancer 137:1457–1466. https://doi.org/10.1002/ijc.29492
31. Koivisto L, Heino J, Häkkinen L, Larjava H (2014) Integrins in wound healing. Adv Wound Care 3:762–783. https://doi.org/10.1089/wound.2013.0436
32. Li Z-B, Chen K, Chen X (2008) 68Ga-labeled multimeric RGD peptides for microPET imaging of integrin αvβ3 expression. Eur J Nucl Med Mol Imaging 35:1100–1108. https://doi.org/10.1007/s00259-007-0692-y

33. Liu S (2009) Radiolabeled cyclic RGD peptides as integrin alpha(v)beta(3)-targeted radiotracers: maximizing binding affinity via bivalency. Bioconjug Chem 20:2199–2213. https://doi.org/10.1021/bc900167c
34. Luo H, England CG, Shi S, Graves SA, Hernandez R, Liu B, Theuer CP, Wong HC, Nickles RJ, Cai W (2016) Dual targeting of tissue factor and CD105 for preclinical PET imaging of pancreatic cancer. Clin Cancer Res 22:3821–3830. https://doi.org/10.1158/1078-0432.CCR-15-2054
35. Massoud TF, Gambhir SS (2003) Molecular imaging in living subjects: seeing fundamental biological processes in a new light. Genes Dev 17:545–580. https://doi.org/10.1101/gad.1047403
36. Matsumura Y (2012) Cancer stromal targeting (CAST) therapy. Adv Drug Deliv Rev 64:710–719. https://doi.org/10.1016/j.addr.2011.12.010
37. McLendon RE, Akabani G, Friedman HS, Reardon DA, Cleveland L, Cokgor I, Herndon JE, Wikstrand C, Boulton ST, Friedman AH, Bigner DD, Zalutsky MR (2007) Tumor resection cavity administered iodine-131-labeled antitenascin 81C6 radioimmunotherapy in patients with malignant glioma: neuropathology aspects. Nucl Med Biol 34:405–413. https://doi.org/10.1016/j.nucmedbio.2007.01.009
38. Midwood KS, Orend G (2009) The role of tenascin-C in tissue injury and tumorigenesis. J Cell Commun Signal 3:287–310. https://doi.org/10.1007/s12079-009-0075-1
39. Morris PE, Steingrub JS, Huang BY, Tang S, Liu PM, Rhode PR, Wong HC (2012) A phase I study evaluating the pharmacokinetics, safety and tolerability of an antibody-based tissue factor antagonist in subjects with acute lung injury or acute respiratory distress syndrome. BMC Pulm Med 12(5). https://doi.org/10.1186/1471-2466-12-5
40. Morris TA, Marsh JJ, Chiles PG, Konopka RG, Pedersen CA, Schmidt PF, Gerometta M (2004a) Single photon emission computed tomography of pulmonary emboli and venous thrombi using anti-D-dimer. Am J Respir Crit Care Med 169:987–993. https://doi.org/10.1164/rccm.200306-735OC
41. Morris TA, Marsh JJ, Konopka R, Pedersen CA, Chiles PG (2004b) Improved imaging of deep venous thrombi during anticoagulation using radiolabelled anti-D-dimer antibodies. Nucl Med Commun 25:917–922
42. Nielsen CH, Erlandsson M, Jeppesen TE, Jensen MM, Kristensen LK, Madsen J, Petersen LC, Kjaer A (2015) Quantitative PET imaging of tissue factor expression using 18F-labled active site inhibited factor VII. J Nucl Med. https://doi.org/10.2967/jnumed.115.154849
43. Nielsen CH, Jeppesen TE, Kristensen LK, Jensen MM, Ali El HH, Madsen J, Wiinberg B, Petersen LC, Kjaer A (2016) PET imaging of tissue factor in pancreatic cancer using 64Cu-labeled active site inhibited factor VII. J Nucl Med 115:170266. https://doi.org/10.2967/jnumed.115.17026
44. Nitori N, Ino Y, Nakanishi Y, Yamada T, Honda K, Yanagihara K, Kosuge T, Kanai Y, Kitajima M, Hirohashi S (2005) Prognostic significance of tissue factor in pancreatic ductal adenocarcinoma. Clin Cancer Res 11:2531–2539. https://doi.org/10.1158/1078-0432.CCR-04-0866
45. Novikov DS, Kiselev VG, Jespersen SN (2018) On modeling. Magn Reson Med 79:3172–3193. https://doi.org/10.1002/mrm.27101
46. Obonai T, Fuchigami H, Furuya F et al (2016) Tumour imaging by the detection of fibrin clots in tumour stroma using an anti-fibrin fab fragment. Sci Rep 6:23613. https://doi.org/10.1038/srep23613
47. Obata T, Kershaw J, Tachibana Y et al (2018) Comparison of diffusion-weighted MRI and anti-stokes Raman scattering (CARS) measurements of the inter-compartmental exchange-time of water in expression-controlled aquaporin-4 cells. Sci Rep 8:13199. https://doi.org/10.1038/s41598-018-36264-9
48. Polasek M, Fuchs BC, Uppal R, Schühle DT, Alford JK, Loving GS, Yamada S, Wei L, Lauwers GY, Guimaraes AR, Tanabe KK, Caravan P (2012) Molecular MR imaging of liver fibrosis: a feasibility study using rat and mouse models. J Hepatol 57:549–555. https://doi.org/10.1016/j.jhep.2012.04.035

49. Polasek M, Yang Y, Schühle DT, Yaseen MA, Kim YR, Sung YS, Guimaraes AR, Caravan P (2017) Molecular MR imaging of fibrosis in a mouse model of pancreatic cancer. Sci Rep 7:8114. https://doi.org/10.1038/s41598-017-08838-6
50. Qing C (2017) The molecular biology in wound healing & non-healing wound. Chin J Traumatol 20:189–193. https://doi.org/10.1016/j.cjtee.2017.06.001
51. Reardon DA, Zalutsky MR, Akabani G, Coleman RE, Friedman AH, Herndon JE, McLendon RE, Pegram CN, Quinn JA, Rich JN, Vredenburgh JJ, Desjardins A, Guruangan S, Boulton S, Raynor RH, Dowell JM, Wong TZ, Zhao X-G, Friedman HS, Bigner DD (2008) A pilot study: 131I-antitenascin monoclonal antibody 81c6 to deliver a 44-Gy resection cavity boost. Neuro-Oncology 10:182–189. https://doi.org/10.1215/15228517-2007-053
52. Reilly RM (2010) The radiochemistry of monoclonal antibodies and peptides. In: Monoclonal antibody and peptide-targeted radiotherapy of Cancer. Wiley, Hoboken, pp 39–100
53. Reulen H-J, Poepperl G, Goetz C, Gildehaus FJ, Schmidt M, Tatsch K, Pietsch T, Kraus T, Rachinger W (2015) Long-term outcome of patients with WHO Grade III and IV gliomas treated by fractionated intracavitary radioimmunotherapy. J Neurosurg 123:1–11. https://doi.org/10.3171/2014.12.JNS142168
54. Saga T, Sakahara H, Nakamoto Y, Sato N, Zhao S, Aoki T, Miyatake S, Namba Y, Konishi J (1999) Radioimmunotherapy of human glioma xenografts in nude mice by indium-111 labelled internalising monoclonal antibody. Eur J Cancer 35:1281–1285
55. Saito Y, Hashimoto Y, Kuroda J-I, Yasunaga M, Koga Y, Takahashi A, Matsumura Y (2011) The inhibition of pancreatic cancer invasion-metastasis cascade in both cellular signal and blood coagulation cascade of tissue factor by its neutralisation antibody. Eur J Cancer 47:2230–2239. https://doi.org/10.1016/j.ejca.2011.04.028
56. Shi S, Hong H, Orbay H, Graves SA, Yang Y, Ohman JD, Liu B, Nickles RJ, Wong HC, Cai W (2015) ImmunoPET of tissue factor expression in triple-negative breast cancer with a radiolabeled antibody Fab fragment. Eur J Nucl Med Mol Imaging 42:1295–1303. https://doi.org/10.1007/s00259-015-3038-1
57. Spuentrup E, Botnar RM, Wiethoff AJ, Ibrahim T, Kelle S, Katoh M, Özgun M, Nagel E, Vymazal J, Graham PB, Günther RW, Maintz D (2008) MR imaging of thrombi using EP-2104R, a fibrin-specific contrast agent: initial results in patients. Eur Radiol 18:1995–2005. https://doi.org/10.1007/s00330-008-0965-2
58. Starmans LWE, van Duijnhoven SMJ, Rossin R, Aime S, Daemen MJAP, Nicolay K, Grüll H (2013) SPECT imaging of fibrin using fibrin-binding peptides. Contrast Media Mol Imaging 8:229–237. https://doi.org/10.1002/cmmi.1521
59. Starmans LWE, van Mourik T, Rossin R, Verel I, Nicolay K, Grüll H (2015) Noninvasive visualization of tumoral fibrin deposition using a peptidic fibrin-binding single photon emission computed tomography tracer. Mol Pharm 12:1921–1928. https://doi.org/10.1021/mp500673u
60. Tachibana Y, Obata T, Tsuchiya H, Omatsu T, Kishimoto R, Kawaguchi H, Nishikori A, Kamagata K, Hori M, Aoki S, Tsuji H, Inoue T (2015) Diffusion-tensor-based method for robust and practical estimation of axial and radial diffusional kurtosis. Eur Radiol 26:2559–2566. https://doi.org/10.1007/s00330-015-4038-z
61. Takashima H, Tsuji AB, Saga T, Yasunaga M, Koga Y, Kuroda J-I, Yano S, Kuratsu J-I, Matsumura Y (2017) Molecular imaging using an anti-human tissue factor monoclonal antibody in an orthotopic glioma xenograft model. Sci Rep 7:12341. https://doi.org/10.1038/s41598-017-12563-5
62. Takayama Y, Ohno T, Kishimoto R, Kato S, Yoneyama R, Kandatsu S, Tsujii H, Obata T (2009) Prediction of early response to radiotherapy of uterine carcinoma with dynamic contrast-enhanced MR imaging using pixel analysis of MR perfusion imaging. Magn Reson Imaging 27:370–376. https://doi.org/10.1016/j.mri.2008.07.007
63. Uppal R, Medarova Z, Farrar CT, Dai G, Moore A, Caravan P (2012) Molecular imaging of fibrin in a breast cancer xenograft mouse model. Investig Radiol 47:553–558. https://doi.org/10.1097/RLI.0b013e31825dddfb

64. Veeravagu A, Liu Z, Niu G, Chen K, Jia B, Cai W, Jin C, Hsu AR, Connolly AJ, Tse V, Wang F, Chen X (2008) Integrin alphavbeta3-targeted radioimmunotherapy of glioblastoma multiforme. Clin Cancer Res 14:7330–7339. https://doi.org/10.1158/1078-0432.CCR-08-0797
65. Vrana JA, Stang MT, Grande JP, Getz MJ (1996) Expression of tissue factor in tumor stroma correlates with progression to invasive human breast cancer: paracrine regulation by carcinoma cell-derived members of the transforming growth factor beta family. Cancer Res 56:5063–5070
66. Yasunaga M, Manabe S, Furuta M, Ogata K, Koga Y, Takashima H, Nishida T, Matsumura Y (2018) Mass spectrometry imaging for early discovery and development of cancer drugs. AIMS Med Sci 5:162–180. https://doi.org/10.3934/medsci.2018.2.162
67. Yasunaga M, Manabe S, Matsumura Y (2011a) New concept of cytotoxic immunoconjugate therapy targeting cancer-induced fibrin clots. Cancer Sci 102:1396–1402. https://doi.org/10.1111/j.1349-7006.2011.01954.x
68. Yasunaga M, Manabe S, Tarin D, Matsumura Y (2011b) Cancer-stroma targeting therapy by cytotoxic immunoconjugate bound to the collagen 4 network in the tumor tissue. Bioconjug Chem 22:1776–1783. https://doi.org/10.1021/bc200158j
69. Yasunaga M, Manabe S, Tarin D, Matsumura Y (2013) Tailored immunoconjugate therapy depending on a quantity of tumor stroma. Cancer Sci 104:231–237. https://doi.org/10.1111/cas.12062
70. Yasunaga M, Manabe S, Tsuji A, Furuta M, Ogata K, Koga Y, Saga T, Matsumura Y (2017) Development of antibody–drug conjugates using DDS and molecular imaging. Bioengineering 4:78. https://doi.org/10.3390/bioengineering4030078
71. Zalutsky MR, Moseley RP, Coakham HB, Coleman RE, Bigner DD (1989) Pharmacokinetics and tumor localization of 131I-labeled anti-tenascin monoclonal antibody 81C6 in patients with gliomas and other intracranial malignancies. Cancer Res 49:2813
72. Zalutsky MR, Narula AS (1987) A method for the radiohalogenation of proteins resulting in decreased thyroid uptake of radioiodine. Int J Rad Appl Instrum A 38:1051–1055
73. Zalutsky MR, Reardon DA, Akabani G, Coleman RE, Friedman AH, Friedman HS, McLendon RE, Wong TZ, Bigner DD (2008) Clinical experience with alpha-particle emitting 211At: treatment of recurrent brain tumor patients with 211At-labeled chimeric antitenascin monoclonal antibody 81C6. J Nucl Med 49:30–38. https://doi.org/10.2967/jnumed.107.046938

Part V
The Current Status of Cancer Drug Delivery Systems and Future Directions

The Current Status of Cancer Drug Delivery Systems and Future Directions

Yasuhiro Matsumura and David Tarin

Abstract In this final chapter we review the sequential developments that have led to the range of choices now available for treating various types of cancer at different stages, summarise the methodological advances described in this book and discuss promising avenues for future progress of drug delivery systems.

Keywords Drug Delivery system · EPR effect · CAST therapy · Liposomes · Micelles · Antibody drug conjugates · Pathophysiology of cancer

The New Drug Delivery System (DDS) Concept

Our survey of the development and current status of the DDS concept includes the following topics, grouped sequentially:

A. The enhanced permeability and retention (EPR) effect: This constitutes the fundamental theoretical basis of DDS for solid cancer treatment.

   There are two fundamental concepts underlying DDS used in modern solid cancer treatment, namely: passive and active targeting of the lesion.

B. Passive targeting takes advantage of the defective vasculature of the tumor to achieve selective accumulation of a drug or other treatment agent in the lesion. This form of drug delivery is achieved by the EPR effect [1–3] alone.

C. Active targeting also exploits the EPR effect to preferentially deliver macromolecular medications to the tumor, but enhances their retention and concentration

Y. Matsumura (✉)
Division of Developmental Therapeutics, Exploratory Oncology Research & Clinical Trial Center, National Cancer Center, Kashiwa, Japan
e-mail: yhmatsum@east.ncc.go.jp

D. Tarin
Moores UCSD Cancer Center and Department of Pathology, University of California San Diego, La Jolla, CA, USA

Y. Matsumura, D. Tarin (eds.), *Cancer Drug Delivery Systems Based on the Tumor Microenvironment*, https://doi.org/10.1007/978-4-431-56880-3_14

within the tumor by coupling EPR with mechanisms to bind the medication to living or inert components of the tumor. The purpose of this strategy is to hinder rapid clearance of the therapeutic agent from the tumor via fluid seepage into the blood and achieve its slow sustained release in the tissues. It utilizes

- the specific binding properties of molecules such as antibodies to antigens, receptors and ligands on the cancer cells or the intercellular components of the tumor and
- linker molecules attaching the antibodies to the chosen drug. The linkers are specifically designed to be cleaved, slowly, by location-specific conditions such as pH, specific enzymes or by chemical reactions resulting in hydrolysis or chemical recombinations with surrounding tissue molecules or administered chemicals
- The released drug or other treatment agent (DNA, siRNA etc) may also be encapsulated in envelopes designed to facilitate passage through cell, endoplasmic or nuclear membranes.

The term DDS can be used flexibly to refer to any of the above items individually or to the whole concept, depending on the context.

The first paper on the EPR effect was published by Matsumura and Maeda in 1986 [1], and the number of citations for the paper is now over 5500 (as of August 2018 from Google Scholar). At the animal experiment level, applications of EPR are accepted worldwide as providing proof of concept support for delivering anticancer drugs, genes and nucleic acids to cancer tissues with high molecular polymers and nanoparticles such as liposomes and micelles [2–3].

Recent DDS Formulations Approved or Under Development for Clinical Use

DDS was first developed in the 1960's mainly in the United States as a controlled release formulation. A microcapsule-type sustained-release preparation of a luteinizing hormone-releasing hormone (LH-RH) derivative from Japan was approved in the United States in 1989, and is still used worldwide for treatment of prostate cancer and breast cancer [4]. Later, the concept was refined and a nanoparticle formulation, which is a targeting type DDS modality, based on the EPR effect, was produced. Subsequently, Doxil, a doxorubicin-containing liposome conjugated with PEG, was approved in the United States in 1995, for ovarian cancer treatment, and was also approved later in other countries including in Japan in 2007 [5]. More recently, the progress of antibody production technology, has led to the creation of payload (small molecule anticancer drug) -added antibodies, otherwise known as, antibody drug conjugates (ADCs). The following section introduces DDS formulations that have been approved or that are undergoing clinical development. It is noteworthy that there is still a big gap between the promising results in basic research and those obtained in clinical trials. This does not signify that DDS is ineffective in humans. Instead it indicates that human cancers are so much more varied and complicated than the types of cancers used to test therapeutic initiatives in animals that current DDS technology is not yet sufficiently optimized to achieve its full potential in human clinical settings. Under such circumstances, it is important to know the development status of various DDS preparations and to understand the dosage ranges so far approved for different types of cancer to achieve specific objectives.

Liposomal Formulations

Irinotecan-Containing Liposomes (PEPO2)

A randomized phase 3 trial was globally conducted for patients with metastatic pancreatic cancer who had been treated with gemcitabine in order to see whether addition of the DDS formulation PEPO2 to standard treatment fluorouracil plus leucovorin (F/L) would be beneficial. The result showed that combination of PEPO2 and F/L significantly prolonged survival time compared to PEPO2 alone or F/L alone [6]. This result is very interesting because this DDS formulation (PEP02) was thought to be minimally effective in pancreatic cancer with abundant cancer stroma that may hinder the efficient distribution of the DDS within the cancer tissue. The clear reason for this success is unknown. However, it is presumed that the combination of the high molecular weight DDS and low molecular weight anticancer agents compensates for operational deficiencies of each agent when used alone. PEPO2 was approved for the pancreatic cancer as a second line therapy in the United States in 2018.

Dual-Drug Liposomal Encapsulation of Cytarabine and Daunorubicin

CPX-351 is a liposome preparation in which two anticancer agents, daunorubicin and cytarabine, are encapsulated in the same vehicle. The FDA approved the liposomal formulation in 2017 because phase 3 results for acute myeloid leukemia resulted in significant prolongation of survival compared to standard treatment [7]. This result is very interesting because the improved clinical outcome has nothing to do with the EPR effect, as the target disease is leukemia, in which the malignant cells circulate and do not form a focal tumor. The conclusion to be drawn from this observation is that, a DDS formulation can enhance treatment results in malignant diseases where there is no solid tumor. This result may be due to the fact that the retention of the low molecular weight anticancer drug in the blood has been enhanced by liposome encapsulation and the exposure time of circulating leukemia cells to the drug is thus extended. At the same time, because of the molecular size of the liposome preparation, the distribution of the drug to normal tissues is diminished, causing the therapeutic window to be expanded.

Micellar Formulations

Paclitaxel-Incorporating Micelles (NK105)

NK105 contains paclitaxel (PTX) in micelles [8]. As phase 1 trial of NK105 appeared to be well tolerated [9] and the phase 2 study showed that NK105 exhibited less neurotoxicity than PTX, a phase 3 non-inferiority trial, to compare the efficacy and safety of NK105 and PTX in metastatic breast cancer was conducted [10].

In this trial, NK105 had a better peripheral sensory neuropathy toxicity profile than paclitaxel [10]. Many of the oncologists involved in this phase 3 claimed that the dose of NK105 was too low. Currently, a randomized phase 2 trial is being conducted for the same metastatic breast cancer subject, reflecting that the NK105 dose was too low.

Epirubicin-Incorporating Micelles (NC-6300)

NC-6300 is an epirubicin-incorporating polymeric micelle developed by Kataoka et al. [11]. The size is from 40 to 80 nm in diameter. In this formulation, epirubicin is bound to a poly-aspartic hydrophobic chain via an azomethine bond that can be hydrolysed only in acidic conditions. We hypothesize that NC-6300 selectively accumulates in tumor tissue due to the EPR effect and that the formulation disintegrates spontaneously during the stay within the tumor tissue. Disintegrated epirubicin-bound unimers, being small molecules, diffuse rapidly through the interstitial space, reach the cancer cells, are integrated into the cell and ferried in endosomes to the lysosomes. Here the acidic conditions liberate free epirubicin which kills the cancer cells. NC-6300 showed a significantly higher antitumor effect than an equivalent dose of epirubicin in vivo [12].

Echocardiograms in mice showed that NC-6300 markedly reduces cardiotoxicity of epirubicin [13]. A phase 1 trial showed that the human pharmacokinetic profile of NC-6300 was consistent with that observed in studies in rats and monkeys. Partial response was observed in one patient with breast cancer and stable disease was observed in 10 patients [14]. A phase 2 study is now underway.

Antibody Drug Conjugates (ADC)

ADC was a treatment proposed in the 1970s and was once referred to as missile therapy. At that time, the antibody used was a mouse antibody, immunogenic in humans, and, because the linker technology to combine the antibody and the anticancer drug was immature, it was not clinically implemented. Despite such past failures, recent advances in antibody production technology, such as humanized antibodies and improvements in linker technology have resulted in the rapid development of drugs known as antibody drug conjugates (ADCs). Currently, more than 70 ADCs have been tested in clinical trials, and six have progressed to phase 3. This large expansion of the ADC repertoire is largely due to the clinical success of two ADCs, as well as to further technological innovations discussed below,. One is Brentuximab vedotin, which was approved in the United States in 2011 for refractory lymphomas with CD30 expression [15]. The other is trastuzumab emtansine, which was approved for HER2-positive breast cancer in the United States in 2013 [16].

Among the ADCs currently undergoing phase 3, the two ADCs introduced below contain new ADC concepts and technologies.

IMMU-132 That Is a SN-38 Conjugated Anti-TROP2 Antibody

The linker moiety is a carbonate bond, and SN-38 is released by carboxyesterase [17]. Generally, after binding to the surface of cancer cells, ADCs internalize into cancer cells and the linker is cleaved by intracellular lysosomal enzymes to release the anticancer drug. However, since the linker of this ADC is cleaved by the carboxylesterase outside the cancer cell, the anticancer agent is excised by the carboxylesterase present in the tumor stroma when the ADC is bound to the cancer cell surface. It is a mechanism to attack from the outside of cancer cells. Since the carboxylesterase is also present in cells, the anticancer drug is released also from any ADC that has entered cells. Therefore, SN-38 is released and exerts a cell-killing effect both inside and outside cancer cells. In the case of such a linker, even if the internalization is weak, binding to cancer cells is considered to bring about an effect.

DS-8201a Is an ADC in Which the Topoisomerase I Inhibitor, Exatecan Derivative Is Bound to an Antibody [18]

It shows remarkable effects on HER2-positive stomach cancer and breast cancer. Preclinical and early clinical studies showed that DS-8201a is effective HER2 positive metastatic breast cancer resistant to T-DM1. The ADC is described in detail in another chapter (See Part II, 3).

The Future of DDS

Importance of DDS Development Based on Pathophysiology of Cancer

Medication for cancer is on the rise. However, as is apparent from the side effects and drug resistance of anticancer agents and molecular targeting agents, their limitations have also been seen. So what is the future role of DDS in cancer treatment? It is certainly true, that at least at the present time, it is not in the mainstream of drug treatment protocols for cancer. As discussed in this book, the wide range of antibodies, linkers and envelopes now available to assemble into drug delivery vehicles offers great promise for the future. The findings described in its chapters indicate that we stand on the threshold of a new era of highly effective cancer therapy. They also highlight that, in order to discover optimal combinations for specific human cancer types, DDS research will not only need to take into account the characteristics of their constituent cancer cells, but also the characteristics of the extracellular constituents of each solid cancer tissue [19]. This includes understanding of macrodynamic events, such as blood clotting and histopathological changes, such as fibrosis in cancer tissue, that occur within the tumor microenvironment, but outside the cancer cells..

More importantly, the overall objective in clinical cancer treatment is to find systemic drug regimens that will target not only primary tumors but also their metastases. We consider that the range of therapeutic modalities now available in the DDS repertoire and described in various chapters of this book offers realistic promise to reach this goal. Accordingly, the mechanisms of metastasis as well as the characteristics of the environments in the secondary tumors in different organs will need to be taken into consideration [20, 21] in designing effective DDS combinations. Knowledge of the cellular and molecular biological characteristics as well as the pathophysiological properties of specific cancer tissue types is needed for achieving these objectives, which makes DDS research and development difficult. Since various aspects of these issues are already described in other chapters, details are omitted in this chapter.

In summary, it will take more time to identify and perfect optimal DDS preparations for individual tumor types based on their cellular, molecular and pathophysiological phenotypes.

The Contribution of CAST Therapy to the Future of DDS Cancer Research

To further the aims described above, CAST therapy has been developed as a new form of DDS aiming to provide *systemic* cancer treatment *focused* upon destroying both primary and secondary tumors, without damaging healthy normal organs of the patient. It may seem a contradiction in terms to discuss *systemic* and *focused* delivery of treatment in the same sentence, but this is what exploitation of the EPR effect, coupled with modern pharmaceutical chemistry described in other chapters in this book, can achieve in this latest phase of DDS cancer research. Focused effects of systemic drug therapy have occasionally been used for specific purposes in other branches of medicine such as, for example, the administration of parenteral sildenafil to relieve pulmonary hypertension in neonates [22], but the local amplification of drug effect achieved by the EPR phenomenon in neoplasms, coupled with mechanisms (ADCs, micelles, liposomes etc) to hold the therapeutic agent in the disease site is unique.

Additionally, the counter-intuitive development of antibodies to non-malignant, acellular components (collagen, insoluble fibrin) of the tumor stroma, previously considered inert impediments to the diffusion of cytotoxic macromolecules, has opened an unexpected new arena of therapeutic opportunities for human cancer therapy. The formulations described by Matsumura and colleagues in other chapters of this book, use ADCs, but other adaptations using liposomes and micelles in combination with ADCs are obvious avenues to explore for further, amplified, effects against neoplastic tissues. It has not escaped our notice that such new concepts and vehicles can also lead to the development of focused systemic therapy for other, non-neoplastic, diseases.

Incorporation of Findings from Many Disciplines Advances the Development of DDS

The goal of modern medicine is to use advancing knowledge of molecular mechanisms of disease to formulate new therapeutic agents. However, what has emerged from such efforts is that the mechanisms by which cancer cells survive and proliferate are very diverse. Additionally, tumor cells are often difficult to distinguish from their normal counterparts and their genomic instability results in increasing cellular and molecular heterogeneity in the neoplastic population, rendering them hard to eliminate. We feel that this factual observation is contrary to the directionality of current cancer research, which seeks drugs to alter or neutralize the functionality of specific molecules deemed to be a "drivers" of the neoplastic process. Thus, as we human beings still do not have "the hand of God" to distinguish normal cells from cancer cells unambiguously, there is a need for a DDS to selectively carry highly toxic payloads into the immediate environment of cells within the cancer and destroy them indiscriminately, without damaging normal cells of other organs. This can be viewed as a highly localized "scorched earth" policy, by which the cancer is denied support from adjacent stromal resources and poisoned, but the destructive agent is caged within the zones chosen by the physician.

As is evident in this book, DDS formulations of increasing potency are being designed by incorporating principles from many different scientific disciplines including organic and physical chemistry, biochemistry, molecular biology and pathology. For example, the above-mentioned CAST therapy is based on pathophysiology-based medicine, antibody chemistry such as protein production / modification, and synthetic organic chemistry. Other promising formulations based upon liposomes and micelles discussed in this book also result from multidisciplinary collaborations. The promising clinical results recently obtained with the introduction of immune checkpoint inhibitory antibodies into cancer treatment also offer new possibilities for multidisciplinary therapy using DDS. Evidence is accumulating that the anti-tumor effect of these antibodies is enhanced when used in combination with conventional anti-cancer agents [23]. Since CAST therapy is a method of selectively delivering therapeutic agents to the cancer stroma and it is thought that the effect of CAST on whole body lymphocytes may be small, it is rational to test the proposition that the combination of DDS agents with immune checkpoint inhibiting antibodies could enhance the therapeutic effects of both.

Basic research on cancer treatment is progressing rapidly. On the other hand, advances in cancer treatment in the clinic only progress slowly. Partly, this is because our biological test systems are not as closely representative of conditions in living human bodies as we believe, but also our conclusions from experiments are sometimes incorrect. Therefore, we will frequently need to review the concepts that we assume to be true, so that direct observations in clinical practice guide theoretical principles of therapeutic design. By doing so, the effort to achieve control of cancer continues with further DDS research, based upon principles described in this book, that bring together the knowledge obtained from many disciplines to deliver better medicines to cancer patients.

References

1. Matsumura Y, Maeda H (1986) A new concept for macromolecular therapeutics in cancer chemotherapy: mechanism of tumoritropic accumulation of proteins and the antitumor agent smancs. Cancer Res 46:6387–6392
2. Duncan R (2003) The dawning era of polymer therapeutics. Nat Rev 2:347–360
3. Peer D, Karp JM, Hong S, Farokhzad OC, Margalit R, Langer R (2007) Nanocarriers as an emerging platform for cancer therapy. Nat Nanotech 2:751–760
4. Okada H (1997) One- and three-month release injectable microspheres of the LH-RH superagonist leuprorelin acetate. Adv Drug Deliv Rev 28(1):43–70
5. Muggia FM, Hainsworth JD, Jeffers S, Miller P, Groshen S, Tan M, Roman L, Uziely B, Muderspach L, Garcia A, Burnett A, Greco FA, Morrow CP, Paradiso LJ, Liang LJ Phase II study of liposomal doxorubicin in refractory ovarian cancer: antitumor activity and toxicity modification by liposomal encapsulation. J Clin Oncol 15:987–993
6. Wang-Gillam A, Li CP, Bodoky G, Dean A, Shan YS, Jameson G, Macarulla T, Lee KH, Cunningham D, Blanc JF, Hubner RA, Chiu CF, Schwartsmann G, Siveke JT, Braiteh F, Moyo V, Belanger B, Dhindsa N, Bayever E, Von Hoff DD, Chen LT, NAPOLI-1 Study Group (2016) Nanoliposomal irinotecan with fluorouracil and folinic acid in metastatic pancreatic cancer after previous gemcitabine-based therapy (NAPOLI-1): a global, randomised, open-label, phase 3 trial. Lancet 387:545–557
7. Lancet JE, Uy GL, Cortes JE, Newell LF, Lin TL, Ritchie EK, Stuart RK, Strickland SA, Hogge D, Solomon SR, Stone RM, Bixby DL, Kolitz JE, Schiller GJ, Wieduwilt MJ, Ryan DH, Hoering A, Banerjee K, Chiarella M, Louie AC, Medeiros BC (2018) CPX-351 (cytarabine and daunorubicin) liposome for injection versus conventional Cytarabine Plus Daunorubicin in older patients with newly diagnosed secondary Acute Myeloid Leukemia. J Clin Oncol 36(26):2684–2692
8. Hamaguchi T, Kato K, Yasui H, Morizane C, Ikeda M, Ueno H, Muro K, Yamada Y, Okusaka T, Shirao K, Shimada Y, Nakahama H, Matsumura Y (2007) lA phase I and pharmacokinetic study of NK105, a paclitaxel-incorporating micellar nanoparticle formulation. Br J Cancer 97(2):170–176
9. Kato K, Chin K, Yoshikawa T, Yamaguchi K, Tsuji Y, Esaki T, Sakai K, Kimura M, Hamaguchi T, Shimada Y, Matsumura Y, Ikeda R (2012) Phase II study of NK105, a paclitaxel-incorporating

micellar nanoparticle, for previously treated advanced or recurrent gastric cancer. Investig New Drugs 30(4):1621–1627
10. Fujiwara Y, Mukai H, Saeki T, Ro J, Lin YC, Nagai SE, Lee KS, Watanabe J, Ohtani S, Kim SB, Kuroi K, Tsugawa K, Tokuda Y, Iwata H, Park YH, Yang Y, Nambu Y (2019) A multinational, randomised, open-label, parallel, phase III non-inferiority study comparing NK105 and paclitaxel in metastatic or recurrent breast cancer patients. Br J Cancer 120(5):475–480
11. Bae Y, Jang WD, Nishiyama N, Fukushima S, Kataoka K (2005) Multifunctional polymeric micelles with folate-mediated cancer cell targeting and pH-triggered drug releasing properties for active intracellular drug delivery. Mol Biosyst 1(3):242–250
12. Harada M, Bobe I, Saito H, Shibata N, Tanaka R, Hayashi T, Kato Y (2011) Improved antitumor activity of stabilized anthracycline polymeric micelle formulation, NC-6300. Cancer Sci 102(1):192–199
13. Takahashi A, Yamamoto Y, Yasunaga M, Koga Y, Kuroda J, Takigahira M, Harada M, Saito H, Hayashi T, Kato Y, Kinoshita T, Ohkohchi N, Hyodo I, Matsumura Y (2013) NC-6300, an epirubicin-incorporating micelle, extends the antitumor effect and reduces the cardiotoxicity of epirubicin. Cancer Sci 104(7):920–925
14. Mukai H, Kogawa T, Matsubara N, Naito Y, Sasaki M, Hosono A (2017) A first-in-human phase 1 study of epirubicin-conjugated polymer micelles (K-912/NC-6300) in patients with advanced or recurrent solid tumors. Invest New Drugs 35(3):307–314
15. Younes A, Bartlett NL, Leonard JP, Kennedy DA, Lynch CM, Sievers EL, Forero-Torres A (2010) Brentuximab vedotin (SGN-35) for relapsed CD30-positive lymphomas. N Engl J Med 363(19):1812–1821
16. Verma S, Miles D, Gianni L, Krop IE, Welslau M, Baselga J, Pegram M, Oh DY, Diéras V, Guardino E, Fang L, Lu MW, Olsen S, Blackwell K, EMILIA Study Group (2012) Trastuzumab Emtansine for HER2-positive advanced breast cancer. N Engl J Med 367:1783–1791
17. Bardia A, Mayer IA, Diamond JR, Moroose RL, Isakoff SJ, Starodub AN, Shah NC, O'Shaughnessy J, Kalinsky K, Guarino M, Abramson V, Juric D, Tolaney SM, Berlin J, Messersmith WA, Ocean AJ, Wegener WA, Maliakal P, Sharkey RM, Govindan SV, Goldenberg DM, Vahdat LT (2017) Efficacy and safety of anti-Trop-2 antibody drug conjugate Sacituzumab Govitecan (IMMU-132) in heavily pretreated patients with metastatic triple-negative breast Cancer. J Clin Oncol 35(19):2141–2148
18. Doi T, Shitara K, Naito Y, Shimomura A, Fujiwara Y, Yonemori K, Shimizu C, Shimoi T, Kuboki Y, Matsubara N, Kitano A, Jikoh T, Lee C, Fujisaki Y, Ogitani Y, Yver A, Tamura K (2017) Safety, pharmacokinetics, and antitumour activity of trastuzumab deruxtecan (DS-8201), a HER2-targeting antibody-drug conjugate, in patients with advanced breast and gastric or gastro-oesophageal tumours: a phase 1 dose-escalation study. Lancet Oncol 18(11):1512–1522
19. Matsumura Y (2012) Cancer stromal targeting (CAST) therapy. Adv Drug Deliv Rev 64:710–719
20. Tarin D, Price JE, Kettlewell MG, Souter RG, Vass AC, Crossley B (1984) Mechanisms of human tumor metastasis studied in patients with peritoneovenous shunts. Cancer Res 44(8):3584–3592
21. Tarin D (2011) Cell and tissue interactions in carcinogenesis and metastasis and their clinical significance. Semin Cancer Biol 21(2):72–82
22. Kelly LE, Ohlsson A, Shah PS (2017) Sildenafil for pulmonary hypertension in neonates. Cochrane Database Syst Rev 8:CD005494. https://doi.org/10.1002/14651858.CD005494.pub4
23. Gandhi L, Rodríguez-Abreu D, Gadgeel S, Esteban E, Felip E, De Angelis F, Domine M, Clingan P, Hochmair MJ, Powell SF, Cheng SY, Bischoff HG, Peled N, Grossi F, Jennens RR, Reck M, Hui R, Garon EB, Boyer M, Rubio-Viqueira B, Novello S, Kurata T, Gray JE, Vida J, Wei Z, Yang J, Raftopoulos H, Pietanza MC, Garassino MC, KEYNOTE-189 Investigators (2018) Pembrolizumab plus chemotherapy in metastatic non-small-cell lung Cancer. N Engl J Med 378(22):2078–2092

Index

Y. Matsumura, D. Tarin (eds.), *Cancer Drug Delivery Systems Based on the Tumor Microenvironment*, https://doi.org/10.1007/978-4-431-56880-3

The manufacturer's authorised representative in the EU is Springer Nature Customer Service Centre GmbH, Europaplatz 3, 69115 Heidelberg, Germany. If you have any concerns regarding our products, please contact ProductSafety@springernature.com

Printed and bound by CPI Group (UK) Ltd, Croydon, CR0 4YY
22/09/2026
02221057-0014